Methods in Plant Molecular Biology

Methods in Plant Molecular Biology

A LABORATORY COURSE MANUAL

Pal Maliga Waksman Institute, Rutgers University

Daniel F. Klessig Waksman Institute, Rutgers University

Anthony R. Cashmore University of Pennsylvania, Philadelphia

Wilhelm Gruissem University of California, Berkeley

Joseph E. Varner Washington University, St. Louis

COLD SPRING HARBOR LABORATORY PRESS 1995

Methods in Plant Molecular Biology
A LABORATORY COURSE MANUAL

Printed in the United States of America

Cover and text design by Emily Harste

Front cover: A chimeric tobacco plant with transformed and wild-type plastid genome copies. In the yellow sectors the photosynthetic *psbA* gene was replaced with the bacterial *aadA* gene by targeted gene insertion. Plant growth is supported by the photosynthetically competent green sectors. (Photograph by Zora Svab and Pal Maliga [Waksman Institute, Rutgers University].)

Back cover: In situ hybridization was used to detect the activation of the pathogenesis-related PR-1 genes in a tobacco plant responding to an infection by tobacco mosaic virus. Darkfield microscopy reveals the presence of silver grains over the leaf section indicating the accumulation of PR-1 mRNA within the cells of this infected leaf. (Photograph by David C. Dixon [U.S. Department of Agriculture/ARS/SRRC] and Daniel F. Klessig [Waksman Institute, Rutgers University].)

Library of Congress Cataloging-in-Publication Data

Methods in plant molecular biology : a laboratory course manual/ Pal Maliga ... [et al.].
p. cm.
Includes bibliographical references (p.) and index.
ISBN 0-87969-450-5 (cloth bound). -- ISBN 0-87969-386-X (comb bound)
1. Plant molecular biology--Laboratory manuals. I. Maliga, P.
QK728.M482 1995
581.8'8'078--dc20 94-36570
CIP

All Cold Spring Harbor Laboratory Press publications may be ordered directly from Cold Spring Harbor Laboratory Press, 10 Skyline Drive, Plainview, New York 11803-2500. Phone: 1-800-843-4388 in Continental U.S. and Canada. All other locations: (516) 349-1930. FAX: (516) 349-1946.

Preface

This manual began as a collection of laboratory protocols handed out to participants of the three-week course Molecular and Developmental Biology of Plants, held every summer since 1981 at Cold Spring Harbor Laboratory. The editors of this book were instructors on the course with an overlapping tenure for a total of four years: Pal Maliga (1989–1992), Joseph Varner (1989–1991), Wilhelm Gruissem (1989–1990), Daniel Klessig (1991–1992), and Anthony Cashmore (1992). The "plant course" was established by Frederick Ausubel (1981–1982) and John Bedbrook (1981–1982), and subsequently organized by Ian Sussex (1983–1988), Russel Malmberg (1983), Joachim Messing (1984–1988), and Robert Horsch (1984, 1986, 1987). Since its inception, the course has been supported by the National Science Foundation.

The course program consisted of a series of daily lectures to present a comprehensive overview of the frontiers in plant sciences and hands-on teaching of molecular techniques in laboratory sessions. The laboratory sessions were designed to demonstrate a diverse set of principal methods in plant biology. Since the experiments were sometimes performed with participants who had limited experience in molecular biology, we tried to prepare as detailed and specific a protocol as feasible. The manual reflects the philosophy of the training course: it is a collection of detailed protocols, each designed to both demonstrate a principal method and maximize the probability of success.

Each chapter of the laboratory manual is authored by the instructor responsible for teaching the protocol at the course. As a consequence, we had to make difficult choices in selecting among similar protocols taught by different scientists during the years. As a principle, in the case of two essentially identical procedures, we chose the protocol of the instructor who was participating in the course at the time the decision was made to publish the course manual. This is why chapters by Ken Keegstra, Chris Sommerville, and Sue Gibson are missing, although they proposed and implemented laboratory exercises at the course that were later taught by others. We would like to take this opportunity to thank them and all the other invited instructors and speakers for their contributions to the course, as well as for their dedicated service in training a new generation of plant scientists. We are also indebted to our colleagues at Cold Spring Harbor Laboratory who have supported the course over the years, and who have often found their own research programs grind to a halt when course participants suddenly needed their equipment.

We also extend our gratitude to the staff of Cold Spring Harbor Laboratory Press, particularly Nancy Ford and John Inglis, who kept the project of assembling this laboratory manual alive, and Maryliz Dickerson, Dorothy Brown, Lee Martin, and Joan Ebert for their excellent editorial and technical assistance. A great part of the credit should go to Catriona Simpson, whose careful editing made this manual more accessible, and who executed her work with extreme patience and understanding.

The Editors

Contributors

Bell, Callum J., Plant Science Institute, Department of Biology, University of Pennsylvania, Philadelphia. *Present Address:* Division of Genetics, Children's Hospital of Philadelphia, Pennsylvania

Carrasco, Pedro, Department of Plant Biology, University of California, Berkeley. *Present Address:* Departament de Bioquímica i Biologia Molecular, Universitat de València, Spain

Cashmore, Anthony R., Plant Science Institute, Department of Biology, University of Pennsylvania, Philadelphia

Cutt, John R., Schering-Plough Research, Kenilworth, New Jersey

Deng, Xing-Wang, Department of Plant Biology, University of California, Berkeley. *Present Address:* Department of Biology, Yale University, New Haven, Connecticut

Dixon, David C., U.S. Department of Agriculture, Agricultural Research Service, Southern Regional Research Center, New Orleans, Louisiana

Dunn, Patrick J., Plant Science Institute, Department of Biology, University of Pennsylvania, Philadelphia

Ecker, Joseph R., Plant Science Institute, Department of Biology, University of Pennsylvania, Philadelphia

Gruissem, Wilhelm, Department of Plant Biology, University of California, Berkeley

Hajdukiewicz, Peter, Waksman Institute, Rutgers, The State University of New Jersey, Piscataway

Klessig, Daniel F., Waksman Institute, Rutgers, The State University of New Jersey, Piscataway

Klimczak, Leszek J., Plant Science Institute, Department of Biology, University of Pennsylvania, Philadelphia

Lamppa, Gayle K., Department of Molecular Genetics and Cell Biology, The University of Chicago, Illinois

Lu, Meiqing, Plant Science Institute, Department of Biology, University of Pennsylvania, Philadelphia

Lucas, William J., Section of Plant Biology/DBS, University of California, Davis

Maliga, Pal, Waksman Institute, Rutgers, The State University of New Jersey, Piscataway

Manzara, Thianda, Laboratory for Molecular Biology, Department of Biological Sciences, University of Illinois at Chicago

Matallana, Emilia, Plant Science Institute, Department of Biology, University of Pennsylvania, Philadelphia. *Present Address:* Departamento de Bioquimicay Biology Molecular, Facultades de Ciencias, Universitat de Valencia, Spain

McCarty, Donald R., Horticultural Sciences Department, University of Florida, Gainesville

Morgan, Michael K., Plant Gene Expression Center, U.S. Department of Agriculture and University of California at Berkeley, Albany

Ow, David W., Plant Gene Expression Center, U.S. Department of Agriculture and University of California at Berkeley, Albany

Rosenkrans, Leonard, Horticultural Sciences Department, University of Florida, Gainesville

Schindler, Ulrike, Plant Science Institute, Department of Biology, University of Pennsylvania, Philadelphia. *Present Address:* Tularik, Inc., South San Francisco, California

Schuster, Gadi, Department of Plant Biology, University of California, Berkeley. *Present Address:* Department of Biology, Technion-Israel Institute of Technology, Haifa, Israel

Stark, Kenneth, Plant Science Institute, Department of Biology, University of Pennsylvania, Philadelphia. *Present Address:* Pennsylvania College of Optometry, Philadelphia

Svab, Zora, Waksman Institute, Rutgers, The State University of New Jersey, Piscataway

van der Schoot, Chris, Section of Plant Biology/DBS, University of California, Davis. *Present address:* Agrotechnological Research Institute (ATO-DLO), Wageningen, The Netherlands

Varner, Joseph E., Department of Biology, Washington University, St. Louis, Missouri

Vasil, Indra K., Horticultural Sciences Department, University of Florida, Gainesville

Vasil, Vimla, Horticultural Sciences Department, University of Florida, Gainesville

Ye, Zheng-Hua, Department of Biology, Washington University, St. Louis, Missouri

Contents

• *Section 1*

Polyethylene Glycol-mediated Transformation of Tobacco Leaf Mesophyll Protoplasts: An Experiment in the Study of Cre-*lox* Recombination

This section describes a protocol for the transformation of plant leaf mesophyll protoplasts with nucleic acids and the transient expression of the encoded proteins. Although the use of stably transformed materials may be preferable in most instances, the transient introduction of macromolecules into plant protoplasts permits a more rapid analysis of the biological activity of the introduced material. The transient assay approach has facilitated, for example, the identification of *cis*-acting regulatory sequences of promoter regions (see, e.g., Ebert et al. 1987; Ellis et al. 1987; Ow et al. 1987), the comparison of transcriptional strength among different promoters (see, e.g., Fromm et al. 1985; Boston et al. 1987; Hauptmann et al. 1987), and the study of environmental factors that affect gene expression (see, e.g., Howard et al. 1987; Marcotte et al. 1988). In experiments in which cells are transformed with RNA molecules, notable progress has been made in defining the *cis* requirements for efficient maturation, stability, and translation of messenger and viral RNAs (see, e.g., Callis et al. 1987; Gallie et al. 1989). More recently, proteins have been introduced transiently into plant cells in the analysis of transcription factors (Katagiri and Chua 1992). In this section, the biological activity of a DNA molecule is examined in transgenic plant cells, exemplifying the combined use of stably transformed and transiently introduced gene constructs.

POLYETHYLENE GLYCOL-MEDIATED TRANSFORMATION

A large variety of methods have been developed to transfer DNA into plant cells (for review, see Potrykus 1991). Among these, polyethylene glycol (PEG)-mediated transformation has been used to establish both stable transformation and transient gene expression (Krens et al. 1982; Paszkowski et al. 1984; Shillito et al. 1985). PEG-mediated transformation offers several attractive features. First, conditions that maximize transient gene expression are associated with high survival and division rates in the transformed cells (Negrutiu et al. 1990). Second, PEG-mediated transformations utilize common, inexpensive supplies and equipment. Finally, extensive studies have demonstrated that this simple and reliable method can be adapted to a wide range of plant species and tissue sources used for protoplast preparation. The protocol presented here was devised to study the site-specific recombination of a transgenic construct within the genome of tobacco (*Nicotiana tabacum*) leaf mesophyll protoplasts.

EXPERIMENTAL DESIGN

The objective of this experiment is to detect an inversion event in the tobacco genome mediated by transforming DNA. The site-specific recombinase is the 38.5-kD product of the bacteriophage P1 *cre* (*c*ontrol of *re*combination) gene, which catalyzes recombination between 34-bp sequences known as *lox* (*lo*cus of cross[*x*]over) sites. As has been shown previously, this site-specific recombination system can mediate the excision, inversion, and cointegration of DNA in plant cells (Dale and Ow 1990, 1991; Odell et al. 1990). The target of the recombinase in this experiment is a construct, pED32 (Dale and Ow 1990), that is stably incorporated into the *Nicotiana tabacum* genome by *Agrobacterium*-mediated transformation. This construct consists of a cauliflower mosaic virus *35S* RNA promoter (*35S*) transcriptionally fused to a fragment containing a firefly luciferase cDNA-nopaline synthase poly(A) region (*luc-nos3′*). However, as shown in Figure 1, the *luc-nos3′* fragment is in an inverted orientation relative to the *35S* promoter. This fragment is also flanked by a *lox* site on each end, but the two sites are of opposing orientations. Transcription of the *luc* cDNA would require inversion of the *luc-nos3′* fragment via recombination at the flanking *lox* sequences. The recombinase required for this inversion event is provided by the transient expression of *cre* from pMM23, a *35S-cre-nos3′* construct. This chimeric *cre* gene is similar to the previously described pED23 (Dale and Ow 1990) except that the prokaryotic ribosome binding site has been removed. The Cre recombinase produced by pMM23 must enter the plant nucleus to catalyze the site-specific inversion event, which can then be scored as luciferase activity (light production). The efficiency of DNA transformation can be assessed using a positive control construct such as pDO432, a *luc* expression plasmid (Ow et al. 1986). For a recent review of the *luc* marker gene, see Millar et al. (1992).

Table 1 illustrates the data from a typical experiment. Without the addition of a *cre* expression plasmid, leaf mesophyll protoplasts

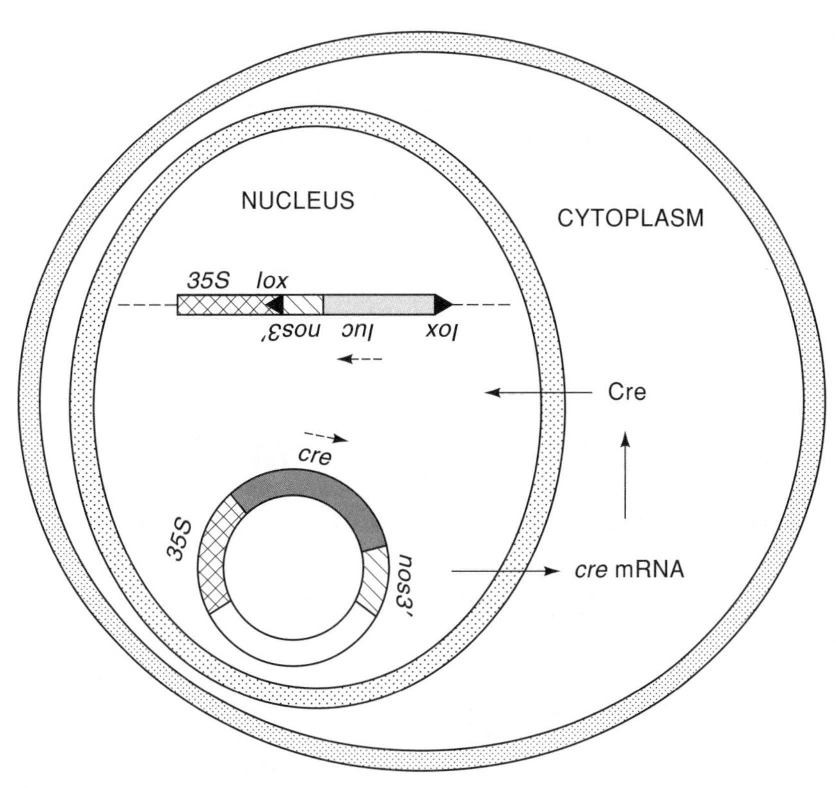

Figure 1
Cre-mediated site-specific inversion of DNA in the plant genome. The Cre recombinase, expressed from the transforming DNA, flips the coding region of the *luc* transgene which results in luciferase activity.

Table 1 **Cre-induced Inversion of Plant Chromosomal DNA**

Transgenic *luc* Plants	Transforming DNA[1]	Light Units[2]	Relative Activity
nt35.9	carrier only	820	1
nt35.9	carrier + 10 μg of pMM23	1.14×10^5	139
nt35.10	carrier only	1190	1
nt35.10	carrier + 10 μg of pMM23	1.57×10^5	132

[1]Transformation of approximately 2×10^6 protoplasts with 100 μg of calf thymus carrier DNA and 10 μg of pMM23, where indicated.
[2]Extracts were prepared 26 hours after transformation. Each extract was adjusted to 1 ml total volume with luciferase extraction buffer. 100 μl of extract were combined with 100 μl of luciferase reaction buffer followed by injection of 100 μl of 0.5 mM D-luciferin. Light emission was counted for 30 seconds in integrated mode. An average of three readings was taken. The baseline count of a blank (extraction buffer only) was 279 light units.

from two independent transgenic tobacco lines (nt35.9 and nt35.10) that harbor the inverted luciferase construct produced very little luciferase activity. These baseline counts were only three- to fourfold higher than the background level of the extraction buffer alone or values that would be observed in protoplasts from wild-type tobacco plants without the construct. Addition of the *cre* expression plasmid to the transformation mix resulted in over a 100-fold increase in luciferase activity. This demonstrates that the Cre recombinase can penetrate the plant nucleus to catalyze recombination of plant chromosomal DNA. In its original intent, this experiment was conducted to test the nuclear entry capability of this prokaryotic protein prior to proceeding with stable transformation experiments (Dale and Ow 1991). Since this experiment demonstrated that the assay is effective, a possible future use of this transient assay might be for the analysis of different Cre-encoding sequences that were modified in vitro.

ADAPTATION OF PROTOCOL TO OTHER SYSTEMS

To adapt the protocol described here to other systems, it may be necessary to modify some of the steps.

- The tissue source and the method of protoplast preparation may affect the transformation process (Negrutiu et al. 1987, 1990). It has been suggested that the results of transformation experiments are superior if the protoplasts are isolated by methods in which the cell wall is digested under mild conditions for a long period of time (e.g., overnight) (Negrutiu et al. 1990). For advice on alternative methods of protoplast isolation, see Potrykus and Shillito (1986).
- Researchers have noted that heat shock increases levels of transient expression in some systems (Oliveiria et al. 1991) but decreases levels of transient expression in others (Negrutiu et al. 1990). It may be useful to test the transformation with and without the brief heat treatment.
- If DNA is being used in the transformation process, the topology of the plasmid template (circular versus linear) may affect expression levels (Ballas et al. 1988). If the plasmid is linearized, cleavage at a site distant from the 3′ end of the gene may enhance levels of gene expression (Negrutiu et al. 1990).
- The divalent cation in the PEG solution is another variable worth testing. Negrutiu et al. (1990) have suggested that the use of Ca^{++} instead of Mg^{++} results in higher transient expression levels and that a divalent cation concentration of 5–15 mM results in a good balance between cell survival and DNA uptake.
- Finally, different PEG concentrations and treatment times should be tested. In addition, the source and molecular weight of the PEG

may be important. The use of PEG with a high molecular weight such as 4000 or especially 8000 may be most appropriate for the isolation of fragile protoplast preparations (Negrutiu et al. 1987).

ACKNOWLEDGMENTS

We thank R. Shillito and H. Albert for useful comments on this manuscript.

REFERENCES

Ballas, N., N. Zakai, D. Friedberg, and A. Loyter. 1988. Linear forms of plasmid DNA are superior to supercoiled structures as active templates for gene expression in plant protoplasts. *Plant Mol. Biol.* **11**: 517–527.

Boston, R.S., M.R. Becwar, R.D. Ryan, P.B. Goldsbrough, B.A. Larkins, and T.K. Hodges. 1987. Expression from heterologous promoters in electroporated carrot protoplasts. *Plant Physiol.* **83**: 742–746.

Callis, J., M. Fromm, and V. Walbot. 1987. Introns increase gene expression in cultured maize cells. *Genes Dev.* **1**: 1183–1200.

Dale, E.C. and D.W. Ow. 1990. Intra- and intermolecular site-specific recombination in plant cells mediated by bacteriophage P1 recombinase. *Gene* **91**: 79–85.

———. 1991. Gene transfer with subsequent removal of the selection gene from the host genome. *Proc. Natl. Acad. Sci.* **88**: 10558–10562.

Ebert, P.R., S.B. Ha, and G. An. 1987. Identification of an essential upstream element in the nopaline synthase promoter by stable and transient assays. *Proc. Natl. Acad. Sci.* **84**: 5745–5749.

Ellis, J.G., D.J. Llewellyn, J.C. Walker, E.S. Dennis, and W.J. Peacock. 1987. The *ocs* element: A 16 base pair palindrome essential for activity of the octopine synthase enhancer. *EMBO J.* **6**: 3203–3208.

Fromm, M., L.P. Taylor, and V. Walbot. 1985. Expression of genes transferred into monocot and dicot plant cells by electroporation. *Proc. Natl. Acad. Sci.* **82**: 5824–5828.

Gallie, D.R., W.J. Lucas, and V. Walbot. 1989. Visualizing mRNA expression in plant protoplasts: Factors influencing efficient mRNA uptake and translation. *Plant Cell* **1**: 301–311.

Hauptmann, R.M., P.O. Akins, V. Vasil, Z. Tabaeiuzadeh, S.G. Rogers, R.B. Horsch, I.K. Vasil, and R.T. Fraley. 1987. Transient expression of electroporated DNA in monocotyledonous and dicotyledonous species. *Plant Cell Rep.* **6**: 265–270.

Howard, E.A., J.C. Walker, E.S. Dennis, and W.J. Peacock. 1987. Regulated expression of an alcohol dehydrogenase 1 chimeric gene introduced into maize protoplasts. *Planta* **170**: 535–540.

Katagiri, F. and N.-H. Chua. 1992. Plant transcription factors: Present knowledge and future challenges. *Trends Genet.* **8**: 22–27.

Krens, F.A., L. Molendijk, G.J. Wullems, and R.A. Schilperoort. 1982. *In vitro* transformation of plant protoplasts with Ti-plasmid DNA. *Nature* **296**: 72–74.

Marcotte, W.R., Jr., C.C. Bayley, and R.S. Quatrano. 1988. Regulation of a wheat promoter by abscisic acid in rice protoplasts. *Nature* **335**: 454–457.

Millar, A.J., S.R. Short, K. Hiratsuka, N.-H. Chua, and S.A. Kay. 1992. Firefly luciferase as a reporter of regulated gene expression in higher plants. *Plant Mol. Biol. Rep.* **10**: 324–337.

Negrutiu, I., R. Shillito, I. Potrykus, G. Biasini, and F. Sala. 1987. Hybrid genes in the analysis of transformation conditions. I. Setting up a simple method for direct gene transfer in plant protoplasts. *Plant Mol. Biol.* **8**: 363–373.

Negrutiu, I., J. Dewulf, M. Pietrzak, J. Botterman, E. Rietveld, E.M. Wurzer-Figurelli, D. Ye, and M. Jacobs. 1990. Hybrid genes in the analysis of transformation conditions. II. Transient expression vs. stable transformation—Analysis of parameters influencing gene expression levels and transformation efficiency. *Physiol. Plant* **79**: 197–205.

Odell, J., P.G. Caimi, B. Sauer, and S.H. Russell. 1990. Site-directed recombination in the genome of transgenic tobacco. *Mol. Gen. Genet.* **223**: 369–378.

Oliveiria, M.M., J. Barroso, and M.S.S. Pais. 1991. Direct gene transfer into *Actinidia deliciosa* protoplasts: Analysis of transient expression of the CAT gene using TLC autoradiography and a GC-MS-based method. *Plant Mol. Biol.* **17**: 235–242.

Ow, D.W., J.D. Jacobs, and S.H. Howell. 1987. Functional regions of the cauliflower mosaic virus 35S RNA promoter determined by the use of the firefly luciferase gene as a reporter of promoter activity. *Proc. Natl. Acad. Sci.* **84**: 4870–4874.

Ow, D.W., K.V. Wood, M. DeLuca, J.R. de Wet, D.R. Helinski, and S.H. Howell. 1986. Transient and stable expression of the firefly luciferase gene in plant cells and transgenic plants. *Science* **234**: 856–859.

Paszkowski, J., R.D. Shillito, M. Saul, V. Mandak, T. Hohn, B. Hohn, and I. Potrykus. 1984. Direct gene transfer to plants. *EMBO J.* **3**: 2717–2722.

Potrykus, I. 1991. Gene transfer to plants: Assessment of published approaches and results. *Annu. Rev. Plant Physiol. Plant Mol. Biol.* **42**: 205–255

Potrykus, I. and R.D. Shillito. 1986. Protoplasts: Isolation, culture, plant regeneration. *Methods Enzymol.* **118**: 549–578.

Shillito, R.D. and M.W. Saul. 1988. Protoplast isolation and transformation. In *Plant molecular biology: A practical approach* (ed. C.H. Shaw), pp. 161–186. IRL Press, Oxford, United Kingdom.

Shillito, R.D., M.W. Saul, J. Paszkowski, M. Müller, and I. Potrykus. 1985. High efficiency gene transfer to plants. *Biotechnology* **3**: 1099–1103.

• *TIMETABLE*
Polyethylene Glycol-mediated Transformation of Tobacco Leaf Mesophyll Protoplasts

STEPS	*TIME REQUIRED*
Preparation of Protoplasts	
Grow tobacco plants from seed	5–7 weeks
Prepare digestion enzyme solution	1 hour
Prepare leaf digestions for protoplast isolation	1 hour
Incubate leaf digestions overnight	14–18 hours
Continue digestion of leaf strips (if necessary)	1–3 hours
Harvest and wash protoplasts	2 hours
Polyethylene Glycol-mediated Transformation	
Count protoplasts and prepare transformation samples	1 hour
Transform samples	1 hour
Harvest protoplasts and resuspend in K_3AM medium	30 minutes
Incubate protoplasts for transient expression	16–24 hours
Harvest and freeze-thaw lyse protoplasts	1 hour
Luciferase Assay	
Assay luciferase activity to quantify transformation	1 hour

Preparation of Protoplasts

All the steps described below should be performed under aseptic conditions using sterile equipment and reagents. Leaf tissue and protoplasts should always be handled in a sterile flow hood such as a laminar or other type of horizontal flow hood.

MATERIALS AND EQUIPMENT

Tobacco (*Nicotiana tabacum*) plants grown from seed. The tobacco variety used in this experiment, Wisconsin 38, is no longer available commercially. However, small amounts of transgenic Wisconsin 38 seeds (harboring the pED32 construct) and parental, nontransformed seeds can be obtained from the laboratory of Dr. D.W. Ow (see Appendix 2).

Transparent tissue culture boxes (10 x 7.5 x 7.5 cm; GA 7 Magenta boxes, Magenta Corporation)

Sterile filter unit containing a 100-mesh nylon screen (149 μm opening; 104 mesh/inch; part no. K-CMN-149, Small Parts, Inc.)

MEDIA

Digestion enzyme solution in K_3AS medium (prepare fresh)
K_3AS medium
(For recipes, see Preparation of Media, pp. 14–15)

REAGENTS

W_5
(For recipe, see Preparation of Reagents, pp. 16–17)

PROCEDURE

1. Harvest several medium-sized leaves from tobacco plants grown aseptically in transparent tissue culture boxes. These boxes allow plants to be grown aseptically in the presence of light. (For details on the growth of tobacco plants in transparent tissue culture boxes, see Shillito and Saul [1988].) Use the plants before they become crowded in the boxes (i.e., when they are approximately 5–7 weeks old).

Note: The amount of time required for tobacco plants to reach the proper size will vary considerably depending on the variety of plant, the medium used, and the growth conditions. The plants should be harvested before the top of the plant grows up into the top of the tissue culture box or other leaves on the plant become "cramped." The protoplast yield from such plants is usually quite poor.

2. Place three or four leaves in each of eight petri dishes containing 10 ml of digestion enzyme solution in K_3AS medium.

   Note: The speed and extent of digestion can be influenced by the age of the plants and the conditions under which they are grown. It is best to take medium-sized leaves from the middle of the plant, not the oldest or newest leaves.

3. Stack the leaves two layers deep and cut strips 1–2 mm wide perpendicular to the leaf midrib.

4. Seal the dishes with Parafilm and incubate them for 14–18 hours at 24–26°C. Older leaves may require longer digestion time.

5. Check the progress of digestion. Gently agitate the leaf strips to release cells and to expose fresh cells to the digestion enzyme solution. Allow the leaves to digest for an additional 1–3 hours, if necessary.

6. To facilitate the release of protoplasts, gently pass the digestion mixture up and down a wide-bore pipette two or three times.

   Note: Plant protoplasts are quite fragile and are sensitive to slight changes in osmoticum and to rough handling. All protoplast suspensions should be handled gently and transferred using wide-bore pipettes whenever possible.

7. Filter the digestion mixture through a sterile filter unit containing a 100-mesh nylon screen.

8. Place the filtered protoplasts in 50-ml Falcon tubes and spin at 200*g* for 20 minutes to float the protoplasts.

9. Using a wide-bore pipette, remove the top layer of protoplasts in a minimum volume and transfer to a 50-ml tube.

   Note: To remove the floating protoplasts in a minimum volume requires some practice. It is often easiest to tilt the tube and remove the protoplasts from one edge.

10. Dilute the protoplast suspension with four volumes of W_5. Mix gently and pellet the protoplasts by spinning at 200*g* for 5 minutes.

    Note: The protoplasts sediment in W_5, whereas they float to the surface in the digestion enzyme solution in K_3AS medium, which contains sucrose.

11. Resuspend the protoplasts in 30–40 ml of W_5 and pellet them by spinning at 200*g* for 5 minutes.

    Note: Protoplast pellets are often best resuspended by first adding a small volume of the resuspension buffer (<1 ml) and gently resuspending the pellet before adding the remainder of the appropriate buffer.

12. Repeat the W_5 wash and pellet the protoplasts by spinning at 200*g* for 5 minutes.

Polyethylene Glycol-mediated Transformation

The protocol described below was adapted from Negrutiu et al. (1987). All the steps should be performed under aseptic conditions using sterile equipment and reagents. The manipulation of protoplasts and the PEG transformation should be performed in a horizontal flow hood.

MATERIALS

Tobacco leaf protoplasts (see p. 10, step 12)
Transforming DNA (*cre* expression plasmid [pMM23]; can be obtained from the laboratory of Dr. D.W. Ow [see Appendix 2])

MEDIA

K_3AM medium
(For recipe, see Preparation of Media, pp. 14–15)

REAGENTS

MaMg (1x and 2x)
Polyethylene glycol (M.W. 1450) solution (PEG 1450)
W_5
Luciferase extraction buffer
(For recipes, see Preparation of Reagents, pp. 16–17)

PROCEDURE

1. Resuspend the tobacco leaf protoplasts in 10 ml of 1x MaMg. Determine the protoplast concentration by counting the cells in a hemocytometer. Aliquot the protoplasts into 50-ml Falcon tubes to give 1–2 x 10^6 cells per transformation sample. Dilute each sample with 1x MaMg to give a final volume of 10 ml.

2. Pellet the protoplasts by spinning at 200*g* for 5 minutes. Remove the supernatant and resuspend each pellet in 0.5 ml of 1x MaMg.

3. Heat shock the protoplasts by incubating them in a water bath for 5 minutes at 45°C. Remove the samples from the water bath and place them at room temperature for 5–10 minutes.

4. Add the transforming DNA to the samples. In general, for each sample use approximately 10 µg of transforming DNA premixed with approximately 50 µg of carrier DNA/10^6 protoplasts. It is desirable for the DNA to be dissolved in H_2O. However, if it is dissolved in 10 mM Tris/1 mM EDTA, the volume of the DNA solution should be <30 µl.

5. Slowly add 0.5 ml of PEG 1450 to induce DNA uptake. Mix gently but thoroughly.

   Note: Because of the viscosity of the PEG solution, it may take several minutes to completely mix each transformation sample.

6. Allow the transformation to proceed for 25 minutes at room temperature.

7. Dilute each sample slowly with 5 ml of W_5, mix gently, and pellet the protoplasts by spinning at 200*g* for 5 minutes.

8. Resuspend the protoplasts in K_3AM medium at a density of approximately 10^5 protoplasts/ml.

9. To allow transient expression to occur, incubate the protoplasts in 50-ml Falcon tubes in the dark for 16–24 hours at 24–26°C. Slant the tubes to give the maximum surface area.

10. Harvest the protoplasts by spinning at 200*g* for 7 minutes. The protoplasts may pellet better if they are first diluted two- to threefold with W_5.

11. Resuspend each pellet in 500–1000 µl of luciferase extraction buffer and transfer the protoplast suspensions to 1.5-ml microfuge tubes. Freeze-thaw the samples three times by placing them alternately for 3 minutes in a -80°C freezer and for 3 minutes in a room-temperature water bath. Pellet the cell debris by spinning the samples in a microfuge for 5 minutes at 4°C, and assay the supernatants for luciferase activity (and protein concentration, if desired, using an assay such as the Bio-Rad [Bradford] Protein Assay). Extracts can be stored for several weeks at -80°C without loss of activity.

Luciferase Assay

MATERIALS AND EQUIPMENT

Protoplast extract (see p. 12, step 11)
Luminometer (e.g., from Analytical Luminescence Laboratory, Inc. or Turner Designs)

REAGENTS

Reaction buffer
D-Luciferin (0.5 mM)
(For recipes, see Preparation of Reagents, pp. 16–17)

PROCEDURE

1. Mix 100 μl of reaction buffer with 100 μl of protoplast extract in a luminometer cuvette. Depending on the luminometer, the total volume may need to be adjusted for optimal reading.

2. Place the cuvette into the luminometer and start the reaction (outlined below) by injecting 100 μl of 0.5 mM D-luciferin. Record the light emission.

   Luciferase + Mg-ATP + luciferin → luciferase-luciferyl-AMP + PPi

   Luciferase-luciferyl-AMP + O_2 → luciferase-oxyluciferin* + CO_2 + AMP

   Luciferase-oxyluciferin* → luciferase + oxyluciferin + light (560 nm)

• *PREPARATION OF MEDIA*

Digestion enzyme solution in K_3 AS medium

1% Cellulase R1O (Serva 16419)
0.2% Macerozyme R1O (Serva 28302)

Dissolve in K_3AS medium and stir for 30–60 minutes. Adjust the pH to 5.8 with 1 M KOH. Spin at 3000 rpm for 5 minutes to remove undissolved material. Filter sterilize. Prepare fresh before each use.

K_3AM medium

Follow the same recipe as for K_3AS medium, but add 30 g of sucrose/liter (instead of 137 g/liter) and 72.8 g of mannitol/liter. Store at room temperature. K_3AM medium is stable for at least 1 month.

K_3AS medium

(1 liter)

Murashige and Skoog salts (GIBCO 11117-041)	4.3 g
$CaCl_2 \cdot 2H_2O$	0.46 g
Xylose	0.25 g
B5 vitamins stock (1000x)	1 ml
α-Naphthaleneacetic acid stock solution	1 ml
6-Benzylaminopurine stock solution	25 µl
Sucrose	137 g

Dissolve in 900 ml of H_2O and adjust the pH to 5.8 with KOH. Adjust the volume to 1 liter with H_2O and filter sterilize. Store at room temperature. K_3AS medium is stable for at least 1 month.

B5 vitamins stock (1000x)

(100 ml)

Nicotinic acid	0.1 g
Thiamine-HCl	1 g
Pyridoxine-HCl	1 g
Myo-inositol	10 g

Dissolve in H_2O and then adjust the volume to 100 ml. Filter sterilize. Store frozen at –20°C.

α-*Naphthaleneacetic acid (auxin)* (GIBCO 21570-015)
Stock solution: 2 mg/ml in dimethylsulfoxide. Store at –20°C.

6-Benzylaminopurine (cytokinin) (GIBCO 16105-017)
Stock solution: 20 mg/ml in dimethylsulfoxide. Store at –20°C.

• PREPARATION OF REAGENTS

Luciferase extraction buffer

100 mM potassium phosphate buffer (pH 7.5) containing 1 mM dithiothreitol. To prepare 100 mM potassium phosphate buffer, add 16 ml of 0.2 M monobasic salt and 84 ml of 0.2 M dibasic salt to 100 ml of distilled H_2O. Filter sterilize. Store the phosphate buffer at room temperature. Add dithiothreitol to 1 mM immediately before use.

D-Luciferin

Potassium salt (Analytical Luminescence Laboratory, Inc.; M.W. 318)

Stock solution: 10 mM in H_2O

Store at –20°C in the dark. Immediately before use, prepare a 0.5 mM solution by adding 250 µl of D-luciferin stock solution to 4.75 ml of H_2O. Keep the 0.5 mM solution protected from light and on ice.

MaMg (1x)

Dilute 2x MaMg 1:1 with sterile H_2O. Store at room temperature. The solution is stable for at least 1 month.

MaMg (2x)

(100 ml)

0.8 M Mannitol
30 mM $MgCl_2 \cdot 6H_2O$
0.2% 2-(*N*-Morpholino) ethane sulfonic acid (MES)

Dissolve 14.6 g of mannitol, 0.61 g of $MgCl_2 \cdot 6H_2O$, and 0.2 g of MES in 70 ml of H_2O and adjust the pH to 5.6 with KOH. Adjust the volume to 100 ml with H_2O and filter sterilize. Store at room temperature. The solution is stable for at least 1 month.

Polyethylene glycol solution (PEG 1450)

Dissolve PEG (Polysciences, Inc. 1102; M.W. 1450) in 2x MaMg. Adjust the volume with H_2O to make a 40% PEG solution (w/v) in 1x MaMg. Check the pH after the PEG has dissolved. If it is not between 5.6 and 7.0, adjust with dilute KOH or HCl as necessary. Filter sterilize. Prepare this solution on the day of the transformation. Store at room temperature.

Reaction buffer
(10 ml)

500 mM HEPES buffer (pH 7.8)	1 ml
1 M $MgCl_2$	0.2 ml
20 mM ATP	5 ml
50 mg/ml Bovine serum albumin	0.1 ml

Adjust the volume to 10 ml with H_2O. Prepare fresh immediately before the luciferase assay. Store the 20 mM ATP and 50 mg/ml bovine serum albumin stock solutions at –20°C.

W_5
(1 liter)

154 mM NaCl
125 mM $CaCl_2 \cdot 2H_2O$
5 mM KCl
5 mM Glucose

Dissolve 9 g of NaCl, 18.37 g of $CaCl_2 \cdot 2H_2O$, 0.37 g of KCl, and 0.9 g of glucose in H_2O and then adjust the volume to 1 liter with H_2O. Adjust the pH to 5.8 with 1 M KOH. Filter sterilize. Store at room temperature. The solution is stable for at least 1 month.

• *Section 2*

Functional Analysis of a Plant Transcription Factor Using Transient Expression in Maize Protoplasts

Assays for gene expression based on transient expression of cloned DNA constructs introduced into plant cells by polyethylene glycol treatment (Marcotte et al. 1988, 1989), electroporation (McCarty et al. 1991; Hattori et al. 1992), or particle bombardment (Goff et al. 1990, 1991) can have significant advantages for many applications in plant biology over experimental approaches that require stable transformation and regeneration of plants. Notably, transient expression assays allow analyses that are rapid, reproducible, and high in capacity (i.e., a large number of samples can be analyzed simultaneously). Transient expression systems are very flexible, allowing various gene constructs to be tested in combination and allowing the application of a variety of exogenous treatments (e.g., hormones, etc.). These qualities of speed and flexibility have been particularly valuable for functional analysis of plant transcription factors (Goff et al. 1991; McCarty et al. 1991) and for identification and analysis of *cis* regulatory elements in plant promoters (Huttly and Baulcombe 1989; Marcotte et al. 1989; Guiltinan et al. 1990; Hattori et al. 1992).

This section illustrates the application of transient expression to the study of the interaction between a maize transcriptional activator (VP1) and one of its downstream target genes (*C1*). The *Vp1* gene regulates a variety of genes that are expressed during seed maturation. Many maturation-related genes, including the *Em* gene of wheat (Marcotte et al. 1988) and the *C1* anthocyanin regulatory gene in maize (Hattori et al. 1992), are also regulated by the hormone abscisic acid (ABA). In the developing maize seed, VP1 regulates the anthocyanin pathway by activating transcription of the *C1* gene. *C1* encodes a transcription factor that, in turn, activates structural genes in the anthocyanin biosynthetic pathway. An effector gene construct that allows overexpression of VP1 protein in maize protoplasts was prepared by cloning the *Vp1* cDNA sequence downstream of the cauliflower mosaic virus (CaMV) *35S* promoter (McCarty et al. 1991) (Fig. 1). Expression from this construct is further enhanced by incorporation of the first intron sequence of the maize *Sh1* gene (Vasil et al. 1989) into the 5′ untranslated sequence of the *Vp1* cDNA. To mon-

REPORTER PLASMID

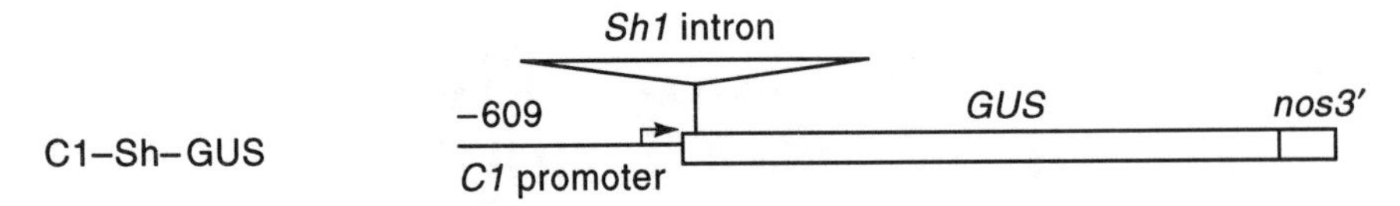

EFFECTOR PLASMID

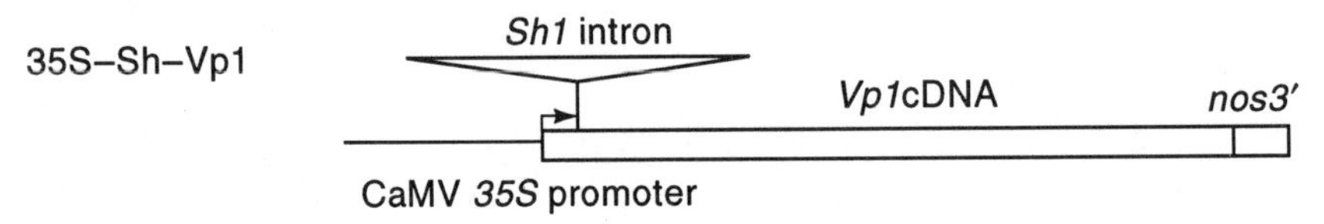

Figure 1
Structure of the effector and reporter gene constructs used in the *trans*-activation experiment.

itor the effect of VP1 overexpression on transcription of a downstream gene, a reporter gene that contains the promoter of the *C1* gene fused to the bacterial β-glucuronidase (*GUS*) gene is used (Fig. 1). The basic experiment is to introduce the C1-Sh-GUS reporter plasmid alone or in combination with the 35S-Sh-Vp1 effector plasmid into protoplasts of maize cell suspension cultures by electroporation. Twenty four hours after electroporation, levels of GUS enzyme activity are determined in protoplast extracts. The expected result is that overexpression of VP1 in cells receiving both plasmids will cause activation of the *C1-Sh-GUS* gene. Therefore, the levels of GUS activity will be higher in these cells than in those receiving only the C1-Sh-GUS plasmid. The influence of ABA on this interaction can be tested by incubating aliquots of each electroporation sample in the presence and absence of 10^{-4} M ABA.

REFERENCES

Chu, C.C., C.C. Wang, C.S. Sun, C. Hsu, K.C. Yin, and C.Y. Chu. 1975. Establishment of an efficient medium for anther culture of rice through comparative experiments on the nitrogen sources. *Sci. Sin.* **16:** 659–688.

Goff, S., K.C. Cone, and M.E. Fromm. 1991. Identification of functional domains in the maize transcriptional activator C1: Comparison of wild-type and dominant inhibitor proteins. *Genes Dev.* **5:** 298–309.

Goff, S., T.M. Klein, B.A. Roth, M.E. Fromm, K.C. Cone, J.P. Radicella, and V.L. Chandler. 1990. Transactivation of anthocyanin biosynthetic genes following transfer of *B* regulatory genes into maize tissues. *EMBO J.* **9:** 2517–2522.

Guiltinan, M.J., W.R. Marcotte, and R.S. Quatrano. 1990. A leucine zipper protein that recognizes an abscisic acid response element. *Science* **250:** 267–270.

Hattori, T., V. Vasil, L. Rosenkrans, L.C. Hannah, D.R. McCarty, and I.K. Vasil. 1992. The *viviparous-1* gene and abscisic acid activate the *C1* regulatory gene for anthocyanin biosynthesis during seed maturation in maize. *Genes Dev.* **6:** 609–618.

Huttly, A.K. and D.C. Baulcombe. 1989. A wheat α-amy2 promoter is regulated by gibberillin in transformed oat aleurone protoplasts. *EMBO J.* **8:** 1907–1913.

Marcotte, W.R., Jr., C.C. Bayley, and R.S. Quatrano. 1988. Regulation of a wheat promoter by abscisic acid in rice protoplasts. *Nature* **335:** 454–457.

Marcotte, W.R., Jr., S.H. Russell, and R.S. Quatrano. 1989. Abscisic acid response sequences from the Em gene of wheat. *Plant Cell* **1:** 523–532.

McCarty, D.R., T. Hattori, C.B. Carson, V. Vasil, and I.K. Vasil. 1991. The *viviparous-1* developmental gene of maize encodes a novel transcriptional activator. *Cell* **66:** 895–905.

Murashige, T. and F. Skoog. 1962. A revised medium for rapid growth and bioassays with tobacco tissue cultures. *Physiol. Plant* **15:** 473–497.

Vasil, V., R.M. Hauptmann, F.M. Morrish, and I. K. Vasil. 1988. Comparative analysis of free DNA delivery and expression into protoplasts of *Panicum maximum* Jacq. (Guinea grass) by electroporation and polyethylene glycol. *Plant Cell Rep.* **7:** 499–503.

Vasil, V., M. Clancy, R.J. Ferl, I.K. Vasil, and L.C. Hannah. 1989. Increased gene expression by the first intron of the maize *sh1* locus in grass species. *Plant Physiol.* **91:** 1575–1579.

• *TIMETABLE*
Transient Expression in Maize Protoplasts

STEPS	*TIME REQUIRED*
Maintenance of Maize Cell Suspension Cultures	
Subculture cells	3–4 days
Preparation of Protoplasts	
Digest cells	4–5 hours
Harvest, wash, and count protoplasts	1 hour
Electroporation	
Electroporate protoplasts	1 hour
Incubate protoplasts for transient expression	24–48 hours
Extract β-glucuronidase	1 hour
Fluorometric β-Glucuronidase Assay	
Assay β-glucuronidase activity	3 hours

Maintenance of Maize Cell Suspension Cultures

The protocol for obtaining protoplasts described in this section was originally developed for a maize cell suspension culture established from immature, embryo-derived type II embryogenic callus (Vasil et al. 1989). Other maize cell suspension cultures can be used, although some adjustment of subculturing interval (for culture growth rate) and digestion time may be desirable for optimal results. The cell line is maintained in modified N6 medium (Chu et al. 1975).

MATERIALS

Maize cell suspension culture (can be obtained from the laboratory of Dr. I.K. Vasil [see Appendix 2])

MEDIA

Modified N6 medium
(For recipe, see Preparation of Media, pp. 31–33)

PROCEDURE

1. Incubate maize cell suspension cultures in the dark on a Gyrotory shaker (120 rpm) at 28°C.

2. Subculture every 4–5 days by transferring 12 ml of the suspension to 25 ml of fresh medium in a 250-ml Erlenmeyer flask.

   Note: Cell transfer is facilitated by the use of 5-ml disposable plastic pipettes with the tips cut off to make a wider opening.

3. To prepare enough protoplasts for 20 electroporation samples, harvest the cells from four flasks 3 days after subculture.

Preparation of Protoplasts

All the steps described below should be performed under aseptic conditions using sterile equipment and reagents. Protoplasts should always be handled in a laminar flow hood.

MATERIALS AND EQUIPMENT

Maize cell suspension culture (see p. 23)
Miracloth (Calbiochem 47855)
Stainless steel filters (100 μm and 50 μm)

REAGENTS

Digestion enzyme solution
Wash solution
Electroporation buffer
(For recipes, see Preparation of Reagents, pp. 34–35)

PROCEDURE

1. Let the maize cell suspension culture settle for 30–40 seconds and transfer 2–3 ml of settled suspension cells into each of 12 disposable petri dishes (15 x 100 mm). Add 16 ml of digestion enzyme solution to each dish and mix. Seal the dishes with Parafilm and incubate them on a Gyrotory shaker at 10–20 rpm for 4–5 hours at room temperature. At the end of the incubation, any partially digested cell clumps can be broken up by passaging them gently through a pipette.

2. Remove debris and undigested cells by passing the protoplast suspensions through a layer of Miracloth and then through 100- and 50-μm stainless steel filters. Transfer the purified protoplasts to 15-ml conical disposable centrifuge tubes and spin in a clinical centrifuge at 70*g* (1000 rpm) for 3 minutes.

3. Decant the supernatants and gently resuspend the protoplast-containing pellets in a small volume of wash solution (~ 1 ml per tube). Combine the suspensions into two centrifuge tubes and wash twice with wash solution using the centrifugation conditions described above.

4. Resuspend the protoplasts in electroporation buffer and adjust the volume in each tube to approximately 11 ml.

5. Dilute aliquots of the resuspensions between 10- and 100-fold with electroporation buffer, and determine the yield and density of the protoplasts using a hemocytometer. Only single spherical protoplasts should be counted. A typical yield from approximately 35 ml of starting material (settled suspension cells) is 9–10 x 10^7 protoplasts. This is sufficient for at least 20 electroporations.

Electroporation

Steps 1–3 described below should be performed under aseptic conditions using sterile equipment and reagents. Protoplasts should always be handled in a laminar flow hood.

MATERIALS AND EQUIPMENT

Maize protoplast suspension (see pp. 24–25)
Electroporation apparatus (e.g., Bio-Rad Gene Pulser® system)
Carborundum or silica grinding powder
Small plastic pestle to fit 1.5-ml microfuge tubes (749515, Kontes Glass, or equivalent)
Drill press
Bradford reagent (Bio-Rad Laboratories)

MEDIA

KM medium containing 0.5 mg of 2,4-dichlorophenoxyacetic acid/liter and 0.5 M glucose
(For recipe, see Preparation of Media, pp. 31–33)

REAGENTS

Plasmid mixtures:
DNA solution A
DNA solution B
Tris/EDTA (pH 8.0) (TE)
Electroporation buffer
Abscisic acid (ABA)
GUS/LUC buffer
(For recipes, see Preparation of Reagents, pp. 34–35)

PROCEDURE

1. Add 0.8 ml of protoplast suspension (i.e., ~ 4 x 10^6 protoplasts) to each of eight sterile microfuge tubes. Add 40 μl of premixed DNA solution A to three tubes (i.e., 20 μg of C1-Sh-GUS reporter plasmid/tube and 20 μg of 35S-CAT plasmid/tube). For the plus effec-

tor treatment, add 40 μl of DNA solution B to three tubes (i.e., 20 μg of C1-Sh-GUS reporter plasmid/tube and 20 μg of 35S-Sh-Vp1 effector plasmid/tube). For control samples, add 50 μl of TE to the two remaining tubes. Bring the total volume in each tube up to 1 ml by adding electroporation buffer. Mix the tubes gently by inverting them several times with the caps closed and then place them on ice for 10 minutes.

2. Adjust the electroporation apparatus to produce an output of 200 V with a capacitance of 950 μFd. Pipette the chilled protoplast samples into electroporation cuvettes, place them individually in the cuvette holder, and electroporate. Transfer the samples back into microfuge tubes and place them on ice for 10 minutes.

   Note: The exact settings that are available will depend upon the instrument model used. The above values pertain to a Bio-Rad Gene Pulser® system. Equally suitable electroporation equipment is available from other manufacturers. Be sure the instrument used is capable of high capacitance (800–1200 μFd) and medium voltage (150–400 V) output and is equipped with 0.4 cm cuvettes.

3. Following electroporation, culture the protoplasts in liquid KM medium containing 0.5 mg of 2,4-dichlorophenoxyacetic acid/liter and 0.5 M glucose. Transfer approximately 0.5 ml of protoplasts from each sample into petri dishes (15 x 100 mm) containing 4.5 ml of the KM medium.

   Note: For hormone treatments, divide each electroporated sample equally between two dishes, one containing medium without ABA and the other containing medium supplemented with 10^{-4} M ABA.

   Seal the dishes with Parafilm and incubate overnight at 28°C in the dark. *Trans*-activation of the *C1-Sh-GUS* gene can first be detected about 8 hours after electroporation and increases linearly for 40 hours. The condition of the protoplasts during the incubation period can be monitored under an inverted microscope, if one is available. After 40 hours, most of the surviving protoplasts have re-formed cell walls and appear oblong.

4. Transfer the cells from each dish into disposable conical centrifuge tubes and place them on ice. Pellet the cells by spinning in a clinical centrifuge at the lowest speed setting for 2 minutes. Discard the supernatant and the wash the cells with 300 μl of GUS/LUC buffer. Transfer the cells to 1.5-ml microfuge tubes, and spin in a microfuge for 5 minutes.

5. Resuspend each pellet in 200 μl of GUS/LUC buffer and add a small amount of carborundum or silica grinding powder (5–10 mg) to each tube. Grind each cell sample individually for 30 seconds using a small plastic pestle attached to a drill press and

vortex vigorously for an additional 30 seconds. To remove the grinding powder and insoluble cell debris, spin the ground samples in a microfuge for 10 minutes. Transfer the supernatant fractions to fresh microfuge tubes.

6. Determine the protein concentrations in the supernatant fractions using a micro-Bradford assay. Up to 20 μl of supernatant is added to the dye reagent for the protein determination. Typical protein concentrations are in the range of 0.3–1.5 mg of protein/ml. Lower values may indicate poor protoplast survival.

Fluorometric β-Glucuronidase (GUS) Assay

MATERIALS AND EQUIPMENT

Cell extract (see p. 27, step 5)
Fluorometer

REAGENTS

4-Methylumbelliferyl β-D-glucuronide (MUG) solution
Stop buffer
4-Methylumbelliferone (MU) stock solution
(For recipes, see Preparation of Reagents, pp. 34–35)

PROCEDURE

1. For each assay, preincubate 112.5 μl of MUG solution in a microfuge tube in a water bath for 5 minutes at 37°C. To initiate the assay, add 75 μl of cell extract, vortex briefly, and return to the water bath. The total volume allows 50-μl samples to be removed at three time points. Initiate the individual assays at 20-second intervals to allow convenient handling.

2. After 15 minutes, remove 50 μl from each assay tube, transfer to a test tube containing 1 ml of stop buffer, and vortex briefly. Repeat this step for each sample at 45-minute and 90-minute time points. The progress of the reaction can be qualitatively assessed by viewing the stopped reactions under ultraviolet (UV) longwave (360 nm) illumination. If the initial GUS activities appear to be low, the second or third time point can be delayed to allow production of more product and increase sensitivity. Reaction rates are linear for several hours.

3. Quantify the concentration of MU produced in the assays in a fluorometer using an excitation wavelength of 350 nm and an emission wavelength of 455 nm. Generate a standard curve by determining the fluorescence of a series of MU standards in the concentration range 0–10 μM prepared by diluting MU stock solu-

tion in stop buffer. It is generally necessary to dilute the assay samples to bring them into this range. The dilution factor may vary widely among treatments and with different protoplast preparations. This may require that several dilutions be made from each sample. However, with experience, it is usually possible to estimate the appropriate dilution factor based on visual assessment of sample fluorescence under a handheld UV source. If GUS activities are extremely high and large dilutions are needed, it may be desirable to reassay the samples at a lower protein concentration to insure linearity of the assay.

4. The concentration of product in the original undiluted reaction is calculated from the fluorescence of each diluted sample. For each sample, the value of the blank (samples electroporated with no DNA) is subtracted. A reaction rate can then be determined from the various time points. Typical results are shown below.

	C1-Sh-GUS	**35S-Sh-Vp1**	**Abscisic acid**	**GUS activity (μmol MU/ minute/mg of protein)**
1	+	–	–	0.26
2	+	–	+	2.75
3	+	+	–	1.56
4	+	+	+	2.82

From Hattori et al. (1992).

• PREPARATION OF MEDIA

KM medium

(as modified in Vasil et al. 1988)

Salts	mg/l
NH_4NO_3	600
KNO_3	1900
KH_2PO_4	170
$MgSO_4 \cdot 7H_2O$	300
$CaCl_2 \cdot 2H_2O$	600
$MnSO_4 \cdot 4H_2O$	10
$ZnSO_4 \cdot 7H_2O$	2.0
$Na_2MoO_4 \cdot 2H_2O$	0.25
$CuSO_4 \cdot 5H_2O$	0.025
KCl	300
H_3BO_3	3.0
KI	0.75
Sequestrene™ 330Fe (Ciba-Geigy)	28

Vitamins	mg/l
Thiamine-HCl	10
Riboflavin	0.2
Ascorbic acid	2.0
Folic acid	0.4
Nicotinamide	1.0
Pyridoxine-HCl	1.0
Inositol	100
Biotin	0.01
Choline chloride	1.0
D-Ca pantothenate	1.0
p-Aminobenzoic acid	0.02
Vitamin A	0.01
Vitamin D_3	0.01
Vitamin B_{12}	0.02

Organic acids (pH 5.5)	mg/l
Sodium pyruvate	20
Malic acid	40
Citric acid	40
Fumaric acid	40

Sugars/sugar alcohol	mg/l
Fructose	250
Ribose	250
Xylose	250
Mannose	250
Rhamnose	250
Cellobiose	250
Sorbitol	250
Mannitol	250

Other	
Casamino acid	250 mg/l
Coconut water	20 ml/l

Premixed KM medium is available commercially from Sigma and Life Technologies, Inc. Add glucose, 2,4-dichlorophenoxyacetic acid, and MES buffer (pH 5.8) to final concentrations of 0.5 M, 0.5 mg/liter, and 3.0 mM, respectively. Autoclave.

Modified N6 medium
(1 liter)

N6-A stock solution (10x)	20 ml
N6-B stock solution (200x)	5 ml
N6-C stock solution (200x)	5 ml
MS-F stock solution (200x)	5 ml
MS-G stock solution (200x)	5 ml
Sucrose	20 g
Casein hydrolysate	100 g
Proline	402 mg
2,4-dichlorophenoxyacetic acid (GIBCO 11215-019)	1.25 mg

Dissolve in H_2O and then adjust the volume to 1 liter with H_2O. Adjust the pH to 5.8 and autoclave. Premixed N6 medium is available commercially from Sigma and Life Technologies, Inc. The recipes for the stock solutions required to prepare N6 medium are described below (see also Chu et al. 1975).

N6-A stock solution (10x)
(1 liter)

$(NH_4)_2SO_4$	23.15 g
KNO_3	141.5 g
KH_2PO_4	20 g
$MgSO_4 \cdot 7H_2O$	9.25 g

Dissolve in H_2O and then adjust the volume to 1 liter with H_2O.

N6-B stock solution (200x)
(1 liter)

Dissolve 8.3 g of $CaCl_2 \cdot 2H_2O$ in H_2O and then adjust the volume to 1 liter with H_2O.

N6-C stock solution (200x)
(1 liter)

$MnSO_4 \cdot 4H_2O$	0.88 g
$ZnSO_4 \cdot 7H_2O$	0.3 g
H_3BO_3	0.32 g
KI	0.16 g

Dissolve in H_2O and then adjust the volume to 1 liter with H_2O.

MS-F stock solution (200x)
(1 liter)

Na_2EDTA	7.45 g
$FeSO_4 \cdot 7H_2O$	5.57 g

Dissolve in H_2O and then adjust the volume to 1 liter with H_2O (Murashige and Skoog 1962). Premixed MS-F medium is available commercially from Sigma and Life Technologies, Inc.

MS-G stock solution (200x)
(1 liter)

Thiamine-HCl	0.02 g
Nicotinic acid	0.1 g
Pyridoxine-HCl	0.1 g
Glycine	0.4 g
Inositol	20 g

Dissolve in H_2O and then adjust the volume to 1 liter with H_2O (Murashige and Skoog 1962). Premixed MS-G medium is available commercially from Sigma and Life Technologies, Inc.

• *PREPARATION OF REAGENTS*

Abscisic acid (ABA)
(Sigma)

Stock solution: 10 mM in H_2O. Store at –20°C.

Digestion enzyme solution

Cellulase RS (Yakult Pharmaceutical Co.)	2 g
Pectinase (Serva)	2 g
Pectolyase Y23 (Seishin Pharmaceutical Co.)	0.2 g

Dissolve in 200 ml of wash solution (for preparation, see p. 35) and stir for 30 minutes. Any insoluble residue can be removed by spinning at 8000 rpm for 8 minutes. Filter sterilize using a disposable 500-ml, 0.2-μm vacuum filter flask. The sterilized digestion enzyme solution can be stored for up to 1 week at 4°C.

Electroporation buffer

10 mM HEPES buffer (pH 7.2)
150 mM NaCl
5 mM $CaCl_2 \cdot 2H_2O$
0.2 M Mannitol

Autoclave. Store at –20°C.

GUS/LUC buffer

0.1 M KPO_4 (pH 7.8)
2 mM Na_2EDTA
2 mM Dithiothreitol
5% Glycerol

Store at –20°C.

4-Methylumbelliferone (MU)

Stock solution: 100 μM in stop buffer (for preparation, see p. 35)

Prepare a series of MU standards in the range 0–10 μM by diluting MU stock solution in stop buffer. MU is available commercially from Sigma.

4-Methylumbelliferyl β-D-glucuronide (MUG) solution

1 mM prepared in GUS/LUC buffer. MUG is available commercially from Sigma.

Plasmid mixtures

DNA Solution A:
0.5 μg/μl C1-Sh-GUS (Hattori et al. 1992)
0.5 μg/μl 35S-CAT

DNA Solution B:
0.5 μg/μl C1-Sh-GUS
0.5 μg/μl 35S-Sh-Vp1 (McCarty et al. 1991)

Precipitate the plasmid mixtures by adding one tenth volume of 3.0 M sodium acetate (pH 5.5) and two volumes of ethanol. Dry the ethanol-rinsed DNA pellets in a vacuum desiccator. Place the chamber under a laminar flow hood prior to release of the vacuum to maintain sterility and resuspend the DNA in the appropriate volume of a sterile solution of 10 mM Tris-HCl (pH 8.0)/2 mM EDTA. The 35S-CAT plasmid is included to keep the total concentration of DNA and, perhaps more importantly, the concentration of the 35S promoter constant in the experiment. Another neutral plasmid containing the 35S promoter may be substituted. Alternatively, we have found that this "dummy" plasmid can be omitted without significantly affecting the results. The plasmids used here can be obtained from the laboratory of Dr. D.R. McCarty (see Appendix 2).

Stop buffer

0.2 M Na_2CO_3

Tris/EDTA (TE)

10 mM Tris-HCl (pH 8.0)
2 mM EDTA

Wash solution

0.4 M Mannitol
0.7 mM $NaPO_4 \cdot H_2O$
5 mM $CaCl_2$
3 mM MES buffer (pH 5.8)

Autoclave.

• *Section 3*

Biolistic Transformation of Tobacco Cells with Nuclear Drug Resistance Genes

The introduction of DNA into plant cells has been achieved by a variety of techniques, including biolistic transformation, polyethylene glycol treatment, microinjection, electroporation, and the use of *Agrobacterium*, plant viruses, and liposomes as DNA carriers (Potrykus 1991; Vasil 1994). In the biolistic process, transforming DNA on the surface of tungsten or gold particles is introduced into plant cells by means of a particle acceleration gun. Both the concept and the gunpowder-charge acceleration gun were first described by John Sanford and collaborators. The early history of the technology is reviewed in Sanford (1988). The more advanced du Pont helium acceleration PDS-1000/He gun is now available from Bio-Rad Laboratories (Sanford et al. 1991). The operation of the helium gun is shown in Figure 1. Biolistic delivery has proved particularly useful for testing transient gene expression in differentiated tissues (Klein et al. 1989), for transforming embryogenic maize (Gordon-Kamm et al. 1990) and wheat cultures (Vasil et al. 1992), and for transforming the genomes of cytoplasmic organelles such as plastids in *Chlamydomonas* (Boynton et al. 1988) and tobacco (Svab et al. 1990a; Svab and Maliga 1993) and mitochondria in yeast (Johnston et al. 1988) and *Chlamydomonas* (Randolph-Anderson et al. 1992). For a review of particle bombardment technology and its applications, see Klein and Fitzpatrick-McElligott (1993), Sanford et al. (1993), and Yang and Christou (1994).

This section describes a protocol for the transformation of tobacco (*Nicotiana tabacum*) cells by bombardment with DNA-coated tungsten particles. The tungsten particles are coated with DNA containing either a chimeric *uidA* gene that encodes β-glucuronidase (GUS) or a chimeric *kan* gene that encodes neomycin phosphotransferase, which confers resistance to kanamycin by converting the drug to an inactive form. The *uidA* and *kan* genes are carried by plasmids pFF19G and pFF19K, respectively. Both genes are expressed in the pFF19 cassette utilizing the cauliflower mosaic virus *35S* gene expression signals (Timmermans et al. 1990). The procedure for coating tungsten particles with DNA is adapted from Gordon-Kamm et al. (1990).

The efficiency of biolistic DNA delivery is assessed by monitoring the transient expression of the *uidA* gene. GUS activity is detected by

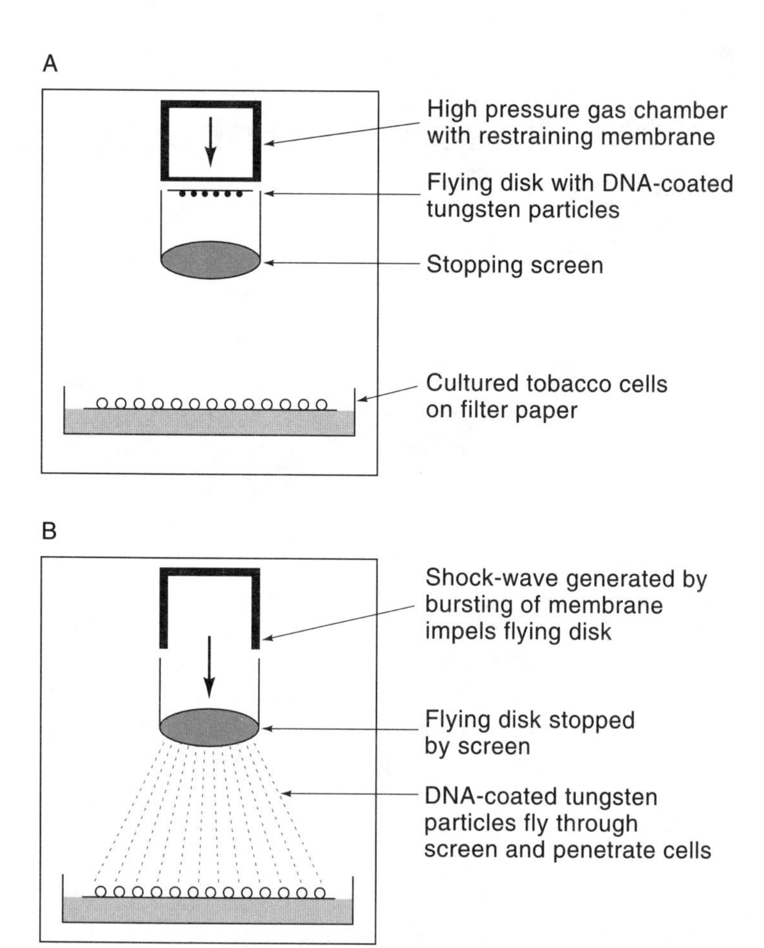

Figure 1
Operation of the helium acceleration PDS-1000/He biolistic gun. (*A*) Helium pressure (1100 psi) builds up behind the restraining membrane (rupture disk). (*B*) The restraining membrane breaks, impelling the flying disk. Membranes of different strengths are available. (This diagram is not drawn to scale.)

histochemical staining 24 hours after the cells are bombarded with pFF19G DNA (Jefferson 1987; Gallagher 1992). Most of the GUS activity is due to expression of non-integrated plasmid DNA.

The number of cells that stably integrate pFF19K DNA introduced by the biolistic process is determined by plating the cells on selective medium 48 hours after bombardment and screening for the growth of kanamycin-resistant colonies 2–8 weeks later (Klein et al. 1988). The number of stable transformants obtained is only a few percent relative to the number of cells that transiently express GUS. A suggested experimental design and the results you can expect to obtain are shown in Table 1. This protocol can be adapted to select for

Table 1 **Transient Gene Expression and Stable Transformation in *Nicotiana tabacum* Cells Following Bombardment with DNA-coated Tungsten Particles**

Plate Number	Plasmid	Gene Introduced	Stain for GUS	Plate on medium[c] Kan_0	Plate on medium[c] Kan_{50}	Experiment (expected results)
1	–	–	–	+	–	to test viability of cells (continuous lawn of calli)
2	–	–	–	–	+	to test kan medium (no kan-resistant calli)
3, 4, 5	pFF19G	*uidA*[a]	+	–	–	to test efficiency of DNA delivery (500–3000 transient GUS-expressing units)
6, 7, 8	pFF19K	*kan*[b]	–	–	+	to select for stably transformed, kan-resistant clones (5–20 stable kan-resistant clones)

[a]*uidA* encodes β-glucuronidase (GUS).
[b]*kan* encodes neomycin phosphotransferase and confers resistance to kanamycin (kan).
[c]Kan_0 and Kan_{50} media contain 0 μg and 50 μg of kanamycin sulfate/ml, respectively.

any of the drug resistance markers reported for plants (see Table 2) using appropriate drug concentrations. Transgenic tobacco shoots can be obtained directly from bombarded leaf sections by culturing the

Table 2 **Selectable Marker Genes for Transformation in Plants**

Selection	References
Bleomycin	Hille at al. (1986); Perez et al. (1989)
Gentamycin	Hayford et al. (1988); Carrer et al. (1991)
Glyphosate	Comai et al. (1985); Shah et al. (1986)
Hygromycin	Van den Elzen et al. (1985); Waldron et al. (1985)
Kanamycin	Bevan et al. (1983); Fraley et al. (1983); Herrera-Estrella et al. (1983); Carrer et al. (1993)
Methotrexate	Eicholtz et al. (1987)
Phleomycin	Perez et al. (1989)
Phosphinotricin	De Block et al. (1987); Wohlleben et al. (1988)
Spectinomycin	Svab et al. (1990b); Svab and Maliga (1993)
Streptomycin	Maliga et al. (1988)
Sulfonamide	Guerineau et al. (1990)
Sulfonylureas	Haughn et al. (1988); Lee et al. (1988)

sections on selective, shoot-inducing RMOP medium (Klein et al. 1988).

Experiments with transgenic plants are subject to state and federal laws in the United States and in most other countries. For information concerning legal constraints on research involving recombinant DNA molecules, see Appendix 1; also consult the safety office at your institution.

REFERENCES

Bevan, M.W., R.B. Flavell, and M.D. Chilton. 1983. A chimeric antibiotic resistance gene as a selectable marker for plant cell transformation. *Nature* **304:** 184–187.

Boynton, J.E., N.W. Gillham, E.H. Harris, J.P. Hosler, A.M. Johnson, A.R. Jones, B.L. Randolph-Anderson, D. Robertson, T.M. Klein, K.B. Shark, and J.C. Sanford. 1988. Chloroplast transformation in *Chlamydomonas* with high velocity microprojectiles. *Science* **240:** 1534–1537.

Carrer, H., J.M. Staub, and P. Maliga. 1991. Gentamycin resistance in *Nicotiana* conferred by AAC(3)-I, a narrow substrate specificity acetyltransferase. *Plant Mol. Biol.* **17:** 301–303.

Carrer, H., T.N. Hockenberry, Z. Svab, and P. Maliga. 1993. Kanamycin resistance as a selectable marker for plastid transformation in tobacco. *Mol. Gen. Genet.* **241:** 49–56.

Comai, L., D. Facciotti, W.R. Hiatt, G. Thompson, E.R. Rose, and D.M. Stalker. 1985. Expression in plants of a mutant *aroA* gene from *Salmonella typhimurium* confers tolerance to glyphosate. *Nature* **317:** 741–744.

De Block, M., J. Botterman, M. Vandewiele, J. Dockx, C. Thoen, V. Gossele, N.R. Movva, C. Thompson, M. Van Montagu, and J. Leemans. 1987. Engineering herbicide resistance in plants by expression of a detoxifying enzyme. *EMBO J.* **6:** 2513-2518.

Eicholtz, D.A., S.G. Rogers, R.B. Horsch, H.J. Klee, M. Hayford, N.L. Hoffmann, S.B. Braford, C. Funk, J. Flick, K.M. O'Connell, and R.T. Fraley. 1987. Expression of mouse dihydrofolate reductase gene confers methotrexate resistance in transgenic *Petunia* plants. *Somatic Cell Mol. Genet.* **13:** 67–76.

Fraley, R.T., S.G. Rogers, R.B. Horsch, P.R. Sanders, J.S. Flick, S.P. Adams, M.L. Bittner, L.A. Brand, C.L. Fink, Y.S. Fry, G.R. Galluppi, S.B. Goldberg, N.L. Hoffmann, and S.C. Woo. 1983. Expression of bacterial genes in plant cells. *Proc. Natl. Acad. Sci.* **80:** 4803–4807.

Gallagher, S.R., ed. 1992. GUS *protocols*. Academic Press, San Diego.

Gordon-Kamm, W.J., T.M. Spencer, M.L. Mangano, T.R. Adams, R.J. Daines, W.G. Start, J.V. O'Brien, S.A. Chambers, W.R. Adams, N.G. Willetts, T.B. Rice, C.J. Mackey, R.W. Krueger, A.P. Kausch, and P.G. Lemaux. 1990. Transformation of maize cells and regeneration of fertile transgenic plants. *Plant Cell* **2:** 603–618.

Guerineau, F., L. Brooks, J. Meadows, A. Lucy, C. Robinson, and P. Mullineaux. 1990. Sulfonamide resistance gene for plant transformation. *Plant. Mol. Biol.* **15:** 127–136.

Haughn, G.W., J. Smith, B. Mazur, and C. Somerville. 1988. Transformation with a mutant *Arabidopsis* acetolactate synthase gene renders tobacco resistant to sulfonylurea herbicides. *Mol. Gen. Genet.* **211:** 266–271.

Hayford, M.B., J.I. Medford, N.L. Hoffman, S.G. Rogers, and H.J. Klee. 1988. Development of a plant transformation selection system based on expression of genes encoding gentamycin acetyltransferases. *Plant Physiol.* **86:** 1216–1222.

Herrera-Estrella, L., M. De Block, E. Messens, J.P. Hernalstens, M. Van Montagu, and J. Schell. 1983. Chimeric genes as dominant selectable markers in plant cells. *EMBO J.* **2:** 987–995.

Hille, J., F. Verheggen, P. Roelvink, H. Franssen, A. van Kammen, and P. Zabel. 1986. Bleomycin resistance, a new dominant selectable marker for plant cell transformation. *Plant Mol. Biol.* **7:** 171–176.

Jefferson, R.A. 1987. Assaying chimeric genes in plants: The *GUS* gene fusion system. *Plant Mol. Biol. Rep.* **5:** 387–405.

Johnston, S.A., R. Butow, K. Shark, and J.C. Sanford. 1988. Transformation of yeast mitochondria by bombardment of cells with microprojectiles. *Science* **240:** 1538–1541.

Klein, T.M. and S. Fitzpatrick-McElligott. 1993. Particle bombardment: A universal approach for gene transfer to cells and tissues. *Curr. Opin. Biotechnol.* **4:** 583–590.

Klein, T.M, B.A. Roth, and M.E. Fromm. 1989. Regulation of anthocyanin biosynthetic genes introduced into intact maize tissues by microprojectiles. *Proc. Natl. Acad. Sci.* **86:** 6681–6685.

Klein, T.M., E.C. Harper, Z. Svab, J.C. Sanford, M.E. Fromm, and P. Maliga. 1988. Stable genetic transformation of intact *Nicotiana* cells by the particle bombardment process. *Proc. Natl. Acad. Sci.* **85:** 8502–8505.

Lee, K.Y., J. Townsend, J. Tepperman, M. Black, C.F. Chi, B. Mazur, P. Dunsmuir, and J. Bedbrook. 1988. The molecular basis of sulfonylurea herbicide resistance in tobacco. *EMBO J.* **7:** 1241–1248.

Maliga, P., Z. Svab, E.C. Harper, J.D.G. Jones. 1988. Improved expression of streptomycin resistance in plants due to a deletion in the streptomycin phosphotransferase coding sequence. *Mol. Gen. Genet.* **214:** 456–459.

Perez, P., G. Tirbay, J. Kallerhoff, and J. Perret. 1989. Phleomycin resistance as a dominant selectable marker for plant cell transformation. *Plant Mol. Biol.* **13:** 365–373.

Potrykus, I. 1991. Gene transfer to plants: Assessment of published approaches and results. *Annu. Rev. Plant Physiol. Plant Mol. Biol.* **42:** 205–225.

Randolph-Anderson, B.L., J.E. Boynton, N.W. Gillham, E.H. Harris, A.M. Johnson, M.P. Dorthu, and R.F. Matagne. 1992. Further characterization of the respiratory deficient *dum-1* mutation of *Chlamydomonas reinhardtii* and its use as a recipient for mitochondrial transformation. *Mol. Gen. Genet.* **236:** 235–244.

Sanford, J.C. 1988. The biolistic process. *Trends Biotechnol.* **6:** 299–302.

Sanford, J.C., F.D. Smith, and J.A. Russel. 1993. Optimizing the biolistic process. *Methods Enzymol.* **217:** 483–509.

Sanford, J.C., M.J. DeVit, J.A. Russel, F.D. Smith, P.R. Harpending, M.K. Roy, and S.A. Johnston. 1991. An improved, helium-driven biolistic device. *Technique* **3:** 3–16.

Shah, D.M., R.B. Horsch, H.J. Klee, G.M. Kishore, J.A. Winter, N.E. Tumer, C.M. Hironaka, P.R. Sanders, C.S. Gasser, S. Aykent, N.R. Siegel, S.G. Rogers, and R.T. Fraley. 1986. Engineering herbicide tolerance in transgenic plants. *Science* **233:** 478–481.

Svab, Z. and P. Maliga. 1993. High-frequency plastid transformation in tobacco by selection for a chimeric *aadA* gene. *Proc. Natl. Acad. Sci.* **90:** 913–917.

Svab, Z., P. Hajdukiewicz, and P. Maliga. 1990a. Stable transformation of plastids in higher plants. *Proc. Natl. Acad. Sci.* **87:** 8526–8530.

Svab, Z., E.C. Harper, J.D.G. Jones, and P. Maliga. 1990b. Aminoglycoside 3′ adenyltransferase confers resistance to spectinomycin and streptomycin in *Nicotiana tabacum. Plant Mol. Biol.* **14:** 197–205.

Timmermans, M.C.P., P. Maliga, J. Vieira, and J. Messing. 1990. The pFF plasmids: Casettes utilising CaMV sequences for expression of foreign genes in plants. *J. Biotechnol.* **14:** 333–344.

Van den Elzen, P.J.M., J. Townsend, K.Y. Lee, and J.R. Bedbrook. 1985. A chimeric hygromycin resistance gene as a selectable marker in plant cells. *Plant. Mol. Biol.* **5:** 299–302.

Vasil, I.K. 1994. Molecular improvement of cereals. *Plant Mol. Biol.* **25:** 925–937.

Vasil, V., A.M. Castillo, M.E. Fromm, and I.K. Vasil. 1992. Herbicide resistant fertile transgenic wheat plants obtained by microprojectile bombardment of regenerable embryogenic callus. *Biotechnology* **10:** 667–674.

Waldron, C., E.B. Murphy, J.L. Roberts, G.D. Gustafson, S.L. Armour, and S.K. Malcolm. 1985. Resistance to hygromycin B. A new marker for plant transformation. *Plant Mol. Biol.* **5:** 103–108.

Wohlleben, W., W. Arnold, I. Broer, D. Hillemann, E. Strauch, and A. Puhler. 1988. Nucleotide sequence of the phosphinotricin *N* acetyltransferase gene from *Streptomyces viridochromogenes* Tu494 and its expression in *Nicotiana tabacum*. *Gene* **70:** 25–37.

Yang, N.S. and P. Christou, eds. 1994. *Particle bombardment technology for gene transfer.* Oxford University Press, England.

• *TIMETABLE*
Biolistic Transformation of Tobacco Cells

STEPS	***TIME REQUIRED***
Preparation of Tobacco XD Cells for Transformation	
Dilute suspension culture cells and incubate	3 days
Cover eight filter paper disks with suspension culture cells and transfer disks with cells onto RMS medium	1 hour
Coating of Tungsten Particles with DNA	
Prepare tungsten particles coated with pFF19G and pFF19K plasmid DNA	2 hours
Introduction of DNA into Tobacco XD Cells with the PDS-1000/He Particle Gun	
Bombard eight samples	1 hour
Incubate cells bombarded with pFF19G plasmid DNA	24 hours
Incubate cells bombarded with pFF19K plasmid DNA on nonselective medium	48 hours
Assay of Transient β-Glucuronidase (GUS) Expression	
Stain for GUS	15 minutes
Monitor development of blue color	2–12 hours
Selection of Kanamycin-resistant Lines	
Plate cells on kanamycin-containing medium	2 hours
Incubate cells until all kanamycin-resistant calli appear	8 weeks

Preparation of Tobacco XD Cells for Transformation

MATERIALS

Tobacco (*Nicotiana tabacum*) XD cells (Klein et al. 1988), or other fast-growing cell culture line

MEDIA

Liquid:
RMS medium
Plates:
RMS medium solidified with 0.7% agar
(For recipe, see Preparation of Media, pp. 52–53)

PROCEDURE

1. Filter 5 ml of tobacco XD suspension culture cells onto each of eight sterile Whatman No. 4 filter paper disks (5.5 cm in diameter) using a Buchner funnel and vacuum. Cover the disks uniformly with cells.

   Note: The tobacco XD suspension cultures are maintained in 250-ml Erlenmeyer flasks containing 50 ml of RMS medium and agitated on a Gyrotory shaker at 180 rpm under continuous light (1000 lux) at 28°C. The cells are subcultured approximately every 7 days by a 5–10-fold dilution. The cells are harvested for transformation 3–5 days after subculture.

2. Transfer the filter paper disks with cells onto RMS medium solidified with agar. The cells can be stored like this indefinitely. However, it is practical to transform the cells a few hours.

Coating of Tungsten Particles with DNA

MATERIALS

Sterile distilled H_2O
pFF19G plasmid stock solution (pFF19G plasmid DNA carries the *uidA* gene; Timmermans et al. 1990) (1 μg/ml in 10 mM Tris-HCl/ 1 mM EDTA [pH 8.0])
pFF19K plasmid stock solution (pFF19K plasmid DNA carries the *kan* gene; Timmermans et al. 1990) (1 μg/ml in 10 mM Tris-HCl/ 1 mM EDTA [pH 8.0])
$CaCl_2$ (2.5 M; filter sterilized)
Spermidine free base (0.1 M; filter sterilized) (Sigma S 0266)
Ethanol (100%)

REAGENTS

Tungsten suspension
(For recipe, see Preparation of Reagents, p. 54)

PROCEDURE

1. Place 0.5 ml of the tungsten suspension in a microfuge tube and sediment the tungsten particles by spinning in a microfuge at 14,000 rpm for 5 seconds. Decant the ethanol, and wash the tungsten particles in 0.5 ml of sterile distilled H_2O.

2. Repeat the wash three times.

3. Resuspend the tungsten particles in 0.5 ml of sterile distilled H_2O.

4. Transfer 50-μl aliquots of the aqueous tungsten suspension into three microfuge tubes labeled A, B, and C. (Only 150 μl of the aqueous tungsten suspension is used in this experiment.)

5. Add 10 μl of pFF19G and pFF19K plasmid stock solutions to tubes B and C, respectively. Tube A is the minus-plasmid control.

6. Add 50 µl of 2.5 M $CaCl_2$ and 20 µl of 0.1 M spermidine free base to each tube.

7. Shake or vortex each suspension for 20 minutes in a cold room at 4°C.

8. Add 200 µl of 100% ethanol to each tube. Spin in a microfuge at 8000 rpm for 2 seconds at room temperature.

9. Rinse each sample of DNA-coated tungsten particles four times in 100% ethanol. Sediment the particles by spinning briefly as in step 8.

10. Resuspend each sample of DNA-coated and control tungsten particles in 30 µl of 100% ethanol. Store the samples on ice and use within 2 hours.

Introduction of DNA into Tobacco XD Cells with the PDS-1000/He Particle Gun

MATERIALS AND EQUIPMENT

PDS-1000/He particle gun (Bio-Rad Laboratories)
Flying disks (Bio-Rad Laboratories)
Rupture disks (restraining membranes) (1100 psi; Bio-Rad Laboratories)
Ethanol (70% and 100%)
Tungsten particle suspensions (see p. 46, step 10)
  A — non-coated control
  B — coated with pFF19G plasmid DNA
  C — coated with pFF19K plasmid DNA
Anhydrous $CaSO_4$ (optional; see note to step 3 below)
Eight petri dishes containing tobacco XD cells on filter paper disks (see p. 44)

PROCEDURE

1. Sterilize the gun chamber and the gas acceleration tube with 70% ethanol.

   Note: The operation of the PDS-1000/He particle gun is shown in Figure 1 (p. 38).

2. Sterilize the rupture disks and the flying disks by soaking them in 100% ethanol for 5 minutes. Air dry in a laminar flow hood.

3. Pipette 5 µl-aliquots of the tungsten particle suspensions onto flying disks (placed in the holder) and air dry in a laminar flow hood for 5 minutes. Prepare two disks (5 µl/disk) from the control tungsten particle suspension in tube A (for plates 1 and 2) (see Table 1, p. 39). Prepare three disks each (5 µl/disk) from the DNA-coated tungsten particle suspensions in tubes B and C (for plates 3–5 and 6–8, respectively). Depending on the design of your experiment, five disks can be prepared from each suspension at one time.

   Note: The flying disks and the DNA-coated tungsten particles must be completely dry for efficient DNA delivery. To reduce moisture, samples can be stored in a small box above anhydrous $CaSO_4$ if the air humidity is above 50%.

4. Turn on the helium tank. Set the delivery pressure in the regulator (distal to the tank) to 1300 psi (i.e., 200–300 psi above the rupture disk value).

5. Turn on the vacuum pump.

6. Place a rupture disk in the retaining cap and screw tightly into place.

7. Put a flying disk in place (face down), and place the assembly in the chamber at level two from the top.

8. Place a petri dish containing tobacco XD cells on the sample holder inserted into the chamber at level five. Close the door.

9. Open the vacuum valve. When the vacuum reaches 25–28 inches of Hg, press the fire button. Keep pressing until you hear the noise of the gas breaking the rupture disk.

10. Release the vacuum. Remove the sample.

11. Repeat from step 6 for continuous operation.

    Note: When you have finished using the particle gun, turn off the helium tank, and release the pressure by holding down the fire button while the vacuum is still on.

12. Incubate the cells bombarded with non-coated tungsten (plates 1 and 2) on nonselective medium in the dark for 48 hours at 28°C.

13. Incubate the cells bombarded with pFF19G plasmid DNA (plates 3, 4, and 5) in the dark for 24 hours at 28°C.

14. Incubate the cells bombarded with pFF19K plasmid DNA (plates 6, 7, and 8) on nonselective medium in the dark for 48 hours at 28°C.

Assay of Transient β-Glucuronidase (GUS) Expression

MATERIALS

Plates 3, 4, and 5 (see p. 48, step 13)

REAGENTS

X-Gluc solution
(For recipe, see Preparation of Reagents, p. 54)

PROCEDURE

1. Twenty-four hours after the introduction of pFF19G plasmid DNA into tobacco XD cells, transfer the filter paper disks from plates 3, 4, and 5 into sterile petri dishes.

2. Assay GUS expression in these cells by histochemical staining with X-Gluc solution. Pipette 500 μl of X-Gluc solution one drop at a time onto each filter so that all of the cells are covered with the stain.

3. Seal the petri dishes with Parafilm and incubate them overnight at 37°C. Blue spots are visible within 2–3 hours. The filter paper disks with stained cells can be stored in a refrigerator for at least 2 weeks.

4. Examine the cells under a microscope and determine the number of blue, GUS-expressing units. One unit is an individual blue cell, or a group of adjacent blue cells. Individual GUS-expressing cells are identifiable under a dissecting microscope. Accurate counting is facilitated by laying a transparent disk with a grid over the cells. A typical bombardment yields at least 500–3000 GUS-expressing units.

Selection of Kanamycin-resistant Lines

MATERIALS

Plates 1 and 2 (see p. 48, step 12)
Plates 6, 7, and 8 (see p. 48, step 14)
Kanamycin sulfate stock solution (50 mg/ml in H_2O)

MEDIA

Liquid:
- RMS medium
- RMS medium containing 1% TC agar (cooled to ~48°C)

Plates:
- RMS medium solidified with 0.6% TC agar
- RMS medium solidified with 0.6% TC agar and supplemented with 50 μg of kanamycin/ml

(For recipe, see Preparation of Media, pp. 52–53)

PROCEDURE

In addition to preparing cultures from the cells in plates 6, 7, and 8 (see steps 1–3 below), you should also prepare cultures from the cells bombarded with control tungsten particles (plates 1 and 2). Grow cells from plate 1 on RMS medium solidified with 0.6% agar in the absence of kanamycin to test whether the medium supports cell growth. Grow cells from plate 2 on RMS medium solidified with 0.6% agar and supplemented with 50 μg of kanamycin/ml to verify the absence of kanamycin-resistant cells in the recipient cell population. Note that spontaneous mutation for kanamycin resistance should be extremely low in tobacco cells since none has been reported so far.

1. Forty-eight hours after the introduction of pFF19K plasmid DNA into tobacco XD cells, transfer the filter paper disks from plates 6, 7, and 8 into three sterile 250-ml Erlenmeyer flasks that each con-

tain 20 ml of liquid RMS medium and 40 μl of kanamycin sulfate stock solution. Shake the flasks to wash the cells from the filter paper disks.

2. To each flask, add 20 ml of RMS medium containing 1% TC agar cooled to approximately 48°C. Mix. Prepare four cultures from each flask by spreading 10 ml of the cell suspension onto the surfaces of each of four petri dishes containing 20 ml of RMS medium solidified with 0.6% agar and supplemented with 50 μg of kanamycin/ml. Incubate the plates under continuous light (2000 lux) for 16 hours at 28°C.

   Note: If the density of the XD suspension is high, it may be necessary to plate on as many as 10 dishes. High cell density in selection plates prevents identification of transformants. To dilute the cells more, suspend them in a larger volume and still plate 10 ml of the suspension.

3. The first kanamycin-resistant calli can be recognized after 10–14 days by fast growth and a green color. New resistant calli continue to appear up to approximately 5 weeks. Determine the number of kanamycin-resistant clones per bombardment after 8 weeks. Expect to obtain 5–20 transgenic clones per disk bombarded.

• PREPARATION OF MEDIA

RMS medium
(1 liter)

RM Macro (10x)	100 ml
RM Micro (100x)	10 ml
1% FeEDTA (Sigma E 6760)	5 ml
Sucrose	30 g
RMS vitamins (1000x)	1 ml
α-Naphthaleneacetic acid stock solution	1 ml
6-Benzylaminopurine stock solution	0.1 ml
2,4-Dichlorophenoxyacetic acid stock solution	0.1 ml
Inositol	100 mg
MES	600 mg

Adjust the volume to 1 liter with H_2O, and adjust the pH to 5.8 with 10 M KOH. For solid medium, add the appropriate amount of agar (e.g., TC agar, Hasleton Biologics Co. 91-004-110). Base agar 0.6%; top agar 1% (0.5% after dilution twofold with RMS liquid medium). Sterile medium can be stored for 1 month at room temperature.

RM Macro (10x)
(1 liter)

KNO_3	19 g
$MgSO_4 \cdot 7H_2O$	3.7 g
$CaCl_2 \cdot 2H_2O$	4.4 g
KH_2PO_4	1.7 g
$(NH_4)NO_3$	16.5 g

Dissolve in H_2O and then adjust the volume to 1 liter with H_2O.

RM Micro (100x)
(100 ml)

$MnSO_4 \cdot 2H_2O$	169 mg
H_3BO_3	62 mg
$ZnSO_4 \cdot 7H_2O$	86 mg
KI	8.3 mg
$Na_2MoO_4 \cdot 2H_2O$ (1 mg/ml)	2.5 ml
$CuSO_4 \cdot 5H_2O$ (1 mg/ml)	0.25 ml
$CoCl_2 \cdot 6H_2O$ (1 mg/ml)	0.25 ml

Dissolve in H_2O and then adjust the volume to 100 ml with H_2O.

RMS vitamins (1000x)
(10 ml)

Thiamine-HCl	100 mg
Pyridoxine-HCl	10 mg
Nicotinic acid	10 mg

Dissolve in H_2O and then adjust the volume to 10 ml with H_2O.

α-Naphthaleneacetic acid (auxin) (GIBCO 21570-015)

Stock solution: 1 mg/ml in 0.1 M NaOH

6-Benzylaminopurine (cytokinin) (GIBCO 16105-017)

Stock solution: 1 mg/ml in 0.1 M HCl

2,4-Dichlorophenoxyacetic acid (auxin) (GIBCO 11215-019)

Stock solution: 1 mg/ml in 0.1 M NaOH

• PREPARATION OF REAGENTS

Tungsten suspension

50 mg/ml in 100% ethanol.
Use tungsten with a mean particle size of approximately 1 μm (M–17 = 1.07 μm). Prepare the tungsten suspension by adding tungsten to 100% ethanol and incubating for 1 hour at 95°C. Sonicate at maximum power three times for 5 minutes. Store the tungsten suspension at –20°C for no longer than 2 days.

Note: Biolistic transformation experiments can also be performed with gold particles (1 μm), which are available commercially in a kit from Bio-Rad Laboratories. The procedure for coating gold particles with DNA is essentially the same as that for coating tungsten particles. However, gold particles tend to aggregate and are more difficult to disperse evenly. When washing the gold particles before incubating with DNA, it is best to wait for the gold particles to sediment between washes. Very low-speed centrifugation is also acceptable. In tobacco, we routinely use tungsten for both nuclear and plastid gene transformation. It may be important to use gold in species that are sensitive to tungsten, such as cereals.

X-Gluc (5-bromo-4-chloro-3-indolyl-β-glucuronide) solution

(1 ml)

N,N-Dimethyl formamide	1–2 drops
X-Gluc	1 mg
0.1 M Phosphate buffer (pH 7.0)	980 μl
5 mM Potassium ferricyanide (M.W. 329.3)	10 μl
5 mM Potassium ferrocyanide (M.W. 422.4)	10 μl
Triton X-100	1 μl

Add *N,N*-dimethyl formamide to X-Gluc and mix until dissolved (~ 2 minutes). Add 0.1 M phosphate buffer, 5 mM potassium ferricyanide, and 5 mM potassium ferrocyanide to X-Gluc solution and stir. Add Triton X-100. Prepare fresh before use.

0.1 M Phosphate buffer

0.2 M KH_2PO_4	87 ml
0.2 M K_2HPO_4	122 ml
Distilled H_2O	209 ml

• *Section 4*

Generation of Transgenic Tobacco Plants by Cocultivation of Leaf Disks with *Agrobacterium* pPZP Binary Vectors

The most commonly used method for obtaining transgenic plants involves the introduction of donor DNA into plant cells by the pathogenic bacterium *Agrobacterium tumefaciens. Agrobacterium* introduces a segment of its tumor-inducing (Ti) plasmid, the T-DNA (transferred DNA), into plant nuclei. Transfer of the T-DNA occurs from the 25-bp right border (RB) repeat to the left border (LB) repeat, and is dependent on the virulence (*vir*) functions outside the T-DNA region of the Ti plasmid. In some *Agrobacterium* strains, additional sequences called "overdrive" sequences are required for efficient T-DNA transfer. The overdrive is outside the T-DNA region and adjacent to the RB. For recent reviews on *Agrobacterium* biology, see Cado (1991), Hooykaas and Schilperoort (1992), and Zambryski (1992).

The protocol presented here describes the transformation of tobacco (*Nicotiana tabacum*) cells using an *Agrobacterium* binary vector system. This method exploits the observation that the only *cis*-acting element necessary for transfer of the T-DNA is the RB sequence, and that the T-DNA does not need to be physically linked to the rest of the Ti plasmid (de Framond et al. 1983; Hoekema et al. 1983). The binary vector system consists of two plasmids:

- The binary vector, which carries the T-DNA region that is transferred to plants. This plasmid is small, easy to manipulate in *E. coli*, and replicates both in *E. coli* and *Agrobacterium*

- The large *vir* (virulence) plasmid, which is an engineered Ti plasmid lacking the T-DNA region but carrying all the *vir* functions required for T-DNA transfer.

A third, or "helper", plasmid is required to induce conjugative transfer of the binary vector from *E. coli* into *Agrobacterium*. The helper plasmid is not required if the binary vector is introduced into *Agrobacterium* by transformation using the freeze-thaw method (Holsters et al. 1978) or by electroporation (Cangelosi et al. 1991).

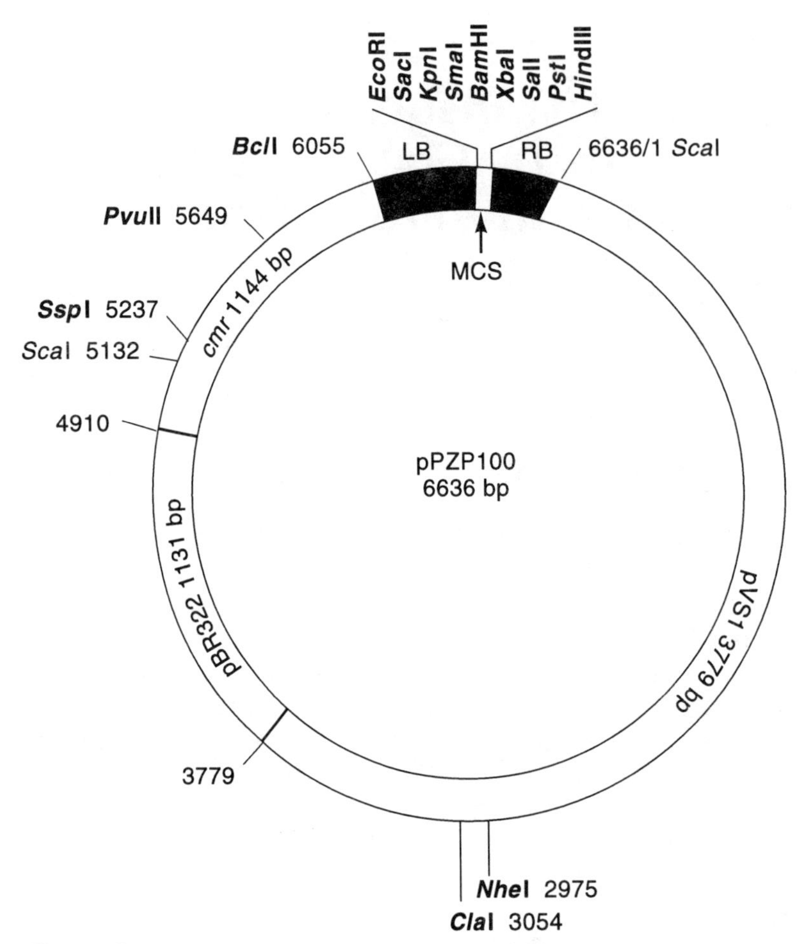

Figure 1
Map of the basic binary vector in the *Agrobacterium* pPZP100 series. In addition to the T-DNA region derived from the *Agrobacterium* plasmid pTiT37 and segments derived from plasmids pBR322 and pVS1, the vector carries a bacterial chloramphenicol resistance (*cmr*) gene. The black bars represent the left border (LB) and right border (RB) sequences of the T-DNA region, and the multiple cloning site (MCS) is located between these sequences. The numbers in light-face type denote nucleotide position at segment borders and restriction sites. Unique restriction sites are shown in bold-face type. (Reprinted, with permission, from Hajdukiewicz et al. 1994, copyright by Kluwer Academic Publishers.)

The pPZP family of binary vectors (Hajdukiewicz et al. 1994) replicate in *E. coli* using the ColE1 replication origin from plasmid pBR322 to form a high copy number (~200 copies per cell). They also replicate in *Agrobacterium* using the broad host-range replication origin of plasmid pVS1 to form 3–5 copies per cell. A map of the basic binary vector in the pPZP100 series is shown in Figure 1. In addition

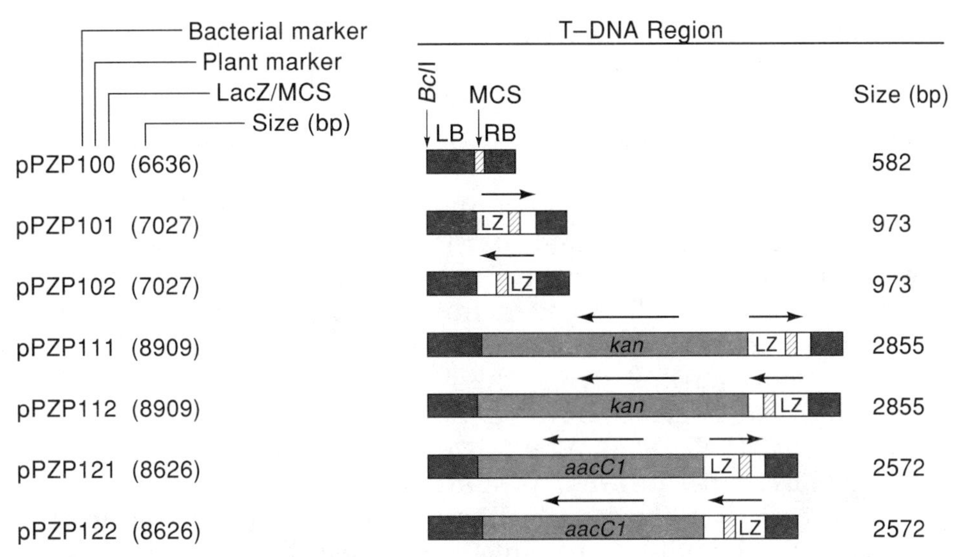

Figure 2
Structure of the T-DNA region in members of the pPZP100 family of binary vectors. Vector nomenclature is explained on the left and maps of the T-DNA regions are shown on the right. The black bars represent the left border (LB) and right border (RB) sequences of the T-DNA. LZ denotes the *Hae*II fragment from plasmid pUC18, which encodes the *LacZ* α-peptide and contains the multiple cloning site (MCS). Unique restriction enzyme cleavage sites for the MCS are shown in Figure 1. *Eco*RI is proximal to and *Hin*dIII is distal to the *LacZ* promoter. *kan* and *aacC1* are chimeric genes that confer resistance to kanamycin sulfate and gentamycin sulfate, respectively, in plants. Horizontal arrows indicate the direction of transcription from the 35S promoter (*kan* and *aacC1*) and from the *LacZ* promoter. The MCS is available in both orientations relative to the plant marker genes. The *Bcl*I site in vector pPZP100 is at nucleotide 6055 (see Fig. 1). (Reprinted, with permission, from Hajdukiewicz et al. 1994, copyright by Kluwer Academic Publishers.)

to the T-DNA region and segments derived from plasmids pBR322 and pVS1, vectors in this series carry a chloramphenicol resistance marker for selection in *E. coli* and *Agrobacterium*. The multiple cloning site (MCS) for the insertion of passenger genes is located between the RB and LB sequences of the T-DNA region and is proximal to the RB sequence.

The structure of the T-DNA region in vectors of the pPZP100 series is shown in Figure 2. In each vector in the series, except pPZP100, the MCS is situated within the LacZ α-peptide coding region (LZ). This feature facilitates identification of *E. coli* carrying binary vector with a passenger gene inserted in the MCS. Insertional inactivation of the α-peptide allows direct identification of recombinant clones by color screening on X-gal indicator plates; interruption of the α-peptide yields white colonies rather than blue (Messing 1991). Some vectors in the pPZP100 family carry *kan* (pPZP111 and pPZP112) or *aacC1* (pPZP121 and pPZP122) chimeric genes that confer resistance to kanamycin sulfate or gentamycin sulfate, respectively, in plants. These plant marker genes are located between LZ and the LB sequence. During transformation, transfer of T-DNA is initiated at the RB sequence and therefore passenger genes in the MCS are transferred to plant nuclei before the plant marker genes (see Fig. 2). Note that the *Hae*II fragment from pUC18 was inserted in both orientations, and therefore the cloning sites in the MCS are available in both orientations relative to the plant marker genes in the vector pairs. The DNA sequence of the pPZP plasmids has been determined (Hajdukiewicz et al. 1994). Alternative binary vectors are available, including pBIN19 (Bevan 1984), pGA482, and pGA492 (An 1986).

In the experiment described here, tobacco cells are transformed by incubating leaf disks with an *Agrobacterium* binary vector system that consists of *Agrobacterium* strain LBA4404, which carries the pLBA4404 *vir* plasmid marked with a streptomycin/spectinomycin resistance gene (Hoekema et al. 1983), and binary vector pPZP111G or pPZP121G. Vectors pPZP111G and pPZP121G are derivatives of pPZP111 and pPZP121, respectively, and both carry a chimeric *uidA* passenger gene that encodes β-glucuronidase (GUS) (Jefferson 1987; Jefferson et al. 1987). The *uidA* gene is expressed in a cauliflower mosaic virus 35S cassette (Timmermans et al. 1990).

Since the binary vectors are not self-transmissible, they are introduced into *Agrobacterium* by triparental mating (Ditta et al. 1980). *Agrobacterium* LBA4404 is incubated with *E. coli* JM83 carrying pPZP111G (JM83[pPZP111G]) or pPZP121G (JM83[pPZP121G]) and *E. coli* JM83 carrying a helper plasmid (pRK2013) (JM83[pRK2013]), which has all the functions required for conjugation. The helper plasmid is transferred into the *E. coli* carrying the binary vector, where it induces the conjugative transfer of the vector into *Agrobacterium*. The products of the triparental mating procedure are *Agrobacterium vir* strains carrying pPZP111G or pPZP121G (LBA4404[pPZP111G] or LBA4404[pPZP121G]), which are identified by genetic markers.

Transformation of tobacco cells is achieved by incubating *Agrobacterium* carrying the binary vectors with leaf disks of *Nicotiana tabacum* (Horsch et al. 1985). We use a tobacco cultivar, cv. Petit Havana, that regenerates readily in tissue culture and has the advantage of a relatively short generation time. Seeds can be obtained from cv. Petit Havana 3–4 months after transfer of transgenic plants to the greenhouse, whereas it takes 6–7 months to obtain seeds from cv. Xanthi or cv. Wisconsin 38, other varieties commonly used in transformation experiments.

In this experiment, transgenic clones are identified by several methods. Kanamycin- and gentamycin-resistant transgenic shoots are identified by incubating the leaf disks on selective medium (Klein et al. 1988; Carrer et al. 1991). Rapid confirmation of transformation is obtained by monitoring expression of the *uidA* gene using a histochemical stain for GUS (Jefferson 1987; Jefferson et al. 1987; Gallagher 1992). Antibiotic resistance in transgenic seedlings is tested using a procedure adapted from Maliga (1984).

Experiments with transgenic plants are subject to state and federal laws in the United States and in most other countries. For information concerning legal constraints on research involving recombinant DNA molecules, see Appendix 1; also consult the safety office at your institution.

REFERENCES

An, G. 1986. Development of plant promoter expression vectors and their use for analysis of differential activity of nopaline synthase promoter in transformed tobacco tissue. *Plant Physiol.* **81:** 86–91.

Bevan, M. 1984. Binary *Agrobacterium* vectors for plant transformation. *Nucleic Acids Res.* **12:** 8711–9721.

Cado, I.C. 1991. Molecular mechanisms of crown gall tumorigenesis. *Crit. Rev. Plant Sci.* **10:** 1–32.

Cangelosi, G.A., E.A. Best, G. Martinetti, and E.W. Nester. 1991. Genetic analysis of *Agrobacterium. Methods Enzymol.* **204:** 384–397.

Carrer, H., J.M. Staub, and P. Maliga. 1991. Gentamycin resistance in *Nicotiana* conferred by AAC(3)-I, a narrow substrate specificity acetyltransferase. *Plant Mol. Biol.* **17:** 301–303.

de Framond, A., K. Barton, and M.-D. Chilton. 1983. Mini-Ti: A vector strategy for plant genetic engineering. *Bio/Technology* **1:** 262–272.

Ditta, G., S. Stanfield, D. Corbin, and D.R. Helinski. 1980. Broad host range cloning system for gram negative bacteria: Construction of a gene bank of *Rhizobium meliloti. Proc. Natl. Acad. Sci.* **77:** 7347–7351.

Gallagher, S.R., ed. 1992. *GUS protocols*. Academic Press, San Diego.

Hajdukiewicz, P., Z. Svab, and P. Maliga. 1994. The pPZP family of *Agrobacterium* binary vectors. *Plant Mol. Biol.* **25:** 989–994.

Hoekema, A., P.R. Hirsch, P.J.J. Hooykaas, and R.A. Schilperoort. 1983. A binary plant vector strategy based on separation of *vir-* and T-region of the *Agrobactrium tumefaciens* Ti plasmid. *Nature* **303:** 179–180.

Holsters, M., D. de Waele, A. Depicker, E. Messens, M. Van Montagu, and J. Schell. 1978. Transfection and transformation of *A. tumefaciens. Mol. Gen. Genet.* **163:** 181–187.

Hooykaas, P.J.J. and R.A. Schilperoort. 1992. *Agrobacterium* and plant genetic engineering. *Plant Mol. Biol.* **19:** 15–38.

Horsch, R.B., J.E. Fry, N.L. Hoffman, D. Eicholtz, S.D. Rogers, and R.T. Fraley. 1985. A simple and general method for transferring genes into plants. *Science* **227:** 1229–1231.

Jefferson, R.A. 1987. Assaying chimeric genes in plants: The GUS gene fusion system. *Plant Mol. Biol. Rep.* **5:** 387–405

Jefferson, R.A., T.A. Kavanagh, and M.W. Bevan. 1987. GUS fusions: Beta-glucuronidase as a sensitive and versatile gene fusion marker in higher plants. *EMBO J.* **6:** 3901–3907.

Klein, T.M., E.C. Harper, Z. Svab, J.C. Sanford, M.E. Fromm, and P. Maliga. 1988. Stable genetic transformation of intact *Nicotiana* cells by the particle bombardment process. *Proc. Natl. Acad. Sci.* **85:** 8502–8505.

Maliga, P. 1984. Cell culture procedures in mutant selection and characterization in *Nicotiana plumbaginifolia*. In *Cell culture and somatic cell genetics of plants* (ed. I.K. Vasil), pp. 552–562. Academic Press, Orlando, Florida.

Messing, J. 1991. Cloning in M13 phage or how to use biology at its best. *Gene* **100:** 3–12.

Murashige, T., and F. Skoog. 1962. A revised medium for the growth and bioassay with tobacco tissue culture. *Physiol. Plant.* **15:** 473–497.

Sambrook, J., E.F. Fritsch, and T. Maniatis. 1989. *Molecular cloning: A laboratory manual.* Cold Spring Harbor Laboratory Press, Cold Spring Harbor, New York.

Timmermans, M.C.P., P. Maliga, J. Vieira, and J. Messing. 1990. The pFF plasmids: Casettes utilising CaMV sequences for expression of foreign genes in plants. *J. Biotechnol.* **14:** 333–344.

Zambryski, P.C. 1992. Chronicles from the *Agrobacterium*-plant cell DNA transfer story. *Annu. Rev. Plant Physiol. Plant Mol. Biol.* **43:** 465–490.

• *TIMETABLE*

Generation of Transgenic Tobacco Plants by *Agrobacterium* Transformation

STEPS	*TIME REQUIRED*
Introduction of pPZP Binary Vectors into *Agrobacterium* by Triparental Mating	
Grow *Agrobacterium* LBA4404	20–24 hours
Grow *E. coli* strains JM83(pPZP111G), JM83(pPZP121G), and JM83(pRK2013)	2–3 hours
Prepare triparental mating mixtures and transfer bacteria onto LB medium	1 hour
Incubate plates at 28°C	20 hours
Prepare dilution series of each culture and transfer undiluted and diluted samples onto selective medium	2 hours
Incubate plates at 28°C until *Agrobacterium* colonies appear	2–4 days
Single-cell purify *Agrobacterium* carrying the binary vectors	2–4 days
Repeat single-cell purification step	2–4 days
Prepare suspension culture of each *Agrobacterium* strain	20–24 hours
Isolate DNA from *Agrobacterium* strains and analyze on gel	8 hours

STEPS	*TIME REQUIRED*
Cocultivation of Tobacco Leaf Disks with *Agrobacterium* and Identification of Antibiotic-resistant Shoots	
Grow tobacco plants from seed	2 months
Grow *Agrobacterium* strains	2 days
Prepare tissue culture media	1 day
Prepare leaf disks and infect with *Agrobacterium*	6 hours
Incubate leaf disks on nonselective RMOP medium	2 days
Transfer leaf disks onto selective RMOP medium	2 hours
Incubate until shoots appear	2–3 weeks
Dissect shoots and insert into selective MS medium	1–2 hours
Incubate until roots form	7–15 days
Plant rooted transgenic shoots in soil	1–2 hours
Histochemical Localization of β-Glucuronidase (GUS)	
Prepare tissue sections and stain for GUS	1 hour
Incubate to develop blue color	2–12 hours
Testing of Antibiotic Resistance Phenotypes in Transgenic Tobacco Seedlings	
Prepare media	6 hours
Surface-sterilize and plate seeds	4 hours
Incubate until seedling phenotype develops	14 days

Introduction of pPZP Binary Vectors into *Agrobacterium* by Triparental Mating

MATERIALS

Agrobacterium LBA4404 (Hoekema et al. 1983)
E. coli JM83 (pRK2013)—carrying helper plasmid (Ditta et al. 1980)
E. coli JM83 (pPZP111G)—carrying binary vector
E. coli JM83 (pPZP121G)—carrying binary vector
Spectinomycin dihydrochloride (100 mg/ml in H_2O)
Kanamycin sulfate (50 mg/ml in H_2O)
Chloramphenicol (25 mg/ml in 100% ethanol)
Rifampicin (100 mg/ml in methanol)
N-Lauroylsarcosine (0.5% in H_2O)

MEDIA

Liquid:
- Minimal A medium
- Minimal A medium containing 100 μg of spectinomycin/ml
- Minimal A medium containing 25 μg of chloramphenicol/ml
- LB medium
- LB medium containing 50 μg of kanamycin sulfate/ml
- LB medium containing 25 μg of chloramphenicol/ml

Plates:
- Minimal A medium containing 1.5% agar and 25 μg of chloramphenicol/ml
- LB medium containing 1.5% agar

(For recipes, see Preparation of Media, pp. 74–76)

PROCEDURE

1. Grow a loopful of *Agrobacterium* LBA4404 in 4 ml of liquid minimal A medium containing 100 μg of spectinomycin/ml for 20–24 hours at 28°C. Allow the culture to attain a density of 1–2 x 10^9 bacteria/ml. Spectinomycin selects for bacteria carrying the pLBA4404 *vir* plasmid.

2. Grow a loopful of *E. coli* JM83 (pRK2013) in 2 ml of liquid LB medium containing 50 µg of kanamycin sulfate/ml for 2–3 hours at 37°C. Allow the culture to attain a density of 1–2 x 10^9 bacteria/ml. Kanamycin sulfate selects for bacteria carrying the helper plasmid.

3. Grow a loopful of *E. coli* JM83 (pPZP111G) and a loopful of *E. coli* JM83 (pPZP121G) each in 2 ml of liquid LB medium containing 25 µg of chloramphenicol/ml for 2–3 hours at 37°C. Allow the cultures to attain densities of 1–2 x 10^9 bacteria/ml. Chloramphenicol selects for bacteria carrying the binary vectors.

4. Sediment the bacteria by spinning each culture in a Beckman GPR centrifuge at 2500 rpm for 5 minutes at room temperature. Decant the antibiotic-containing media, and resuspend each pellet in antibiotic-free, liquid LB medium to give a density of 1–2 x 10^9 bacteria/ml.

5. Prepare triparental mating mixtures (tubes 1 and 2) and control samples (tubes 3–11) in microfuge tubes as follows:

Tube Number	*Agrobacterium* LBA4404	*E. coli* JM83 (pRK2013)	*E. coli* JM83 (pPZP111G)	*E. coli* JM83 (pPZP121G)
1	500 µl	100 µl	100 µl	–
2	500 µl	100 µl	–	100 µl
3	500 µl	–	–	–
4	–	100 µl	–	–
5	–	–	100 µl	–
6	–	–	–	100 µl
7	500 µl	100 µl	–	–
8	500 µl	–	100 µl	–
9	500 µl	–	–	100 µl
10	–	100 µl	100 µl	–
11	–	100 µl	–	100 µl

Mix each sample and then sediment the bacteria by spinning each tube in a microfuge for 1 minute at room temperature. Pour off the supernatants. Resuspend each pellet in the 50–100 µl of liquid LB medium that remains after the removal of the supernatants. Spread 30 µl of each bacterial suspension onto a small area (20 mm x 30 mm) on plates containing agar-solidified LB medium. Leave the plates with the lids open in a laminar flow hood until the patches of bacterial suspension dry. Replace the lids, and incubate the plates for 20 hours at 28°C. It is during this incubation that conjugative transfer occurs.

6. Take a big loopful of each culture and resuspend in 500 μl of liquid minimal A medium. Prepare a 1 in 100 dilution series for each sample. Plate 20 μl of each undiluted suspension and 20 μl of each 10^{-2} and 10^{-4} dilution on agar-solidified minimal A medium containing 25 μg of chloramphenicol/ml. Use a separate plate for each sample. Incubate the plates for 2–4 days at 28°C. Only *Agrobacterium* cells form colonies on the sucrose-containing minimal A medium. (*E. coli* requires glucose as a carbon source.) Chloramphenicol selects for *Agrobacterium* that acquired binary vector (pPZP111G or pPZP121G) by conjugation. Problems with the media or strains can be identified by comparing the experimental and control plates.

   Note: Alternatively, you may select ex-conjugants by plating the samples on agar-solidified LB medium containing 100 μg of rifampicin/ml and 25 μg of chloramphenicol/ml. Incubate the plates for 2–4 days at 28°C. *Agrobacterium* LBA4404 carries a rifampicin resistance gene and the binary vectors carry chloramphenicol resistance genes. The *E. coli* strains are sensitive to rifampicin. Therefore, only *Agrobacterium* carrying binary vector will grow.

7. Using a toothpick, single-cell purify four isolates of each *Agrobacterium* strain carrying binary vector on agar-solidified minimal A medium containing 25 μg of chloramphenicol/ml. Incubate for 2–4 days at 28°C to allow colonies to form.

8. Repeat the single-cell purification step.

9. To confirm the introduction of binary vector, prepare plasmid DNA from the *Agrobacterium* strains. Grow a loopful of each *Agrobacterium* isolate in 2 ml of liquid minimal A medium containing 25 μg of chloramphenicol/ml for 20–24 hours at 28°C. Transfer 1.5 ml of each culture to a sterile microfuge tube and sediment the bacteria by spinning in a microfuge at 14,000 rpm for 5 minutes at room temperature. Resuspend each pellet in 1 ml of 0.5% *N*-lauroylsarcosine and incubate for 2 minutes at room temperature. Sediment the bacteria by centrifugation and decant the *N*-lauroylsarcosine solution. Prepare plasmid DNA from each bacterial pellet using the alkaline lysis procedure (Sambrook et al. 1989). Perform diagnostic digestion of the DNA (Sambrook et al. 1989). The DNA obtained by this procedure is sufficient to run only one or two gels because pPZP vectors replicate in *Agrobacterium* to form only 3–5 copies per cell.

Cocultivation of Tobacco Leaf Disks with *Agrobacterium* and Identification of Antibiotic-resistant Shoots

In this experiment, transformation is carried out with *Agrobacterium* strains LBA4404(pPZP111G) and LBA4404(pPZP121G), which should yield kanamycin-resistant and gentamycin-resistant plants, respectively. Cocultivation of *Agrobacterium* LBA4404 (no binary vector) with leaf disks on kanamycin- and gentamycin-containing plant media tests the efficiency of the antibiotic carbenicillin to eliminate the bacteria and shows kanamycin and gentamycin inhibition of growth and bleaching of nontransformed leaf tissue. Leaf disks are also incubated on drug-free medium without *Agrobacterium* infection to demonstrate callusing and shoot regeneration from tobacco leaf tissue. Resistant transformed tissue is expected to behave on the selective medium as noninfected wild-type tissue does in the absence of drugs.

MATERIALS

Agrobacterium LBA4404 (Hoekema et al. 1983)
Agrobacterium strains LBA4404(pPZP111G) and LBA4404(pPZP121G) (see pp. 63–65; Hajdukiewicz et al. 1994)
Spectinomycin dihydrochloride (100 mg/ml in H_2O)
Chloramphenicol (25 mg/ml in 100% ethanol)
Tobacco (*Nicotiana tabacum*) plants from greenhouse
Ethanol (70%)
Clorox bleach (diluted 1 in 10 with H_2O) or 0.5% sodium hypochlorite
Kanamycin sulfate (50 mg/ml in H_2O)
Gentamycin sulfate (100 mg/ml in H_2O)
Carbenicillin (100 mg/ml in H_2O)

MEDIA

Liquid:
- Minimal A medium
- RMOP medium

Plates:

Minimal A medium containing 1.5% agar and 100 μg of spectinomycin/ml

Minimal A medium containing 1.5% agar, 100 μg of spectinomycin/ml, and 25 μg of chloramphenicol/ml

RMOP medium containing 0.6% agar

RMOP medium containing 0.6% agar, 100 μg of kanamycin sulfate/ml, and 500 μg of carbenicillin/ml

RMOP medium containing 0.6% agar, 100 μg of gentamycin sulfate/ml and 500 μg of carbenicillin/ml

MS medium containing 0.6% agar and 100 μg of kanamycin sulfate/ml

MS medium containing 0.6% agar and 100 μg of gentamycin sulfate/ml

(For recipes, see Preparation of Media, pp. 74–76)

PROCEDURE

Steps 1–8 described below should be performed under aseptic conditions using sterile equipment and reagents. Leaf tissue should always be handled in a laminar flow hood.

1. Grow a lawn of *Agrobacterium* LBA4404 on agar-solidified minimal A medium containing 100 μg of spectinomycin/ml. Grow lawns of LBA4404(pPZP111G) and LBA4404(pPZP121G) on agar-solidified minimal A medium containing 100 μg of spectinomycin/ml and 25 μg of chloramphenicol/ml. Incubate the plates for 2 days at 28°C.

2. Prepare suspensions from each of the three *Agrobacterium* strains by measuring 3 x 10 ml of minimal A medium into sterile, 15-ml plastic screw-cap centrifuge tubes, scraping the bacteria off each plate with a sterile loop, and resuspending them to give a density of 1–2 x 10^9 bacteria/ml.

3. Take the first two fully expanded leaves from two young tobacco plants. Do not use the plants after the flower buds appear. Surface sterilize the leaves by dipping them briefly into 70% ethanol (just long enough to wet the surface of the leaves) and then immersing them in a 1 in 10 dilution of Clorox bleach (or 0.5% sodium hypochlorite) for 5 minutes. Rinse the leaves five times in sterile distilled H_2O.

   Note: You can avoid surface-sterilizing the tobacco leaves if you grow plants from surface-sterilized seeds on agar-solidified MS medium in a Magenta box (Magenta Corporation). These plants can be propagated by cuttings and maintained in culture for several years.

4. Punch 100 disks (5–10 mm in diameter) or cut sections (1 x 1 cm) with a scalpel from the surface-sterilized tobacco leaves. Transfer 20 disks to each of the tubes containing the LBA4404(pPZP111G) and LBA4404(pPZP121G) suspensions and 40 disks to the tube containing the LBA4404 suspension. In addition, place 20 disks in a tube containing sterile RMOP medium. Vortex the samples for 15 seconds to secure contact of the bacteria with the wounded surfaces of the leaf tissue. Remove the leaf disks from the bacterial suspensions and blot them on filter paper. Place the infected leaf disks upside down on Whatman No. 4 filter paper on the surface of agar-solidified RMOP medium (10 disks per 10-cm petri dish). *Agrobacterium* will overgrow the leaf disks if they are placed directly on the RMOP medium.

5. To facilitate transformation during cocultivation, incubate the leaf disks for 2 days at 26°C.

6. Take the leaf disks and place them upside down (10 per 10-cm petri dish) on the following agar-solidifed RMOP media: disks infected with LBA4404(pPZP111G) on medium containing 100 μg of kanamycin sulfate/ml and 500 μg of carbenicillin/ml; disks infected with LBA4404(pPZP121G) on medium containing 100 μg of gentamycin sulfate/ml and 500 μg of carbenicillin/ml; disks infected with LBA4404 on both selective media; and noninfected control disks on drug-free medium. Kanamycin sulfate and gentamycin sulfate select for transformed tobacco cells. Carbenicillin is included in the culture medium to inhibit the growth of *Agrobacterium*. Incubate the plates with 16 hours illumination (2000 lux) per day for 2–3 weeks at 26°C.

7. Noninfected leaf disks form green calli and shoots, whereas leaf disks infected with LBA4404 bleach on the selective media. Look for the appearance of many shoots and green calli at the wounded surfaces of the leaf disks infected and transformed with LBA4404(pPZP111G) or LBA4404(pPZP121G).

   Note: Approximately 10% of the shoots are not transgenic, but form as a result of cross-protection by transgenic tissue. Therefore, after antibiotic selection (step 6), it is important to confirm transformation using another technique. Transformation can be confirmed by a number of methods. (i) Root formation on shoots incubated on selective medium (step 8) is a reliable assay of antibiotic resistance. Moreover, this assay yields plants that are ready to transfer to the greenhouse. (ii) Since the binary vectors used in this experiment carry the *uidA* gene, transformed cells can be identified using a histochemical stain for GUS (see p. 70). (iii) The formation of shoots and calli on leaf disks taken from regenerated shoots and incubated on selective RMOP medium (containing 100 μg of kanamycin sulfate or gentamycin sulfate/ml) can also be used to identify transgenic shoots. (iv) DNA gel blot analysis permits identification of transforming DNA in tobacco cells (Sambrook et al. 1989). (v) PCR amplification of transgenes. However, this analysis must be performed using plants that are free of *Agrobacterium* infection. Since *Agrobacterium* may survive in transformed tissue, verification by methods i–iii is more reliable. (vi) If you are working with a new marker or species, always verify transformation by testing inheritance in the seed progeny (see pp. 71–73).

8. Dissect the shoots from the leaf disks transformed with LBA4404(pPZP111G) or LBA4404(pPZP121G) and insert them into agar-solidified MS medium containing 100 μg of kanamycin sulfate/ml or 100 μg of gentamycin sulfate/ml, respectively. Incubate with 16 hours illumination (2000 lux) per day for 7–15 days at 26°C. Transgenic shoots will form roots, whereas wild-type shoots will bleach and will not form roots.

9. Gently break up the agar-solidified MS medium and wash it from the rooted transgenic shoots. Plant the shoots in soil. Cover the plantlets with transparent plastic for approximately 1 week. Initially (for approximately the first 3 days), the plants should be grown in the shade.

Histochemical Localization of β-Glucuronidase (GUS)

REAGENTS

X-Gluc solution
(For recipe, see Preparation of Reagents, p. 77)

PROCEDURE

Since the binary vectors employed in this experiment carry a chimeric *uidA* gene, transformed cells can be identified using a histochemical stain for GUS. This method allows rapid confirmation of transformation 7–10 days after initiating cocultivation of leaf disks with *Agrobacterium*.

1. Approximately 1 week after the transformed leaf disks have been transferred to selective medium containing carbenicillin (step 6, p. 68), remove one or two disks from each sample for histochemical localization of GUS. Using a razor blade, prepare freehand sections (<0.5 mm thick) from the edge of the leaf disks. The thinner the tissue sections, the better the cellular resolution will be under a microscope.

2. Place a few tissue sections from each sample in microfuge tubes containing 150 μl of X-Gluc solution. Ensure that the tissue sections are covered with the X-Gluc solution. Mix by inverting the tubes several times.

3. Incubate for 2 hours at 37°C. Keep the caps of the microfuge tubes closed to prevent evaporation. The tissue sections can be incubated overnight if necessary.

4. Examine the tissue sections under a dissecting microscope or a compound microscope. GUS-expressing cells are stained blue.

   Note: This procedure can also be used to verify transformation of regenerated shoots. Detection of GUS expression may be improved by removing the chlorophyll from green tissue by repeated ethanol extractions (50% ethanol for 5 minutes; 70% ethanol for 5 minutes; 3 x 100% ethanol for 1 hour each). After extraction, the tissue can be stored in 0.1 M phosphate buffer (pH 7.0). For additional information on tissue cleaning, see Gallagher 1992.

Testing of Antibiotic Resistance Phenotypes in Transgenic Tobacco Seedlings

MATERIALS

Wild-type tobacco seeds
Transgenic tobacco seeds resistant to kanamycin sulfate or gentamycin sulfate (progeny of transgenic plants, see p. 69)
Ethanol (70%)
Clorox bleach (diluted 1 in 10 with H_2O) or 0.5% sodium hypochlorite
Sterile distilled H_2O
Kanamycin sulfate (50 mg/ml in H_2O)
Gentamycin sulfate (100 mg/ml in H_2O)

MEDIA

Liquid:
- MS medium
- MS medium containing 400 μg of kanamycin sulfate/ml
- MS medium containing 1000 μg of gentamycin sulfate/ml
- MS medium containing 1% agar, melted (50°C)

Plates:
- MS medium containing 0.6% agar
- MS medium containing 0.6% agar and 200 μg of kanamycin sulfate/ml
- MS medium containing 0.6% agar and 500 μg of gentamycin sulfate/ml

(For recipe, see Preparation of Media, pp. 74–76)

PROCEDURE

Testing the inheritance of antibiotic resistance phenotypes in the seed progeny of transgenic tobacco plants is an essential part of confirming transformation.

1. Following the experimental design outlined in Table 1, place wild-type, antibiotic-resistant control (if available), and transgenic tobacco seeds into nine 15-ml centrifuge tubes (100–200 seeds/tube).

Table 1 **Experiment to Test the Inheritance of Antibiotic Resistance Phenotypes in Tobacco Seedlings**

Sample number	Type of tobacco seeds	Germination conditions	Experiment (expected results)
1	wild-type	no antibiotic	to test medium (green seedlings)
2[a]	wild-type	+ kanamycin	drug-sensitive control (bleached seedlings)
3[a]	wild-type	+ gentamycin	drug-sensitive control (bleached seedlings)
4[b]	kanamycin-resistant control	+ kanamycin	drug-resistant control (green seedlings)
5[b]	gentamycin-resistant control	+ gentamycin	drug-resistant control (green seedlings)
6	kanamycin-resistant transgenic	no antibiotic	to test viability of seed (green seedlings)
7	kanamycin-resistant transgenic	+ kanamycin	to confirm inheritance of resistance (green and white seedlings)[c]
8	gentamycin-resistant transgenic	no antibiotic	to test viability of seed (green seedlings)
9	gentamycin-resistant transgenic	+ gentamycin	to confirm inheritance of resistance (green and white seedlings)[c]

[a]Control samples 2 and 3 are very important as the degree and speed of bleaching on selective media are dependent on light intensity.
[b]Transgenic, nonsegregating resistant control seeds may not be available.
[c]Transgenes are integrated at one or more loci, but only in one of the homologous chromosomes, therefore progeny segregate for resistance in Mendelian ratios (3:1, 15:1, etc.).

2. Surface-sterilize the seeds by adding a few drops of 70% ethanol (enough to wet the seeds) and 5 ml of a 1 in 10 dilution of Clorox bleach (or 0.5% sodium hypochlorite) to each tube.

3. Incubate the tubes for 3 minutes at room temperature, shaking them occasionally.

4. Remove the bleach from the seeds by pouring the contents of each tube into a funnel containing a 60-μm mesh. The seeds are retained on the mesh.

5. Rinse each group of seeds five times with sterile distilled H_2O to remove all traces of the bleach.

6. For imbibition, transfer each group of seeds into a sterile 15-ml plastic tube containing 5 ml of MS medium with or without selective concentrations of kanamycin sulfate (400 μg/ml) or gentamycin sulfate (1000 μg/ml) and let them stand for 30–60 minutes at room temperature. Check Table 1 for the appropriate conditions for each group of seeds. This step is necessary to achieve uniform, fast germination of the seeds.

   Note: Selective concentrations for other drugs in tobacco seedlings: hygromycin B, 100 μg/ml; lincomycin-HCl, 1000 μg/ml; spectinomycin-HCl, 500 μg/ml; streptomycin sulfate, 500 μg/ml.

7. Add 5 ml of melted MS medium containing 1% agar to each tube. Mix the medium with the seeds. Pour the 10-ml seed suspensions into petri dishes on top of solidified MS medium containing 0.6% agar prepared with or without 200 μg of kanamycin sulfate/ml or 500 μg of gentamycin sulfate/ml. Check Table 1 for the appropriate medium for each group of seeds.

8. Seal the plates with strips of Parafilm or Handi-Wrap. Incubate with 16 hours illumination (5000 lux) per day for 14 days at 28°C.

9. Score for resistance to antibiotics. On selective medium, seedlings that are sensitive to antibiotics are bleached and those that are resistant to antibiotics are green. Count only germinated seedlings.

   Note: Sterilization of a large number of seed samples is time consuming. Seeds in intact tobacco capsules are sterile if the plants are kept free of insects and fungal infection. If your plants were kept free of pests, just break off the tip of the dry capsules and pour the seeds directly onto the surface of sterile MS agar plates (step 7).

• PREPARATION OF MEDIA

LB (Luria-Bertani) medium
(1 liter)

Bacto-tryptone	10 g
Bacto-yeast extract	5 g
NaCl	10 g

Dissolve in H_2O and then adjust the volume to 1 liter with H_2O. Adjust the pH to 7.0 with 5 M NaOH. Autoclave. To prepare agar-solidified LB medium, add 15 g of Bacto agar (Difco)/liter before autoclaving.

Minimal A medium
(1 liter)

Stock solution A	490 ml
Stock solution C	11 ml

To prepare 1 liter of liquid minimal A medium, combine stock solutions A and C with 499 ml of sterile H_2O. To prepare 1 liter of agar-solidified minimal A medium, combine stock solutions A and C with 499 ml of stock solution B. Note that the carbon source required for growing *E. coli* is different from that required for growing *Agrobacterium*.

Stock solution A
(490 ml)

K_2HPO_4	10.5 g
KH_2PO_4	4.5 g
$(NH_4)_2SO_4$	1 g
Sodium citrate	0.5 g

Dissolve in H_2O and then adjust the volume to 490 ml with H_2O. Autoclave.

Stock solution B
(500 ml)

Dissolve 15 g of Bacto agar (Difco) in 500 ml of H_2O. Autoclave.

Stock solution C
(11 ml)

Add 1 ml of 1 M $MgSO_4 \cdot 7H_2O$ (autoclaved) to 10 ml of either 20% glucose (filter sterilized) (for growing *E. coli*) or 40% sucrose (filter sterilized) (for growing *Agrobacterium*).

MS medium
(1 liter)

MS macroelements stock solution (10x)	100 ml
MS microelements stock solution (100x)	10 ml
1% FeEDTA (Sigma E 6760)	5 ml
Sucrose	30 g

Dissolve in H_2O and then adjust the volume to 1 liter with H_2O. Adjust the pH to 5.8 with 1 M KOH. Autoclave. To prepare agar-solidified MS medium for rooting transgenic shoots, add 6 g of TC-agar (Hasleton Biologics Co. 91-004-110)/liter before autoclaving. To prepare agar-solidified MS medium for testing seedling phenotypes, add 6 g of TC-agar/liter for the base layer and 10 g of TC-agar/liter for the top layer. Autoclave after adding the appropriate amount of agar. This recipe is adapted from Murashige and Skoog (1962).

MS macroelements stock solution (10x)
(1 liter)

KNO_3	19 g
$MgSO_4 \cdot 7H_2O$	3.7 g
$CaCl_2 \cdot 2H_2O$	4.4 g
KH_2PO_4	1.7 g
$(NH_4)NO_3$	16.5 g

Dissolve in H_2O and then adjust the volume to 1 liter with H_2O.

MS microelements stock solution (100x)
(100 ml)

$MnSO_4 \cdot 2H_2O$	169 mg
H_3BO_3	62 mg
$ZnSO_4 \cdot 7H_2O$	86 mg
KI	8.3 mg
$Na_2MoO_4 \cdot 2H_2O$ (1 mg/ml)	2.5 ml
$CuSO_4 \cdot 5H_2O$ (1 mg/ml)	0.25 ml
$CoCl_2 \cdot 6H_2O$ (1 mg/ml)	0.25 ml

Dissolve in H_2O and then adjust the volume to 100 ml with H_2O.

RMOP medium
(1 liter)

MS macroelements stock solution (10x)	100 ml
MS microelements stock solution (100x)	10 ml
1% FeEDTA (Sigma E 6760)	5 ml
Thiamine (1 mg/ml)	1 ml
1-Naphthaleneacetic acid (auxin) stock solution	0.1 ml
6-Benzylaminopurine (cytokinin) stock solution	1 ml
Inositol	100 mg
Sucrose	30 g

Dissolve in H_2O and then adjust the volume to 1 liter with H_2O. Adjust the pH to 5.8 with 1 M KOH. Autoclave. To prepare agar-solidified RMOP medium, add 6 g of TC-agar (Hasleton Biologics Co. 91-004-110)/liter before autoclaving.

MS macroelements stock solution (10x)
See recipe for MS medium (p. 75)

MS microelements stock solution (100x)
See recipe for MS medium (p. 75)

1-Naphthaleneacetic acid (auxin) (GIBCO 21570-015)

Stock solution: 1 mg/ml in 0.1 M NaOH

6-Benzylaminopurine (cytokinin) (GIBCO 16105-017)

Stock solution: 1 mg/ml in 0.1 M HCl

• PREPARATION OF REAGENTS

X-Gluc (5-bromo-4-chloro-3-indolyl-β-glucuronide) solution
(1 ml)

N,N-Dimethyl formamide	1–2 drops
X-Gluc	1 mg
0.1 M Phosphate buffer (pH 7.0)	980 μl
5 mM Potassium ferricyanide (M.W. 329.3)	10 μl
5 mM Potassium ferrocyanide (M.W. 422.4)	10 μl
Triton X-100	1 μl

Add 1–2 drops of *N,N*-dimethyl formamide to 1 mg of X-Gluc and mix until dissolved (~2 minutes). Add 980 μl of 0.1 M phosphate buffer (pH 7.0), 10 μl of 5 mM potassium ferricyanide, and 10 μl of 5 mM potassium ferrocyanide to the X-Gluc and stir. Add 1 μl of Triton X-100. Prepare fresh before use.

0.1 M Phosphate buffer (pH 7.0)
(418 ml)

0.2 M KH_2PO_4	87 ml
0.2 M K_2HPO_4	122 ml
Distilled H_2O	209 ml

• *Section 5*
Tissue Printing to Detect Proteins and RNA in Plant Tissues

When a freshly cut section of an organ is pressed gently onto nitrocellulose paper, nylon membrane, or polyvinylidene difluoride membrane, the contents of the cut cells are transferred to the membrane and the macromolecules are immobilized. These molecules constitute a latent print that can be visualized with appropriate probes. When a plant tissue print is made on nitrocellulose paper, the firm cell walls (xylem, phloem, sclerenchyma, and collenchyma) leave a physical print that makes the anatomy visible without further treatment (Varner and Taylor 1989). The simple tissue-printing procedures described here have been made possible by the ready availability of films designed to immobilize proteins and nucleic acids.

Tissue printing on substrate films was pioneered by Daoust (1957), who achieved excellent resolution in the detection of amylase, protease, ribonuclease, and deoxyribonuclease in animal tissues using starch, gelatin, and agar films. These techniques were also applied successfully to plant tissues (Jacobsen and Knox 1973; Yomo and Taylor 1973; Harris and Chrispeels 1975). Advances in membrane technology have increased the range of applications of Daoust's tissue-printing techniques. The first tissue-printing studies performed using modern membranes were the immunocytochemical localization of soybean seed coat extensin on nitrocellulose paper (Cassab and Varner 1987) and the histochemical localization of various plant enzymes on nitrocellulose paper (Spruce et al. 1987).

Examples of applications that use gelatin and starch films are found in Adams and Tuqan (1961), Daoust (1957, 1965), Gaddum and Blandau (1970), Harris and Chrispeels (1975), Jacobsen and Knox (1973), Tremblay (1963), and Yomo and Taylor (1973). The research in all other papers cited here was carried out with new membranes such as nitrocellulose or nylon.

Tissue-printing provides a convenient way to screen at the same time on the same membrane many tissue sections from different plants or different developmental stages. For additional information on the application of the technique, see Cassab et al. (1988), Moore et al. (1988), McClure and Guilfoyle (1989a,b), Pont-Lezica and Varner (1989a,b), Shepherd et al. (1989), Barres et al. (1990), del Campillo et al. (1990), Reid and Pont-Lezica (1992), Reid et al. (1990), and Ye et al. (1991).

REFERENCES

Adams, C.W.M. and N.A. Tuqan. 1961. The histochemical demonstration of protease by a gelatin-silver film substrate. *J. Histochem. Cytochem.* **9:** 469–472.

Barres, B.A., W.J. Koreschetz, L.L. Chun, and D.P. Corey. 1990. Ion channel expression by white matter glia: The type I astrocyte. *Neuron* **5:** 527–544.

Cassab, G.I. and J.E. Varner. 1987. Immunocytolocalization of extensin in developing soybean seed coats by immunogold-silver staining and by tissue printing on nitrocellulose paper. *J. Cell Biol.* **105:** 2581–2588.

Cassab, G.I., J.-J. Lin, L.-S. Lin, and J.E. Varner. 1988. Ethylene effect on extensin and peroxidase distribution in the subapical region of pea epicotyls. *Plant Physiol.* **88:** 522–524.

del Campillo, E., P.D. Reid, R. Sexton, and L.N. Lewis. 1990. Occurrence and localization of 9.5 cellulase in abscising and nonabscising tissues. *Plant Cell* **2:** 245–254.

Daoust, R. 1957. Localization of deoxyribonuclease in tissue sections. A new approach to the histochemistry of enzymes. *Exp. Cell Res.* **12:** 203–211.

———. 1965. Histochemical localization of enzyme activities by substrate film methods: Ribonucleases, deoxyribonucleases, proteases, amylase and hyaluronidase. *Int. Rev. Cytol.* **18:** 191–221.

Gaddum, P. and R.J. Blandau. 1970. Proteolytic reaction of mammalian spermatozoa on gelatin membranes. *Science* **170:** 749–751.

Harris, N. and M.J. Chrispeels. 1975. Histochemical and biochemical observations on storage protein metabolism and protein body autolysis in cotyledons of germinating beans. *Plant Physiol.* **56:** 292–299.

Jacobsen, J.V. and R.B. Knox. 1973. Cytochemical localization and antigenicity of α-amylase in barley aleurone tissue. *Planta* **112:** 213–224.

McClure, B.A. and T. Guilfoyle. 1989a. Rapid redistribution of auxin-regulated RNAs during gravitropism. *Science* **243:** 91–93.

———. 1989b. Tissue print hybridization. A simple technique for detecting organ- and tissue-specific gene expression. *Plant Mol. Biol.* **12:** 517–524.

Moore, B.M., B. Kang, and W.H. Flurkey. 1988. Histochemical and immunochemical localization of tyrosinase in whole tissue sections of mushrooms. *Phytochemistry* **27:** 3735–3737.

Pont-Lezica, R. and J.E. Varner. 1989a. Localization of glycoconjugates and lectins by tissue blotting on solid supports. *J. Cell. Biochem.* (suppl.) **13A:** 130.

———. 1989b. Histochemical localization of cysteine-rich protein by tissue printing on nitrocellulose. *Anal. Biochem.* **182:** 334–337.

Reid, P. and R.F. Pont-Lezica, ed. 1992. *Tissue printing: Tools for the study of anatomy, histochemistry, and gene expression.* Academic Press, San Diego, California.

Reid, P.D., E. Del Campillo, and L.N. Lewis. 1990. Anatomical and immunolocalization of cellulase during abscission as observed on nitrocellulose tissue prints. *Plant Physiol.* **93:** 160–165.

Sambrook, J., E.F. Fritsch, and T. Maniatis. 1989. *Molecular cloning: A laboratory manual,* 2nd edition. Cold Spring Harbor Laboratory Press, Cold Spring Harbor, New York.

Shepherd, G.M.G., B.A. Barres, and D.P. Corey. 1989. "Bundle blot" purification and initial characterization of hair cell stereocilia. *Proc. Natl. Acad. Sci.* **86:** 4973–4977.

Spruce, J., A.M. Mayer, and D.J. Osborne. 1987. A simple histochemical method for locating enzymes in plant tissues using nitrocellulose blotting. *Phytochemistry* **26:** 2901–2903.

Tremblay, G. 1963. The localization of amylase activity in tissue sections by a starch-film method. *J. Histochem. Cytochem.* **11:** 202–206.

Varner, J.E. and R. Taylor. 1989. New ways to look at the architecture of plant cell walls: The localization of polygalaturonate blocks in plant tissues. *Plant Physiol.* **91:** 31–33.

Ye, Z.-H. and J.E. Varner. 1991. Tissue-specific expression of cell wall proteins in developing soybean tissues. *Plant Cell* **3:** 23–37.

Ye, Z.-H., Y.-R. Song, A. Marcus, and J.E. Varner. 1991. Comparative localization of three classes of cell wall proteins. *Plant J.* **1:** 175–183.

Yomo, H. and M.P. Taylor. 1973. Histochemical studies on protease formation in the cotyledons of germinating bean seeds. *Planta* **112:** 35–43.

• *TIMETABLE*

Tissue Printing to Detect Proteins and RNA in Plant Tissues

STEPS	*TIME REQUIRED*
Tissue-print Western Blots	
Prepare tissue section and make tissue prints on nitrocellulose paper	10 minutes
Stain tissue section with toluidine blue and photograph	10 minutes
Wash nitrocellulose paper	15 minutes
Incubate nitrocellulose paper in blocking buffer	3 hours
Incubate nitrocellulose paper with primary antibody	2 hours
Wash nitrocellulose paper	40 minutes
Incubate nitrocellulose paper with secondary antibody	1 hour
Wash nitrocellulose paper and equilibrate in AP buffer	1 hour
Stain tissue prints in AP buffer containing alkaline phosphatase substrates	10–60 minutes
Stop color development reaction and photograph stained tissue prints	20 minutes
Tissue-print Northern Blots	
Prepare ^{35}S-labeled RNA probe	2 hours
Prepare tissue section and make tissue prints on nylon membrane	10 minutes

STEPS	***TIME REQUIRED***
Tissue-print Northern Blots (continued)	
Stain tissue section with toluidine blue and photograph	10 minutes
Wash nylon membrane	4 hours
Prehybridize nylon membrane	4 hours
Incubate nylon membrane with ^{35}S-labeled RNA	20 hours
Wash nylon membrane	2 hours
Expose nylon membrane to T-max 400 film	4 hours–4 days
Develop T-max 400 film and photograph mRNA localization pattern	45 minutes

Tissue-print Western Blots

MATERIALS AND EQUIPMENT

Nitrocellulose paper (Schleicher & Schuell, Inc.)
$CaCl_2$ (0.2 M)
Toluidine blue (0.1% in tap water)
Kodak EKTACHROME 160 (for color slides; used for recording the anatomy of the section)
Primary antibody (diluted 1:15,000 in blocking buffer)
Secondary antibody (alkaline phosphatase-conjugated; diluted 1:20,000 in blocking buffer)
Alkaline phosphatase substrates (nitroblue tetrazolium chloride and 5-bromo-4-chloro-3-indoxylphosphate) (Promega Corporation S3771)
Kodak Technical Pan Film 2415 (for black and white photographs; used for recording the protein localization results)

REAGENTS

Washing buffer I
Blocking buffer
Washing buffer II
AP buffer
Washing buffer III
(For recipes, see Preparation of Reagents, pp. 92–94)

PROCEDURE

Tissue Printing on Nitrocellulose Paper

Nitrocellulose paper is usually used for tissue-print western blots. For the localization of plant cellular proteins, it is not necessary to pretreat the nitrocellulose. However, for the localization of plant cell-wall proteins, the nitrocellulose should be soaked in 0.2 M $CaCl_2$ for 30 minutes and air dried before use. $CaCl_2$ facilitates the release of cell-wall proteins bound via ionic interactions.

1. Place two layers of Whatman No. 1 filter paper on a plastic plate. On top of the filter paper, place one sheet of plain bond paper and then one piece of nitrocellulose paper.

2. Use a double-edge razor blade to prepare a freehand section (1 mm thick) from plant tissue such as soybean stem. If the surface of the tissue section is wet, blot it gently on Kimwipes or wash it in distilled H_2O for a few seconds before blotting it on Kimwipes. Use forceps to transfer the tissue section to the nitrocellulose paper, taking care not to move it once it has been placed on the paper. Place four layers of Kimwipes on top of the tissue section and press gently for 15–20 seconds, applying even pressure with a finger. Remove the Kimwipes and the tissue section carefully with forceps. Retain the tissue section for observation of the anatomy by staining with toluidine blue (see step 4 below).

   Note: Many tissue prints can be made on the same sheet of nitrocellulose paper.

3. Air dry the tissue-printed nitrocellulose paper.

4. Stain the tissue section with toluidine blue by placing it, printed side down, in a drop of the dye solution on a microscope slide. After 2–3 minutes, blot away the excess dye and photograph immediately using oblique lighting. Turn the slide over to photograph the stained side of the tissue section through the glass (the moist tissue will adhere to the slide). Tape small pieces of cardboard to the slide to prevent the tissue section from touching the microscope stage.

Detection of Proteins in Tissue Prints with Specific Antibodies

Perform all the following steps at room temperature.

1. Pour 30–40 ml of washing buffer I into a shallow dish and wash the tissue-printed nitrocellulose paper for 15 minutes.

2. Transfer the nitrocellulose paper to 15 ml of blocking buffer and incubate for 3 hours.

3. Transfer the nitrocellulose paper to 10 ml of a solution of primary antibody diluted in blocking buffer and incubate for 2 hours.

4. Wash the nitrocellulose paper in 40 ml of washing buffer I for 10 minutes. Repeat this step three times with fresh buffer.

5. Transfer the nitrocellulose paper to 10 ml of a solution of alkaline phosphatase-conjugated secondary antibody diluted in blocking buffer and incubate for 1 hour.

6. Wash the nitrocellulose paper in 40 ml of washing buffer I for 10 minutes.

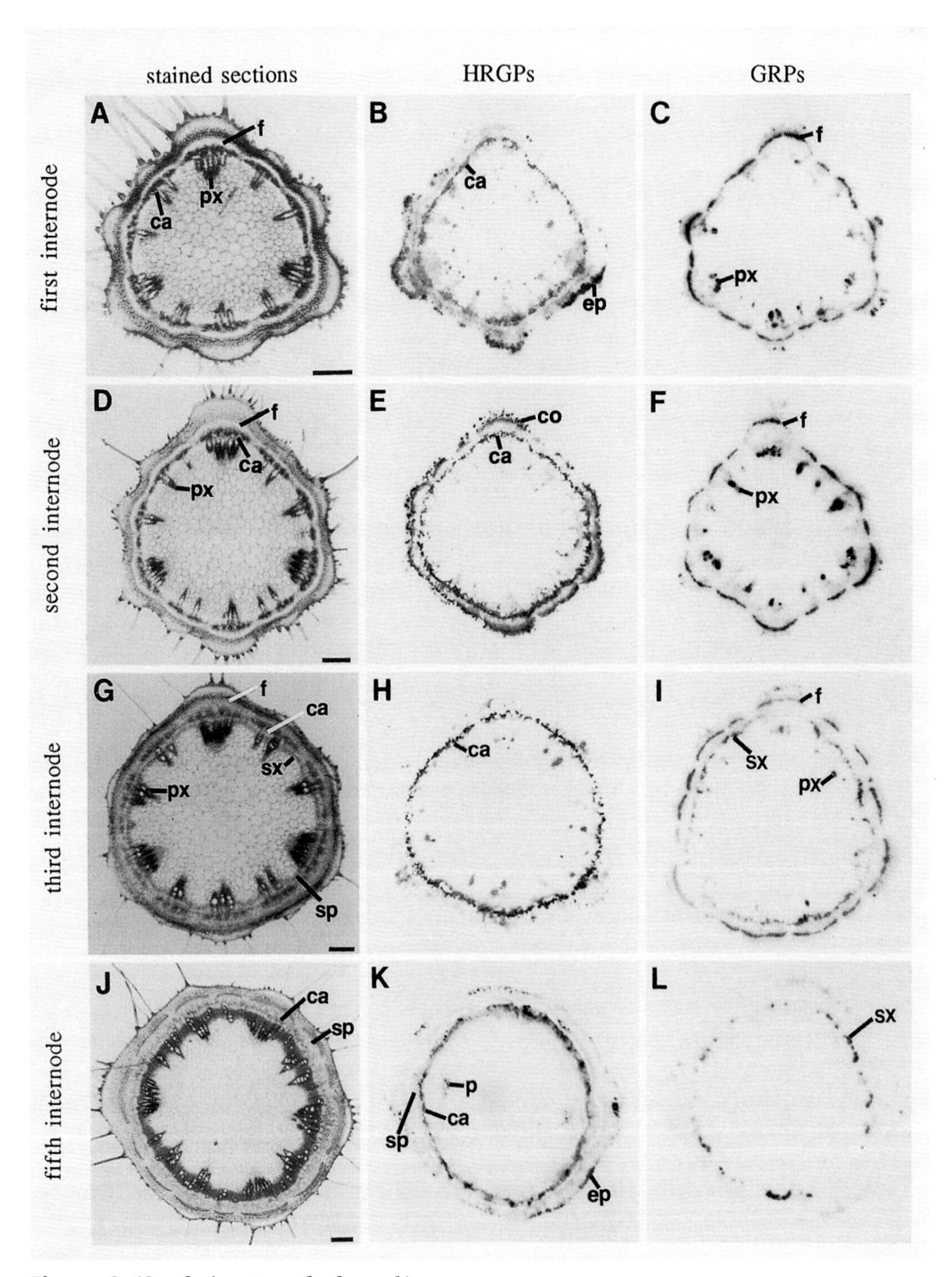

Figure 1 *(See facing page for legend.)*

7. Wash the nitrocellulose paper in 40 ml of washing buffer II for 15 minutes. Repeat this step once with fresh buffer.

8. Wash the nitrocellulose paper in 40 ml of washing buffer I for 10 minutes and equilibrate in 10 ml of AP buffer for 10 minutes.

9. For the color development reaction, place the nitrocellulose paper in 10 ml of AP buffer containing 66 μl of nitroblue tetrazolium chloride and 33 μl of 5-bromo-4-chloro-3-indoxyl-phosphate, both of which are substrates for alkaline phosphatase. A purple color will appear in the places where the protein of interest is localized. Satisfactory color development should occur within 10 minutes. If longer color development is needed, keep the reaction in the dark.

10. Stop the color development reaction by washing the nitrocellulose paper in 20 ml of washing buffer III for 5 minutes. Wash the paper in 40 ml of distilled H_2O for 5 minutes, and air dry.

11. Observe the stained proteins in the tissue print under a dissecting microscope. The protein localization pattern is seen by using transmitted light and incident light in succession. The physical print can be seen easily under incident light. To be certain of the protein localization in the tissue print, it is advisable to compare the stained tissue print directly with the anatomy of the tissue section stained with toluidine blue (see Fig. 1).

Figure 1

Detection of hydroxyproline-rich glycoproteins (HRGPs) and glycine-rich proteins (GRPs) in developing soybean stem by tissue-print western blots. Rabbit polyclonal antibodies raised against soybean seed coat HRGP (dilution 1:15,000) were used for the localization of HRGPs. Rabbit polyclonal antibodies raised against bean GRP1.8 fusion protein (dilution 1:15,000) were used for the localization of GRPs. The same dilution of rabbit preimmune serum was used for the control. Alkaline phosphatase-conjugated goat anti-rabbit antibody (dilution 1:20,000; Sigma) was used as the secondary antibody. HRGPs and GRPs were visualized by a purple color that developed as a result of the metabolism of the alkaline phosphatase substrates, nitroblue tetrazolium chloride and 5-bromo-4-chloro-3-indoxylphosphate (Ye and Varner 1991). The anatomy of the tissue section was observed by staining with toluidine blue. Abbreviations: (ca) cambium; (co) cortex; (f) primary phloem; (ep) epidermis; (p) parenchyma; (px) primary xylem; (sp) secondary phloem; (sx) secondary xylem. Scale bars, 300 μm. (Reprinted, with permission, from Ye and Varner 1991, copyright by American Society of Plant Physiologists.)

Tissue-print Northern Blots

MATERIALS AND EQUIPMENT

^{35}S-UTP (1 mCi/ml; DuPont NEN®)
Nitrocellulose paper (Schleicher & Schuell, Inc.)
Nylon membrane (Zeta-probe membrane, Bio-Rad Laboratories)
UV cross-linker (Bio-Rad Laboratories)
Toluidine blue (0.1% in tap water)
Kodak EKTACHROME 160 (for color slides; used for recording the anatomy of the section)
Kodak T-max 400 film
Kodak X-ray film
Kodak T-max developer
Kodak stop bath
Kodak rapid fixer
Kodak Technical Pan Film 2415

REAGENTS

Washing solution I
Hybridization solution
Washing solution II
Washing solution III
SSC (0.2x)
(For recipes, see Preparation of Reagents, pp. 92–94)

PROCEDURE

Preparation of Radiolabeled Probes

Either ^{32}P- or ^{35}S-labeled DNA and RNA probes can be used for tissue-print northern blots. Higher resolution is obtained with ^{35}S-labeled probes than with ^{32}P-labeled probes. Radiolabeled DNA and RNA probes can be synthesized according to Sambrook et al. (1989).

Tissue Printing on Nitrocellulose Paper or Nylon Membrane

Nitrocellulose paper or nylon membrane can be used for tissue-print northern blots. No pretreatment of either material is necessary. Nylon membrane is generally preferred for this procedure because it is easily handled and often the resolution is higher than that obtained with nitrocellulose.

1. Place two layers of Whatman No. 1 filter paper on a plastic plate. On top of the filter paper, place one sheet of plain bond paper and then one piece of nylon membrane.

2. Use a double-edge razor blade to prepare a freehand section (1 mm thick) from plant tissue such as soybean stem. If the surface of the section is wet, blot it gently on Kimwipes. Use forceps to transfer the tissue section to the nylon membrane, taking care not to move it once is has been placed on the membrane. Place four layers of Kimwipes on top of the tissue section and press gently for 15–20 seconds, applying even pressure with a finger. Remove the Kimwipes and the tissue section carefully with forceps. Retain the tissue section for observation of the anatomy by staining with toluidine blue (see step 4 below).

   Note: Many tissue prints can be made on the same sheet of nylon membrane, so this is a powerful tool for screening organ- and tissue-specific gene expression in different plants and different developmental stages at the same time. The results can be observed in 3 days (see below).

3. Bake the tissue-printed nylon membrane under vacuum for 2 hours at 80°C, or illuminate the membrane in ultraviolet (UV) light using a UV cross-linker (Bio-Rad Laboratories).

4. Stain the tissue section with toluidine blue by placing it, printed side down, in a drop of the dye solution on a microscope slide. After 2–3 minutes, blot away the excess dye and photograph immediately using oblique lighting. Turn the slide over to photograph the stained side of the tissue section through the glass (the moist tissue will adhere to the slide). Tape small pieces of cardboard to the slide to prevent the tissue section from touching the microscope stage.

Detection of mRNA in Tissue Prints with a ^{35}S-labeled RNA Probe

1. Pour 30–40 ml of washing solution I into a Seal-A-Meal™ plastic bag and wash the tissue-printed nylon membrane for 4 hours at 65°C.

2. Prehybridize the membrane by incubating it in 20 ml of hybridization solution for 4 hours at 68°C.

3. For the hybridization reaction, incubate the membrane with ^{35}S-labeled RNA (1 x 10^7 cpm/ml) in 10 ml of hybridization solution for 20 hours at 68°C.

4. Wash the membrane in 100 ml of washing solution II for 20 minutes at 42°C. Repeat this step twice with fresh solution.

5. Wash the membrane in 100 ml of washing solution III at an appropriate temperature (60–68°C). Monitor the amount of radioactivity that remains on the membrane using a hand-held minimonitor. After washing, the membrane hybridized with the sense-RNA probe should not emit a signal, whereas the membrane hybridized with the antisense-RNA probe generally emits a detectable signal with an intensity depending on the abundance of the mRNA of interest. Wash the membrane briefly with 0.2x SSC at room temperature. Air dry the membrane.

6. Expose the membrane to X-ray film or to T-max 400 film. Ensure that the membrane makes good contact with the film by using a flat weight, such as a short stack of books. Generally T-max 400 film gives better resolution than X-ray film. The exposure time depends on the intensity of signal and may vary from 4 hours to 4 days.

7. Develop the X-ray film in an X-ray film processor. Develop the T-max 400 film by soaking in Kodak T-max developer for 11 minutes. Transfer the film to a stop bath for 30 seconds, soak in Kodak rapid fixer for 5 minutes, and then wash in H_2O for 5 minutes.

8. Observe the localization of mRNA in the tissue print under a dissecting microscope. Photograph the mRNA localization pattern using Kodak Technical Pan Film 2415 and compare it with the anatomy of the tissue section stained with toluidine blue (Fig. 2).

Figure 2
Detection of mRNA for hydroxyproline-rich glycoproteins (HRGPs) and glycine-rich proteins (GRPs) in developing soybean stem by tissue-print northern blots. Carrot HRGP genomic DNA and bean GRP1.8 genomic DNA cloned in Bluescript plasmid vector (Stratagene) were used to synthesize ^{35}S-labeled antisense RNA probes by using T7 or T3 RNA polymerase (Ye and Varner 1991). Abbreviations: (ca) cambium; (co) cortex; (ep) epidermis; (f) primary phloem; (p) parenchyma; (pp) pith parenchyma; (px) primary xylem; (sp) secondary phloem; (sx) secondary xylem. Scale bars, 300 μm. (Reprinted, with permission, from Ye and Varner 1991, copyright by American Society of Plant Physiologists.)

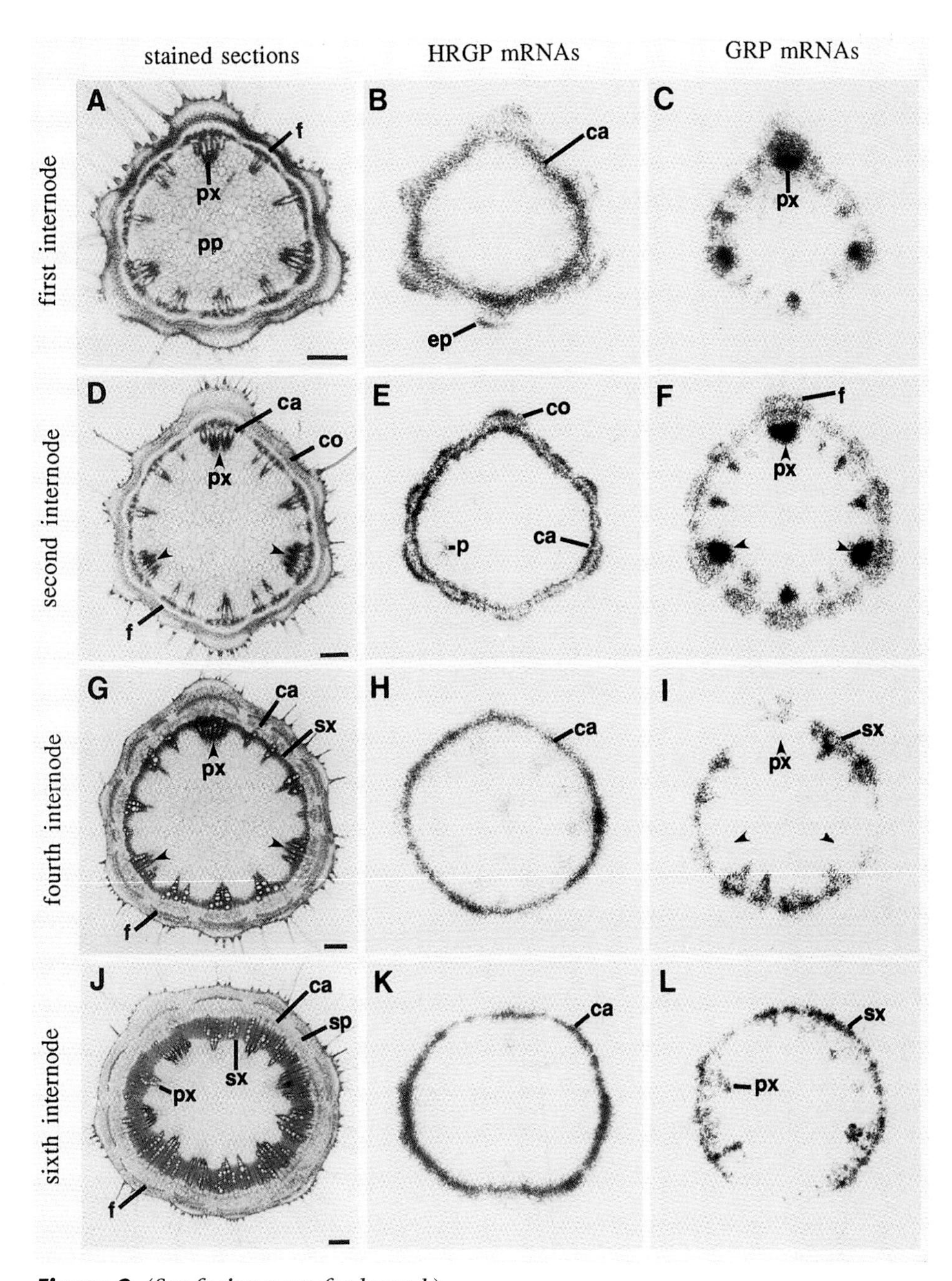

Figure 2 *(See facing page for legend.)*

• PREPARATION OF REAGENTS

TISSUE-PRINT WESTERN BLOTS

AP buffer

0.1 M Tris-HCl (pH 9.5)
0.1 M NaCl
5 mM $MgCl_2$

Store at room temperature.

Blocking buffer

0.1 M Tris-HCl (pH 8.0)
0.05% Sodium azide
0.25% Bovine serum albumin
0.25% Gelatin
0.3% Tween-20

Store at 4°C.

SAFETY NOTE

- Sodium azide is extremely toxic, and gloves should be worn at all times when handling solutions containing this compound.

Washing buffer I

0.1 M Tris-HCl (pH 8.0)
0.05% Sodium azide
0.3% Tween-20

Store at room temperature.

Washing buffer II

0.1 M Tris-HCl (pH 8.0)
0.05% Sodium azide
0.3% Tween-20
0.05% Sodium dodecyl sulfate

Store at room temperature.

Washing buffer III

10 mM Tris-HCl (pH 8.0)
1 mM EDTA

Store at room temperature.

TISSUE-PRINT NORTHERN BLOTS

Hybridization solution

1.5x SSPE
1% Sodium dodecyl sulfate
5x Denhardt's solution
0.1 mg/ml Salmon sperm DNA
100 mM Dithiothreitol

Store at –20°C.

SSPE (1.5x)

0.22 M NaCl
15 mM $NaH_2PO_4 \cdot H_2O$
2 mM EDTA (pH 7.4)

Denhardt's solution (5x)

0.1% Ficoll
0.1% Polyvinylpyrrolidone
0.1% Bovine serum albumin

SSC (0.2x)

0.03 M NaCl
3 mM Sodium citrate (pH 7.0)

Washing solution I

0.2x SSC
1% Sodium dodecyl sulfate

Store at room temperature.

Washing solution II

2x SSC
0.1% Sodium dodecyl sulfate

Store at room temperature.

SSC (2x)

0.3 M NaCl
30 mM Sodium citrate (pH 7.0)

Washing solution III

0.2x SSC
0.1% Sodium dodecyl sulfate

Section 6
Immunolocalization of Proteins in Fixed and Embedded Plant Tissues

Antibodies are powerful tools in biological research. The mammalian immune system is capable of responding to a vast array of foreign substances and, therefore, antibodies are potentially useful for the study of proteins, nucleic acids, polysaccharides, lipids, drugs, and other biological molecules. However, proteins generally elicit a stronger immune response than do the other classes of molecules mentioned and they contain a greater diversity of sites suitable for antibody recognition. For these reasons, antibodies are most commonly used for the study of proteins and they have a wide variety of applications. For example, they can be used for (i) measurement of steady-state levels and rates of synthesis of proteins; (ii) purification of proteins by immunoprecipitation and affinity chromatography; (iii) immunolocalization of proteins; (iv) cloning of genes encoding proteins using expression libraries; and (v) identification of ligand-binding proteins using anti-idiotypic antibodies.

Immunolocalization studies have been performed in a wide range of plants using antibodies specific for a variety of proteins. For example, antibodies specific for red-absorbing and far-red-absorbing forms of phytochrome have been used to characterize the light-induced redistribution of this photosensory protein in oat and soybean seedlings (McCurdy and Pratt 1986; Cope and Pratt 1992). The developmental accumulation and distribution of enzymes involved in C4 photosynthesis have been detailed in the C4 dicot amaranth (Wang et al. 1992). Vegetative storage proteins have been localized within the vacuoles of a variety of cell types in soybean leaves, stems, petioles, and pods (Huang et al. 1991). Extensin proteins have been localized in the cell walls of developing soybean seed coats (Cassab and Varner 1987) and also in the xylem elements of loblolly pine (Bao et al. 1992). Immunolocalization studies in plants are not limited to protein-specific antibodies. For example, antibodies specific for arabinogalactan epitopes have been used for immunolocalization in pea root tips (Pennel et al. 1989) and in tobacco pollen tubes (Li et al. 1992).

In this section, we describe techniques for utilizing antibodies to localize proteins. Successful immunolocalization of specific proteins in biological tissues is dependent upon several important factors, including (i) the abundance of the protein within the tissue; (ii) the specificity of the antibody for the target protein; (iii) the accessibility of the protein to the antibody; and (vi) the inclusion of appropriate controls to distinguish between specific signals and nonspecific background.

The major factor influencing the success or failure of any immunolocalization experiment is the amount of the protein of interest present within the plant tissue. Proteins that accumulate to high levels and are concentrated at specific points within the tissue are generally the easiest to detect. In contrast, a protein that accumulates at low levels and/or is widely dispersed within the tissue may be difficult to detect. Unfortunately, this is a factor over which the scientist has little control. Nevertheless, the sensitivity of immunolocalization protocols should allow for the detection of proteins present at levels ranging from 1000–10,000 molecules per cell, provided they are not too diffuse within the cell (Harlow and Lane 1988).

Both polyclonal and monoclonal antibodies can be used for immunolocalization. Since antiserum contains a number of different antibodies, each capable of binding the target protein at a unique site or epitope, polyclonal antibodies generally provide a stronger signal for immunolocalization studies than do monoclonal antibodies. However, not all the antibodies within the antiserum will be specific for the protein of interest and these nonspecific antibodies will occasionally cause high background in immunolocalization experiments. In contrast, monoclonal antibodies, by definition, are very specific and recognize a single epitope on the protein of interest. However, since these antibodies bind only a single epitope on the target protein, they may give a weaker signal than do polyclonal antibodies when used in immunolocalization experiments. Moreover, because an epitope contains a limited number of amino acids, occasionally a monoclonal antibody will react with an unrelated protein which, by chance, contains the same combination of amino acids that comprise the epitope on the target protein. Before attempting to use an antibody for immunolocalization studies, its specificity should be assessed by immunoblot and/or immunoprecipitation analysis. Obviously antibodies that cross-react with unrelated proteins will make correct interpretation of an immunolocalization experiment difficult, if not impossible.

Inappropriate tissue processing during fixation and embedding may affect the accessibility of the protein of interest to the antibody and may decrease the likelihood of antibody binding. Overfixation of tissue may mask the epitopes on the target protein that are recognized by the antibody, whereas underfixation may result in extraction or redistribution of the protein. Also, the temperatures and/or chemicals involved in some embedding procedures may denature target proteins making them unrecognizable to specific antibodies.

In immunolocalization experiments, appropriate controls are absolutely imperative for distinguishing between specific signals and nonspecific background. These may include preimmune controls, conjugate controls and, if you are working with transformed or infected plants, nontransformed or uninfected controls. Nonspecific signals may be the result of cross-reaction of antibodies with unrelated proteins, autofluorescence of pigments or lignified tissue, or endogenous enzyme activity.

Using the procedure outlined below, we have successfully used a variety of polyclonal and monoclonal antibodies for the immunolocalization of pathogenesis-related (PR) proteins in leaves of tobacco plants infected with tobacco mosaic virus (Dixon et al. 1991) (see Fig. 1). We have also used this procedure to detect PR proteins in floral tissues, extensin in the cell walls of leaves and stems, and the large and

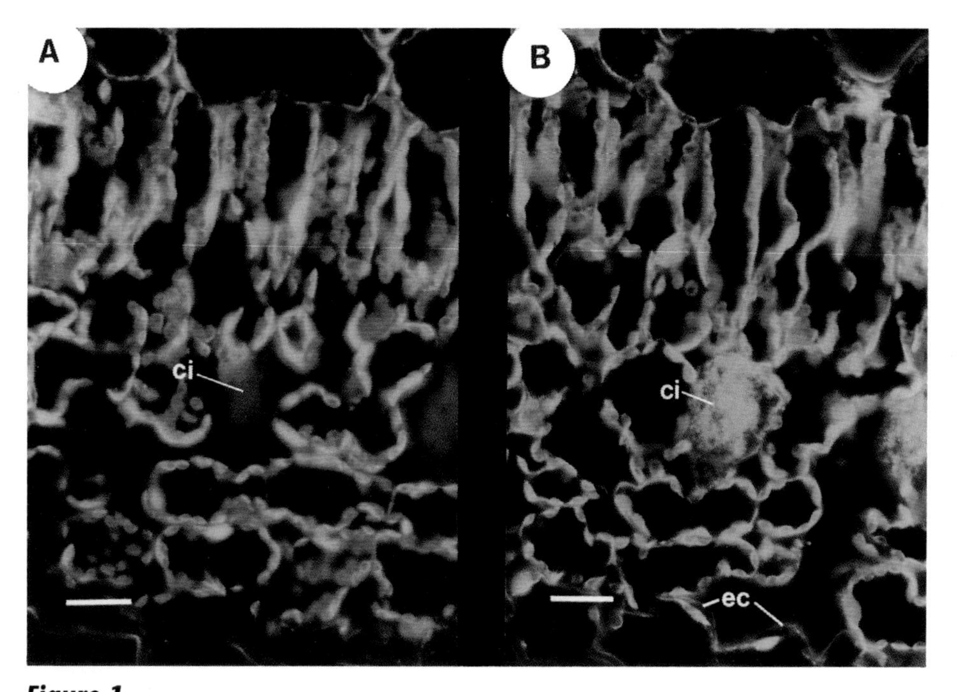

Figure 1
Adjacent, serial cross sections of a tobacco mosaic virus-infected tobacco leaf incubated (A) with a control monoclonal antibody or (B) with a monoclonal antibody specific for pathogenesis-related (PR-1) proteins. PR-1 proteins are synthesized in response to tobacco mosaic virus infection and are detected within crystal idioblast vacuoles (ci) and in the extracellular spaces (ec) of the leaf.

small subunits of ribulose bisphosphate carboxylase in chloroplasts. We feel that this immunolocalization procedure should be generally applicable to a wide variety of proteins and to most plant tissues.

REFERENCES

Bao, W., D.M. O'Malley, and R.R. Sederoff. 1992. Wood contains a cell-wall structural protein. *Proc. Natl. Acad. Sci.* **89:** 6604–6608.

Berlyn, G.P. and J.P. Miksche. 1976. *Botanical microtechnique and cytochemistry.* Iowa State University Press, Ames.

Cassab, G.I. and J.E. Varner. 1987. Immunocytolocalization of extensin in developing soybean seed coats by immunogold-silver staining and by tissue printing on nitrocellulose paper. *J. Cell Biol.* **105:** 2581–2588.

Cope, M. and L.H. Pratt. 1992. Intracellular redistribution of phytochrome in etiolated soybean (*glycine max* L.) seedlings. *Planta* **188:** 115–122.

Dixon, D.C., J.C. Cutt, and D.F. Klessig. 1991. Differential targeting of the tobacco PR-1 pathogenesis-related proteins to the extracellular space and vacuoles of crystal idioblasts. *EMBO J.* **10:** 1317–1324.

Gray, P. 1954. *The microtomist's formulary and guide.* Blakiston, Philadelphia.

Harlow, E. and D. Lane. 1988. *Antibodies: A laboratory manual.* Cold Spring Harbor Laboratory Press, Cold Spring Harbor, New York.

Huang, J.F., D.J. Bantroch, J.S. Greenwood, and P.E. Staswick. 1991. Methyl jasmonate treatment eliminates cell-specific expression of vegetative storage protein genes in soybean leaves. *Plant Physiol.* **97:** 1512–1520.

Li, Y.Q., L. Bruun, E.S. Pierson, and M. Cresti. 1992. Periodic deposition of arabinogalactan epitopes in the cell wall of pollen tubes of *Nicotiana tabacum* L. *Planta* **188:** 532–538.

McCurdy, D.W. and L.H. Pratt. 1986. Kinetics of intracellular redistribution of phytochrome in *Avena* coleptiles after its photoconversion to the active, far-red-absorbing form. *Planta* **167:** 330–336.

Pennel, R.J., J.P. Knox, G.N. Scofield, R.R. Selvendran, and K. Roberts. 1989. A family of abundant plasma membrane-associated glycoproteins related to the arabinogalactan proteins is unique to flowering plants. *J. Cell Biol.* **108:** 1967–1977.

Seagull, R.W. 1986. Changes in microtubule organization and wall microfibril orientation during *in vitro* cotton fiber development: An immunofluorescent study. *Can. J. Bot.* **64:** 1373–1381.

Wang, J.L., D.F. Klessig, and J.O. Berry. 1992. Regulation of C4 gene expression in developing amaranth leaves. *Plant Cell* **4:** 173–184.

• *TIMETABLE*
Immunolocalization of Proteins in Fixed and Embedded Plant Tissues

STEPS	***TIME REQUIRED***
Formaldehyde Fixation and Embedding of Plant Tissues[a]	
Fix tissue pieces in formaldehyde	2–3 hours
Wash tissue pieces	20 minutes
Dehydrate tissue pieces by incubating in a series of dilutions of ethanol in H_2O	1.5 hours[b]
Incubate in 100% ethanol	overnight
Infiltrate tissue pieces with xylene by incubating in a series of dilutions of xylene in ethanol	5 hours[b]
Incubate tissue pieces in a series of dilutions of Paraplast Plus™ in xylene	8 hours[b]
Incubate in 100% Paraplast Plus™	overnight
Embed the tissue pieces in blocks of Paraplast Plus™	1–2 hours
Sectioning and Mounting of Embedded Plant Tissue[a]	
Cut and mount tissue sections	1–3 hours
Dry slides in an incubator	overnight
Immunostaining	
Remove Paraplast Plus™ from mounted tissue sections and rehydrate	30 minutes

STEPS	*TIME REQUIRED*
Immunostaining (continued)	
Wash slides in TSBSA	5 minutes
Incubate slides with primary antibody	30–45 minutes
Wash slides in TSBSA	5 minutes
Incubate slides with secondary antibody	30–45 minutes
Wash slides in TSBSA	5 minutes
Apply Gel/Mount™	1 minute per slide

[a]Fixation, embedding, and sectioning of tissue can be done in advance. The embedded tissue sections mounted on slides are stable for several months.
[b]The times listed for these procedures are general guidelines and can be varied to meet specific needs. We have had no problems extending these times to meet laboratory schedules. However, incubation times shorter than those listed may not give adequate dehydration and/or infiltration.

Formaldehyde Fixation and Embedding of Plant Tissues

MATERIALS AND EQUIPMENT

PIPES buffer (50 mM; pH 6.8)
Ethanol (100%, 75%, 50%, and 25% in H_2O)
Xylene (100%, 75%, 50%, and 25% in ethanol)
Paraplast Plus™ (Fisher Scientific 12-646-111, or similar) (100%, 75%, 50%, and 25% in xylene)
Histological base molds (Fisher Scientific 15-182-505B, or similar)
Slide warmer
Microtome

REAGENTS

Fixative solution
(For recipe, see Preparation of Reagents, p. 110)

SAFETY NOTES

- The fixative solution contains formaldehyde which is toxic and a carcinogen. It is readily absorbed through the skin and is destructive to the skin, eyes, mucous membranes, and upper respiratory tract. Formaldehyde should only be used in a chemical fume hood. Gloves and safety glasses should be worn.
- Xylene should always be handled in a fume hood. It is flammable and may be narcotic at high concentrations.

PROCEDURE

Several fixatives can be used successfully for the preservation of plant tissues. Each has advantages and disadvantages. We have found formaldehyde fixation to be adequate for all immunolocalization procedures performed in our laboratory (Dixon et al. 1991). Formaldehyde fixation gives good tissue morphology for light microscopy and is compatible with the use of fluorescently tagged antibodies. Penetration of this fixative into the plant tissue can be facilitated by the application of a vacuum. Procedures using other fixatives are mentioned briefly at the end of this section.

1. Cut the plant tissue into small pieces (e.g., cut leaf tissue into strips approximately 2 x 10 mm). Immediately transfer the tissue pieces into a beaker containing 10–50 ml of fixative solution. Place the beaker in a vacuum desiccator; apply a gentle vacuum and release it slowly. Apply and release the vacuum as necessary to facilitate the movement of fixative into the tissue. After the tissue pieces are well infiltrated, allow them to fix for 2–3 hours at room temperature.

   Note: (i) The volume of fixative required depends on the amount of tissue being fixed. (ii) Tissue pieces that are well infiltrated with fixative have a "water-soaked" appearance and generally sink to the bottom of the container as liquid replaces air in the tissue.

2. Wash the tissue pieces in 50–100 ml of 50 mM PIPES buffer (pH 6.8) for 20 minutes at room temperature.

3. Dehydrate the tissue pieces by incubating them in 50–100 ml of each of a graded series of dilutions of ethanol in H_2O (25%, 50%, 75%, and 100%) for 20 minutes at room temperature. Incubate the tissue pieces in 50–100 ml of 100% ethanol overnight at 4°C. Exchange of solvent in the tissue can be facilitated by gentle stirring or agitation.

4. In the morning, transfer the tissue pieces to 50–100 ml of fresh 100% ethanol and incubate for an additional 30 minutes.

   Note: If the plant tissue is green, all the chlorophyll should be extracted by this stage. Excess chlorophyll or other pigments remaining in the tissue may produce background fluorescence when fluorescently tagged secondary antibodies are used.

5. Infiltrate the tissue pieces with xylene by incubating them in 50–100 ml of each of a graded series of dilutions of xylene in ethanol (25%, 50%, 75%, and 100%) for 60 minutes at room temperature. Repeat the incubation in 100% xylene twice.

6. Infiltrate the tissue pieces with increasing concentrations of Paraplast Plus™ by incubating them in 50–100 ml of each of a graded series of dilutions of Paraplast Plus™ in xylene (25%, 50%, 75%, and 100%) for 2 hours at 60°C. Incubate the tissue pieces in 100% Paraplast Plus™ overnight at 60°C. During the incubations at 60°C, we have found that beakers in a water bath are better at holding the correct temperature than are ovens, but you need to be sure each beaker is well covered to prevent water from condensing into the embedding medium.

   Note: The manufacturer of Paraplast Plus™ cautions against heating above 62°C. Temperatures in excess of 62°C may adversely effect sectioning of the embedded tissue.

7. The tissue pieces can now be embedded in blocks of Paraplast Plus™. To make the blocks, place histological base molds on a slide warmer set at approximately 60°C (a hot plate set at its lowest temperature should do if you do not have a slide warmer). Avoid overheating. Fill the bottom of the mold with melted Paraplast Plus™. Transfer the tissue pieces to the mold and arrange them using a bacterial inoculating needle. When the tissue is arranged according to your needs, add more melted Paraplast Plus™, and place into the mold the appropriate adaptor for fitting your block onto the microtome. If necessary, add more melted Paraplast Plus™ to insure firm attachment of the adaptor when the Paraplast Plus™ cools. Transfer the mold to the bench top and allow it to cool. The cooled blocks can be left in the mold until you are ready to section them.

   Note: It is useful to have a Bunsen burner nearby to warm equipment (pipettes, forceps, inoculating needles, etc.) that will be inserted into the melted Paraplast Plus™. Put as many pieces of tissue into the mold as will conveniently fit. Placing as much tissue as possible in the mold will improve your chances of finding what you are looking for. Arrange the tissue pieces in an orientation that meets your sectioning needs. For example, it is impossible to cut cross sections from tissue pieces that are placed transversely into the mold.

8. To remove the blocks from the mold, place the mold firmly in ice, allow 5–10 seconds for the metal mold to contract, then pry the wax block out with a spatula.

   Note: Embedded tissue is stable for several months or longer when stored at room temperature.

ALTERNATIVE FIXATIVES

Glutaraldehyde is probably the best fixative for the preservation of tissue and cellular morphology and should be used when preservation of cellular structure is a high priority. The procedure for fixing plant tissues with glutaraldehyde is essentially the same as that described above for formaldehyde, except that the fixative solution contains 1–2% glutaraldehyde in 50 mM PIPES buffer (pH 6.8).

SAFETY NOTE

- Glutaraldehyde is toxic. It is readily absorbed through the skin and is extremely destructive to the skin, eyes, mucous membranes, and upper respiratory tract. Glutaraldehyde should only be used in a chemical fume hood. Gloves and safety glasses should be worn.

Tissue pieces fixed with glutaraldehyde fluoresce brightly and cannot be used in immunolocalization experiments with fluorescently labeled antibodies, unless they are pretreated with sodium borohydride to reduce free aldehydes. This is achieved by washing the tissue pieces three times in a solution of sodium borohydride in H_2O (1 mg/ml; prepared immediately before use) for 10 minutes at room temperature (Seagull 1986). The most convenient stage at which to treat the tissue slices with sodium borohydride is after the embedded tissue has been sectioned, mounted, and rehydrated and before it is treated with Tris-buffered saline containing bovine serum albumin (see p. 107).

SAFETY NOTE

- Hydrogen is evolved when sodium borohydride comes into contact with water. Care should be taken to avoid open flames.

Ethanol/acetic acid (3:1) easily penetrates plant tissues and gives rapid fixation. However, the resultant morphology in tissues fixed by this method is not quite as good as that in tissues fixed with formaldehyde or glutaraldehyde. The procedure for fixing plant tissue with ethanol/acetic acid (3:1) is similar to that described for formaldehyde and glutaraldehyde, with the following modifications. This fixative penetrates the tissue rapidly and hence application of a vacuum is not essential. When dehydrating tissue fixed with ethanol/acetic acid (3:1), it is not necessary to wash with 25% and 50% ethanol in H_2O. Dehydration can begin with a wash in 75% ethanol in H_2O.

Sectioning and Mounting of Embedded Plant Tissue

EQUIPMENT

Corning frosted microscope slides (75 x 25 mm) (Fisher Scientific 12-568-15, or similar)
Microtome

REAGENTS

Tissue section adhesive solution
(For recipe, see Preparation of Reagents, p. 110)

PROCEDURE

1. Pour approximately 1 liter of tissue section adhesive solution into a Pyrex dish and place it in a water bath at 45°C.

   Note: A wide variety of preparations have been described for the adhesion of tissue sections to microscope slides (Gray 1954). The most reliable of these involve the use of egg albumin, gelatin, or polylysine. We use a modification of a method described in Berlyn and Miksche (1976) that is inexpensive, requires minimal slide preparation, and has consistently given good results with leaf, stem, root, and floral tissue from tobacco plants.

2. Using a sharp, single-edge razor blade, trim the excess wax from the edges of the embedded tissue block, leaving a cleanly cut, square edge rather than the rounded edges produced by the mold. Make sure the top and bottom edges of the block are parallel. Lock the block into the microtome chuck so that the trimmed parallel edges are at the top and bottom of the block.

SAFETY NOTE

- Microtome blades are **extremely** sharp! Use care when sectioning. If you are unfamiliar with the use of a microtome, it may be appropriate to have someone demonstrate its use.

---

3. Cut a ribbon of sections 8–10 μm in thickness from the block. Using a moistened fingertip and a small artist's paintbrush, trans-

fer the ribbon (shiny side down) to the tissue section adhesive solution in the Pyrex dish. Allow the ribbon to float on the solution for 2–3 minutes, then pick up the sections with a microscope slide. To do this, hold the slide in a near vertical orientation by the frosted end and dip it into the solution. Align the edge of the slide at the joint between the first and second sections in the ribbon (see Fig. 2). Move the slide toward the section until it touches, then gently lift. With a little practice, a single section should separate from the ribbon as you lift the slide. The paintbrush can be used to hold the ribbon steady while you separate and lift the individual sections from the ribbon. To reposition the section on the slide, lower the slide into the adhesive solution to refloat the single section. With the slide still held in a near vertical orientation, use the paintbrush to position the section at the face of the slide. Then gently lift the slide from the adhesive solution.

Note: No preparation of slides is necessary for immunolocalization procedures if good quality, precleaned slides are used. We have found Corning 75 x 25 mm single frosted slides to be acceptable for most immunolocalization work in our laboratory. Mount at least two adjacent sections of tissue on each slide, making sure they are well separated and in approximately the same orientation. This facilitates microscopic examination and interpretation of the data since one section can be stained with antibody specific for the protein of interest and the other can be stained with control or preimmune antibody.

4. Stand the slides on edge and dry them in an incubator overnight at 42°C.

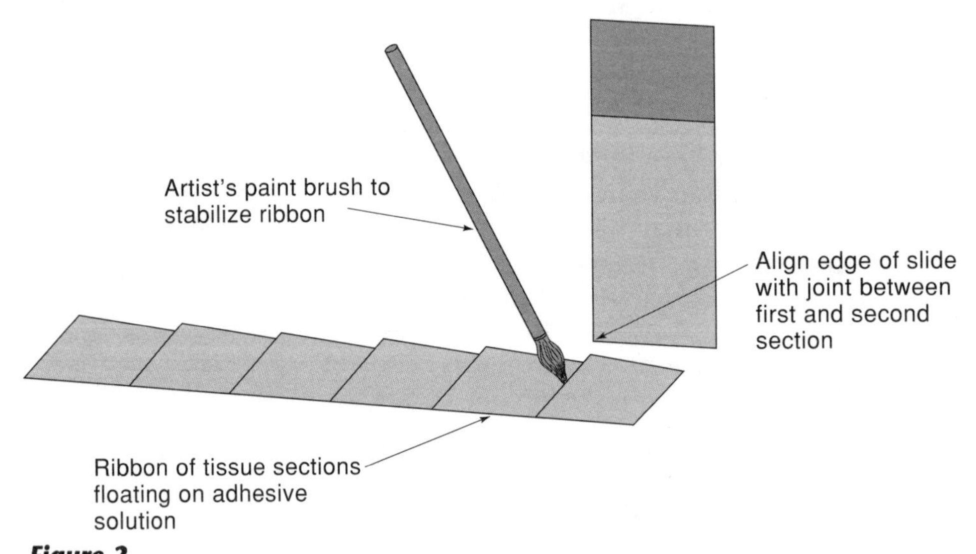

Figure 2
Diagram illustrating the technique for mounting sections of embedded tissue on microscope slides.

Immunostaining

MATERIALS

Xylene (100%)
Ethanol (100%, 95%, 75%, 50%, and 25% in H_2O)
Primary antibody against protein of interest
Secondary antibody conjugate (Sigma F 6257, or similar)
Gel/Mount™ coverslip mounting medium (Fisher Scientific BM-MOI, or similar)

REAGENTS

Tris-buffered saline with bovine serum albumin (TSBSA)
(For recipe, see Preparation of Reagents, p. 110)

SAFETY NOTE

- Xylene should always be handled in a fume hood. It is flammable and may be narcotic at high concentrations.

PROCEDURE

1. Remove the Paraplast Plus™ from the mounted tissue sections by washing each slide twice in 100% xylene for 10 minutes. Transfer the slides to 100% ethanol and wash them by repeatedly dipping until all streaks disappear.

2. Rehydrate the mounted tissue sections by washing the slides in each of a graded series of dilutions of ethanol in H_20 (95%, 75%, 50%, and 25%) at room temperature. At each stage, wash the slides by repeatedly dipping until all streaks disappear.

3. Transfer the slides to TSBSA and incubate them for 5 minutes at room temperature.

   Note: In this experiment the salt concentration in the TSBSA is 150 mM, but it can be increased to 300 mM to reduce nonspecific binding of the primary and secondary antibodies.

4. Take a plastic container with a sealable lid, line the bottom with sponges or paper towels, and saturate them with water.

5. Remove *one* slide from the TSBSA and drain it. Wipe the excess buffer from the back and edges of the slide with a Kimwipe, taking care not to disturb the sections. With a cotton swab, thoroughly dry the edges of the slide and the areas between the sections. Place the slide on top of the wet sponge or towel in the plastic container. Work quickly so the tissue sections do not dry.

6. Pipette approximately 100 μl of primary antibody diluted with TSBSA onto each tissue section mounted on the slide.

   Note: The dilution of primary antibody required for immunolocalization procedures is determined empirically for each antibody used. Antibody dilutions similar to those used for immunoblots seem to be a good starting point. For example, we have found that the primary antibody dilution used for immunoblots probed with ^{125}I-labeled *Staphylococcus aureas* protein A is appropriate for immunolocalization experiments using fluorescently labeled secondary antibodies. However, immunoblots in which the protein is visualized by chemiluminescence (Amersham ECL, or similar) appear to be approximately fivefold more sensitive than those probed with ^{125}I-labeled protein A. Therefore, if a primary antibody is diluted 1:5000 for immunoblots probed by chemiluminescence, it should be diluted approximately 1:1000 for immunolocalization with a fluorescently labeled antibody. If necessary, nonspecific staining can be reduced by including normal serum with the diluted primary antibody. Normal serum (diluted 1:10 to 1:100 in the primary antibody solution) should be derived from the same species used to produce the secondary antibody. As mentioned in the previous section, if two adjacent tissue sections are mounted on each slide, one section can be incubated with specific antibody and the other can be incubated with a control or preimmune antibody.

7. Repeat steps 5 and 6 until all the slides have been treated.

8. Incubate the slides in the sealed plastic container for 30–45 minutes at 37°C.

9. Inspect each slide to check that the antibody solutions on the tissue sections have not mixed (this should not happen if the areas surrounding the sections are well dried with the cotton swab). Remove the slides one at a time and rinse the antibody solution off each section with a gentle stream of TSBSA. Place the slides in a staining rack submerged in TSBSA.

10. Wash the slides in TSBSA with occasional gentle agitation for 5 minutes.

    Note: Additional washes may be necessary to reduce nonspecific antibody binding.

11. Repeat steps 5–7, using approximately 100 µl of affinity-purified secondary antibody diluted in TSBSA.

    Note: Follow the manufacturer's recommendations for initial dilutions of the secondary antibody conjugate. The procedure described here was originally devised for use with an affinity-purified, sheep anti-mouse antibody conjugated to FITC (Sigma F 6257). However, the procedure can be readily adapted for use with other antibody conjugates. When selecting an antibody conjugate for the visualization of primary antibody bound specifically to the tissue section, keep the following general points in mind. Fluorescent antibody conjugates offer good sensitivity and high resolution, but may not be appropriate if the tissue has a high level of autofluorescence. Enzyme conjugates have higher sensitivity than fluorescent or gold-labeled conjugates, but they do not provide the same level of resolution. Moreover, endogenous enzyme activities in the tissue may cause nonspecific background. Colloidal gold conjugates provide high resolution and can offer high sensitivity if they are silver enhanced. Gold-labeled antibodies may be particularly useful with tissue that has high autofluorescence and high endogenous enzyme activities. Although the use of all of these conjugates can be readily adapted to this protocol, detailed use of each is beyond the scope of this section. Therefore, the investigator may wish to refer to the chapter on cell staining in Harlow and Lane (1988) for more specific information on the use of each of these conjugates.

12. Incubate the slides in the sealed plastic container for 30–45 minutes at 37°C.

13. Remove the slides one at a time and rinse the antibody solution off each section with a gentle stream of TSBSA. Place the slides in a staining rack submerged in TSBSA.

14. Wash the slides in TSBSA with occasional gentle agitation for 5 minutes.

    Note: Additional washes may be necessary to reduce nonspecific antibody binding.

15. Remove the slides one at a time from the TSBSA and shake off the excess buffer. Apply approximately 60 µl of Gel/Mount™ (or other mounting medium) near the frosted end of each slide. Touch the edge of a 22 x 50-mm coverslip to the drop of mounting medium and slowly lower the coverslip to spread the mounting medium uniformly over the tissue sections. If bubbles form over the tissue section, return the slide to TSBSA, allow the coverslip to drop off, and remount the slide with fresh Gel/Mount™ and a clean coverslip.

16. Allow the Gel/Mount™ to dry at least partially (i.e., for 10–15 minutes) before examining the slide under the microscope.

• PREPARATION OF REAGENTS

Fixative solution (4% formaldehyde in 50 mM PIPES buffer [pH 6.8])
(50 ml)

10% Formaldehyde	20 ml
500 mM PIPES buffer (pH 6.8)	5 ml

Adjust the volume to 50 ml with H_2O. If necessary, adjust the pH to 6.8. Make fresh immediately before use. Do not store.

Formaldehyde (10%)
(20 ml)

In the fume hood, add 2 g of *p*-formaldehyde to 10 ml of H_2O and mix well. Heat to 60°C while stirring. Add 5 M NaOH drop by drop until all the *p*-formaldehyde has dissolved. Adjust the volume to 20 ml with H_2O. Adjust the pH to near neutral. Alternatively, prepared formaldehyde stock solutions can be purchased in sealed glass vials from microscopy suppliers.

SAFETY NOTE

- Formaldehyde is toxic and a carcinogen. It is readily absorbed through the skin and is destructive to the skin, eyes, mucous membranes, and upper respiratory tract. Formaldehyde should only be used in a chemical fume hood. Gloves and safety glasses should be worn.

Tissue section adhesive solution
(1 liter)

Gelatin (Sigma G 2500, 300 Bloom)	0.1%
Chromium potassium sulfate	0.02%

Sprinkle 1 g of gelatin on top of 1 liter of H_2O. Heat in a microwave until all the gelatin has dissolved. Cool the gelatin solution to 45°C and add 0.2 g of chromium potassium sulfate. Store at 4°C.

Tris-buffered saline with bovine serum albumin (TSBSA)
(1 liter)

Dissolve 1 g of bovine serum albumin (fraction V) in 1 liter of 50 mM Tris-HCl (pH 7.4) containing 150 mM NaCl. Store at 4°C.

• *Section 7*

In Situ Hybridization for the Detection of RNA in Plant Tissues

In situ hybridization is not a new technique—it has been used in animal systems for over 30 years and more recently has been adapted and optimized for use in plant systems. The basic principle of this technique is that a labeled, single-stranded nucleic acid probe will bind to a complementary strand of cellular DNA or RNA and, under the appropriate conditions, these molecules will form a stable hybrid. When nucleic acid hybridization is combined with histological techniques, specific DNA or RNA sequences can be identified within a single cell. Thus, the advantage of in situ hybridization over more conventional techniques for the detection of nucleic acids (northern and Southern blots) is that it allows an investigator to determine the precise spatial distribution of a nucleic acid sequence in a heterogeneous cell population. In situ hybridization can be used to map the chromosomal locations of genes and it can also serve as a diagnostic tool for the detection of viral (pathogen) nucleic acid sequences (Wu and Davidson 1981; McDougall et al. 1986). However, in situ hybridization is most often used to localize specific RNA sequences in cells (Angerer et al. 1984, 1985; Dixon et al. 1991). It is a highly sensitive technique and, under optimal conditions, RNA sequences present in as few as 5–10 copies per cell can be detected (Hardin et al. 1989). Typically, in situ hybridization is used to measure the steady-state level of RNA accumulation. However, in some studies, the technique has been adapted to examine the de novo transcription of specific genes (Raikhel et al. 1988).

In situ hybridization experiments involve a number of basic steps—tissue preparation, synthesis of probes, hybridization, washing, and data analysis—and each procedure is discussed below.

TISSUE PREPARATION

The procedures described in Section 6 for the fixation and embedding of plant tissue for immunolocalization are appropriate for the preparation of tissue for in situ hybridization studies. Only minor modifications of the procedures are needed to minimize degradation of target

RNA by RNases. Some researchers have obtained good results using cryostat sections for in situ hybridization (Meyerowitz 1987; Nasrallah et al. 1988; Raikhel et al. 1988). Cryostat sections are rapidly prepared since the tissue requires minimal fixation and embedding is unnecessary. However, the tissue preservation and cellular morphology is generally not as good in cryostat sections compared with sections that have been cut from fixed and embedded plant tissues.

For in situ hybridization studies, the tissue sections usually require prehybridization treatment to facilitate access of the probe to the target nucleic acid and to minimize nonspecific background. Depending on the protocol, prehybridization treatment may include exposure to heat or incubation with dilute HCl, proteinase K, bovine serum albumin, or acetic anhydride. Concentrations of reagents and pretreatment times (particularly proteinase K treatment) may have to be optimized for each tissue studied.

SYNTHESIS OF PROBES

Both DNA and RNA probes have been used successfully for in situ hybridization (Aoyagi and Chua 1988; Barker et al. 1988). However, the probes most commonly used are radiolabeled, single-stranded RNA molecules generated by in vitro transcription. The advantages of these molecules over other types of probes are that (i) both sense and antisense transcripts can be readily synthesized, and (ii) low background can be obtained by digesting with RNase after hybridization and by washing at higher temperatures (Cox et al. 1984; Angerer et al. 1987). ^{3}H, ^{35}S, and ^{32}P can be used to label probes, and each radioisotope has advantages and disadvantages. ^{3}H-labeled probes provide excellent resolution, but often require prolonged exposure time. In contrast, ^{32}P-labeled probes minimize exposure time, but provide low resolution. ^{35}S-labeled probes seem to offer a good compromise between resolution and exposure time. Other, nonradioactive methods for labeling probes, such as the use of biotin and digoxigenin, are becoming available and these appear to offer a good alternative to radiolabeling (Johns et al. 1992). The length of the probe used for in situ hybridization can also be an important factor. The optimal size for probes is approximately 150–200 bases. Oligonucleotides can be used, but they offer limited sensitivity due to their low complexity (Angerer and Angerer 1989).

HYBRIDIZATION AND WASHING

A number of protocols have been devised for in situ hybridization, each with individualized hybridization and washing conditions (see,

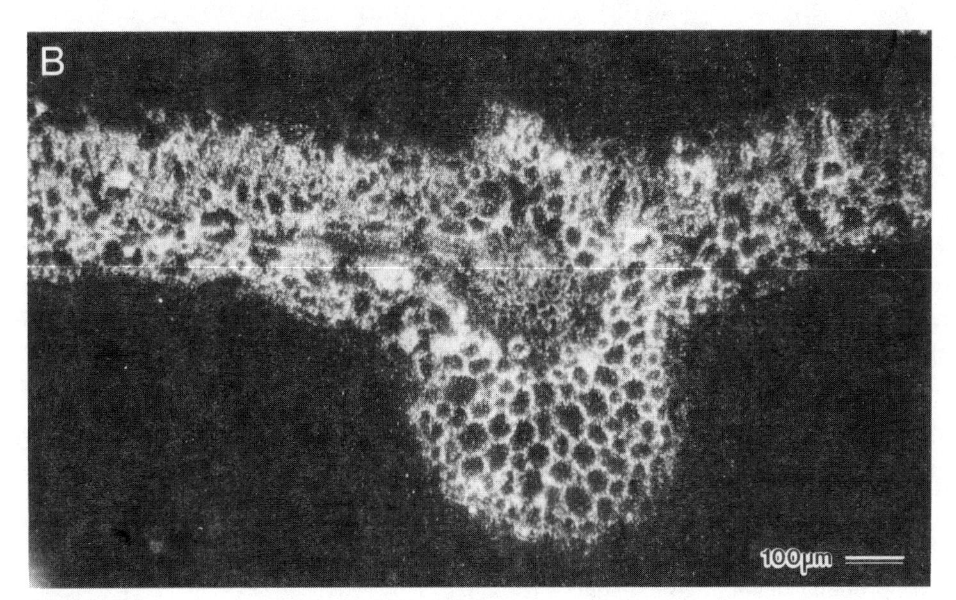

Figure 1
Adjacent serial cross sections of a tobacco mosaic virus-infected tobacco leaf viewed with darkfield microscopy after in situ hybridization to detect induction of RNA in the host plant in response to viral infection. The section hybridized with the sense riboprobe (A) shows little or no hybridization of the probe. In contrast, the high concentration of silver grains over the leaf section hybridized with the antisense riboprobe (B) indicates the presence mRNA coding for the pathogenesis-related protein PR-1.

e.g., Meyerowitz 1987; Cox and Goldberg 1988; Raikhel et al. 1989). The experimental conditions in the protocol described here are intended as basic guidelines. They have been assimilated from a number of different in situ hybridization procedures and have worked well in our laboratory. However, the reader may wish to refer to the references listed to optimize the protocol to his/her specific requirements.

DATA ANALYSIS

The probes employed for in situ hybridization studies are usually radiolabeled and, therefore, autoradiography is the standard method for data analysis. Following hybridization, the microscope slides with mounted tissue sections are washed, coated with Kodak NTB2 photoemulsion, and exposed for an appropriate length of time. The slides are developed and fixed in a manner similar to that used for standard photographic or X-ray film. After counterstaining the tissue sections and mounting the coverslips, the slides are examined for the presence of silver grains in the emulsion overlaying the tissue. Silver grains indicating the presence of bound probe are most easily detected by darkfield microscopy (see Fig. 1). Silver grains can also be detected by brightfield and phase microscopy, but these methods are not as sensitive. If nonradiolabeled probes are used, the visualization of bound probe is somewhat simplified since autoradiography, which requires a good deal of work in a *dimly* lit darkroom, is unnecessary. For example, following hybridization and washing, biotin-labeled probes hybridized to RNA sequences in the tissue can be detected "immunologically" by any one of a number of avidin-labeled conjugates (fluorescein, horseradish peroxidase, alkaline phosphatase, etc.). After washing to remove excess avidin conjugate and incubating with an appropriate enzyme substrate (if enzyme conjugates are used), the tissue is essentially ready for microscopic examination.

REFERENCES

Angerer, L.M. and R.C. Angerer. 1989. *In situ* hybridization with ^{35}S-labeled RNA probes. *DuPont Biotech. Update* **4** (no. 5).

Angerer, L.M., K.H. Cox, and R.C. Angerer. 1987. Demonstration of tissue specific gene expression by *in situ* hybridization. *Methods Enzymol.* **152:** 649–661.

Angerer, L.M., D.V. DeLeon, R.C. Angerer, R.M. Showman, D.E. Wells, and R.A. Raff. 1984. Delayed accumulation of maternal histone mRNA during sea urchin oogenesis. *Dev. Biol.* **101:** 477–484.

Angerer, L.M., D.V. DeLeon, K.H. Cox, R. Maxson, L. Kedes, J. Kaumeyer, E. Wingber, and R.C. Angerer. 1985. Simultaneous expression of early and late histone messenger RNAs in individual cells during development of the sea urchin embryo. *Dev. Biol.* **112:** 157–166.

Aoyagi, K. and N. Chua. 1988. Cell-specific expression of pyruvate Pi dikinase. *Plant Physiol.* **86:** 364–368.

Barker, S.J., J.J. Harada, and R.B. Goldberg. 1988. Cellular localization of soybean storage protein mRNA in transformed tobacco seeds. *Proc. Natl. Acad. Sci.* **85:** 458–462.

Cox, K.H. and R.B. Goldberg. 1988. Analysis of plant gene expression. In *Plant molecular biology: A practical approach* (ed. C.H. Shaw), pp. 1–35. IRL Press, Oxford.

Cox, K.H., D.V. DeLeon, L.M. Angerer, and R.C. Angerer. 1984. Detection of mRNAs in sea urchin embryos by *in situ* hybridization using asymmetric RNA probes. *Dev. Biol.* **101:** 485–502.

Dixon, D.C., J.C. Cutt, and D.F. Klessig. 1991. Differential targeting of the tobacco PR-1 pathogenesis-related proteins to the extracellular space and vacuoles of crystal idioblasts. *EMBO J.* **10:** 1317–1324.

Gottlieb, D.I. and L. Glaser. 1975. A novel assay of neuronal cell adhesion. *Biochem. Biophys. Res. Commun.* **63:** 815–821.

Hardin, P.E., L.M. Angerer, S.H. Hardin, R.C. Angerer, and W.H. Klein. 1989. Spec2 genes of *Strongylocentrotus purpuratus* structure and differential expression in embryonic aboral ectoderm cells. *J. Mol. Biol.* **202:** 417–431.

Johns, C., L. Meiqing, A. Lyznic, and S. Mackenzie. 1992. A mitochondrial DNA sequence is associated with abnormal pollen development in cytoplasmic male sterile bean plants. *Plant Cell* **4:** 435–449.

McDougall, J.K., D. Myerson, and A.M. Beckmann. 1986. Detection of viral DNA and RNA by in situ hybridization. *J. Histochem. Cytochem.* **34:** 33–38.

Meyerowitz, E.M. 1987. *In situ* hybridization to RNA in plant tissue. *Plant Mol. Biol. Rep.* **5:** 242–250.

Nasrallah, J.B., S. Yu, and M.E. Nasrallah. 1988. Self-incompatibility genes of *Brassica oleracea:* Expression, isolation, and structure. *Proc. Natl. Acad. Sci.* **85:** 5551–5555.

Raikhel, N.V., S.Y. Bednarek, and D.R. Lerner. 1989. *In situ* RNA hybridization in plant tissues. In *Plant molecular biology manual,* vol. B9, pp. 1–32. Kluwer Academic Publisher, Dordrecht, Belgium.

Raikhel, N.V., S.Y. Bednarek, and T. A. Wilkins 1988. Cell-type specific expression of a wheat-germ agglutinin gene in embryos and young seedlings of *Triticum aestivum. Planta* **176:** 406–414.

Sambrook, J., E.F. Fritsch, and T. Maniatis. 1989. *Molecular cloning: A laboratory manual,* 2nd edition, Cold Spring Harbor Laboratory Press, Cold Spring Harbor, New York.

Wu, M. and N. Davidson. 1981. Transmission electron microscopic method for gene mapping on polytene chromosomes by *in situ* hybridization. *Proc. Natl. Acad. Sci.* **78:** 7059–7063.

• *TIMETABLE*
In Situ Hybridization for the Detection of RNA in Plant Tissues[a]

STEPS	*TIME REQUIRED*
Tissue Preparation[b]	
Fix and embed plant tissue (see Section 6, pp. 101–104)	2–3 days
Soak microscope slides in Chromerge™	overnight
Rinse slides in H_2O	3 hours
Bake slides at 180°C	overnight
Coat slides with TESPA	1 hour
Cut and mount tissue sections (see Section 6, pp. 105–106)	1–3 hours
Synthesis of RNA Probe	
Linearize DNA template	2 hours
Transcribe DNA template to yield high-specific-activity RNA probe	2 hours
Calculate amount of RNA synthesized	30 minutes
Shear RNA	variable (depending on starting length of RNA)
Prehybridization Treatment of Tissue	
Remove Paraplast Plus™ from mounted tissue sections and rehydrate	30 minutes

STEPS	***TIME REQUIRED***
Prehybridization Treatment of Tissue (continued)	
Incubate slides in proteinase K	30 minutes
Acetylate with acetic anhydride	10 minutes
Wash slides in 2x SSC	15 minutes
Dehydrate mounted tissue sections	5 minutes
Dry under vacuum	1 hour
Hybridization	
Apply probe to mounted tissue sections	1 hour
Incubate	overnight
Washing After Hybridization	
Wash slides in 4x SSC/5mM DTT	1 hour
Incubate slides with RNase A	3 minutes
Remove RNase A	1 hour
Perform low-stringency wash	1 hour
Perform high-stringency wash	1 hour
Dehydrate mounted tissue sections	5 minutes
Dry under vacuum	1 hour
Autoradiography	
Coat slides with emulsion	3–4 hours
Expose slides	1–2 days

STEPS	***TIME REQUIRED***
Autoradiography (continued)	
Develop slides	<1 hour
Staining	
Stain slides with toluidine blue	<30 minutes

[a]Many of the steps can be performed concurrently since they do not require constant attention.
[b]Fixation, embedding, and sectioning of tissue can be done in advance. The embedded tissue sections mounted on slides are stable for several months.

Tissue Preparation

When preparing plant tissue for in situ hybridization, it is essential to protect the RNA from digestion. To minimize the contamination of samples with RNases, always wear gloves, treat solutions with diethylpyrocarbonate (DEPC), and bake all glassware (Sambrook et al. 1989).

MATERIALS AND EQUIPMENT

Corning frosted microscope slides (75 x 25 mm) (Fisher Scientific 12-568-15, or similar)
Chromerge™ (Fisher Scientific)
3-Aminopropyltriethoxysilane (TESPA; Sigma A 3648) (2% in dry acetone)
Acetone
Diethylpyrocarbonate (DEPC)
DEPC-treated (RNase-free) H_2O (see Sambrook et al. 1989)
Microtome

SAFETY NOTES

- DEPC is suspected to be a carcinogen and should be handled with care.
- Microtome blades are **extremely** sharp! Use care when sectioning. If you are unfamiliar with the use of a microtome, it may be appropriate to have someone demonstrate its use.

PROCEDURE

Fixation and Embedding of Plant Tissue

For a description of this procedure, including a list of materials, equipment, and reagents, see Section 6, pp. 101–104. To adapt this procedure for RNA detection in plant tissues, take care to minimize the potential for RNase contamination by using the precautions described above. Fresh solvents such as ethanol and xylene can be assumed to be RNase free.

Preparation of Microscope Slides

1. Place the required number of microscope slides in a slide holder and soak them in Chromerge™ overnight at room temperature.

2. Remove the slides from the Chromerge™ and rinse them in running H_2O for 3 hours. Rinse the slides in distilled H_2O, wrap them in aluminum foil, and bake them overnight at 180°C.

   Note: In addition to thoroughly cleaning the slides, this treatment inactivates RNases, which can destroy RNA probes as well as RNA within the tissue sections.

3. To coat the slides for tissue adhesion, dip them in 2% TESPA in dry acetone for 5–10 seconds. Quickly rinse them twice in acetone and once in distilled H_2O, then air dry (Gottlieb and Glaser 1975; Angerer and Angerer 1989). Slides coated with TESPA can be kept for several months.

Sectioning and Mounting of Embedded Plant Tissue

The procedure for sectioning and mounting the embedded plant tissue is as described in Section 6 (pp. 105–106), except that the ribbon of sections should be floated on sterile, distilled, RNase-free (DEPC-treated) H_2O instead of on tissue section adhesive solution. Care should also be taken not to contaminate sections with RNase from ungloved hands or "dirty" equipment.

Note: We routinely mount two adjacent sections in similar orientations on each slide. One section can be hybridized with the antisense probe and the other with the sense probe, thereby facilitating final analysis of the data.

Synthesis of RNA Probe

When synthesizing an RNA probe, it is essential to protect the RNA from digestion. To minimize the contamination of samples with RNases, always wear gloves, treat solutions with diethylpyrocarbonate (DEPC), and bake all glassware. For details of basic molecular biology techniques, see Sambrook et al. 1989.

MATERIALS AND EQUIPMENT

DNA template
Restriction enzyme (appropriate for DNA template used)
Phenol/chloroform (1:1)
Chloroform
Sodium acetate (3 M; pH 6.0)
Ethanol (100%, 95%, and 70%)
Transcription kit (e.g., Promega Corporation P2020), which contains:
- SP6 and T7 RNA polymerases (specific activity 15–20 units/ml)
- pGEM-3Z and/or pGEM-4Z plasmids
- Transcription buffer (5x) (200 mM Tris-HCl [pH 7.5], 30 mM $MgCl_2$, 10 mM spermidine, 50 mM NaCl)
- Dithiothreitol (DTT) (0.1 M)
- RNasin
- ATP (10 mM)
- CTP (10 mM)
- GTP (10 mM)
- UTP (500 μM)
- RNase-free DNase

^{35}S-UTP (1200 Ci/mmol) (DuPont NEN®)
tRNA carrier (10 μg/μl)
DE-81 ion exchange cellulose paper (Whatman 3658B323)
Na_2HPO_4 (500 mM; pH 7.0)
Glacial acetic acid (10%)

REAGENTS

Tris/EDTA (pH 7.6)
Proteinase K solution (1 mg/ml)

Na_2CO_3 (0.2 M)
$NaHCO_3$ (0.2 M)
Hybridization stock solution B (4x)
(For recipes, see Preparation of Reagents, pp. 137–139)

SAFETY NOTES

- Phenol is highly corrosive and can cause severe burns. Wear gloves, protective clothing, and safety glasses when handling it. All manipulations should be carried out in a chemical fume hood. Any areas of skin that come in contact with phenol should be rinsed with a large volume of water or polyethylene glycol 400 and washed with soap and water; do not use ethanol!
- Chloroform is irritating to the skin, eyes, mucous membranes, and respiratory tract. It should only be used in a chemical fume hood. Gloves and safety glasses should be worn. Chloroform is a carcinogen and may damage the liver and kidneys.
- Wear gloves when handling radioactive substances. Consult the local safety office for further guidance in the appropriate use of radioactive materials.
- Glacial acetic acid is volatile and should be used in a chemical fume hood. Concentrated acids should be handled with great care; gloves and a face protector should be worn.

PROCEDURE

Linearization of DNA Template

The DNA fragment used for generating antisense and sense RNA probes should be subcloned into the pGEM plasmid or another appropriate transcription vector. For details of the subcloning procedure, see Sambrook et al. 1989. RNA synthesis is probably more reliable from the T7 promoter than from the SP6 promoter. Therefore, subclone so that the antisense RNA strand is synthesized from the T7 promoter. This antisense probe is complementary to the mRNA and will hybridize to it. The sense probe (synthesized from the SP6 promoter) will not hybridize to the mRNA, and it can be used as a negative control to assess the level of background, nonspecific binding. Once the DNA fragment has been cloned into the transcription vector, the plasmid must be linearized for synthesis of a run-off transcript.

1. Using a five- to tenfold excess of an appropriate restriction enzyme (50–100 units), digest 10-20 μg of the plasmid in a total volume of approximately 200 μl for 2 hours at 37°C. (This should provide enough template for several rounds of probe synthesis.)

   Note: It is extremely important that the digestion is complete. Promega Corporation recommends avoiding the use of enzymes that produce 3′ overhangs (e.g., *Pst*I). It has been their observation that extraneous RNA is produced when these enzymes are used to linearize the template.

2. Extract the linearized DNA by adding an equal volume (200 μl) of phenol/chloroform (1:1) to the digested plasmid preparation. Mix by gently vortexing the tube. Separate the phases by centrifugation. Transfer the aqueous (upper) phase to a clean tube and perform a second phenol/chloroform extraction.

3. Remove the aqueous phase and extract twice with an equal volume of chloroform.

4. Precipitate the DNA fragments in the aqueous phase by adding 1/10 volume of 3 M sodium acetate and two volumes of ice-cold 100% ethanol. Incubate for 1 hour at -20°C. Spin at 12,000*g* for 10 minutes.

5. Resuspend the DNA fragments in Tris/EDTA (pH 7.6) to give a final concentration of approximately 1.0 μg/μl.

6. Determine the DNA concentration by reading the optical density at 260 nm and 280 nm. One OD_{260} unit corresponds to approximately 40 μg/ml of double-stranded DNA and the OD_{260}/OD_{280} ratio should be approximately 1.8.

7. Run 0.5 μg of the DNA on a minigel to check the completeness of the digestion (Sambrook et al. 1989).

   Note: If the plasmid DNA is contaminated with RNases, it may be necessary to treat the DNA template preparation with proteinase K (Sambrook et al. 1989).

Transcription of DNA Template to Yield a High-Specific-Activity RNA Probe

1. Set up the following reaction mixture:

H_2O	4 μl
Transcription buffer (5x)	5 μl
DTT (0.1 M)	2.5 μl
RNasin	1.25 μl
ATP (10 mM)	1.25 μl
CTP (10 mM)	1.25 μl
GTP (10 mM)	1.25 μl
UTP (500 μM)	1 μl
DNA template (1 μg/μl)	2.5 μl
^{35}S-UTP (1200 Ci/mmol)	5 μl
Total volume:	25 μl

To avoid the possibility of spermidine precipitation of the DNA, add each reagent (except the polymerase, see step 3) at room temperature.

Note: (i) The reaction can be scaled up or down as necessary. (ii) A high-quality, commercial transcription kit (e.g., from Promega Corporation) can be used.

2. Remove 1 μl from the reaction mixture and add it to 1 ml of H_2O. Label this sample t_0 and use it for the calculation of the amount of RNA synthesized. Keep the t_0 sample on ice until you are ready to count it.

3. Add 1 μl of polymerase T7 or SP6 to the reaction mixture (final volume 25 μl) and incubate for 45 minutes at 40°C.

4. Remove 1 μl from the reaction mixture and add it to 1 ml of H_2O. Label this sample t_{45} and use it for the calculation of the amount of RNA synthesized. Keep the t_{45} sample on ice until you are ready to count it.

5. To the remaining reaction mixture (24 μl), add 1 unit of RNase-free DNase per μg of DNA template. Incubate for 10 minutes at 37°C.

6. Add 2 μl of 10 μg/μl tRNA carrier. Perform one phenol/chloroform extraction and one chloroform extraction (see p. 123, steps 2 and 3).

7. Precipitate the RNA by adding 3 M sodium acetate (pH 6.0) to give a final concentration of 0.3 M and two volumes of ice-cold 100% ethanol. Incubate for 1 hour at –20°C.

Calculation of the Amount of RNA Synthesized

1. Spot 10 μl of the t_0 and t_{45} samples in duplicate onto DE-81 filters. Wash one of each set of filters four times in 500 mM Na_2HPO_4 (pH 7.0) for 5 minutes each, twice in distilled H_2O for 5 minutes each, and then briefly in 95% ethanol. Use a volume of approximately 200 ml for each wash step. Dry the washed filters. Count all the filters using the ^3H + ^{14}C channel.

2. The unwashed filters represent the total number of ^{35}S counts in each sample and the washed filters represent the number of ^{35}S counts incorporated into RNA. After a reaction time of 45 minutes, approximately 70% of the ^{35}S counts should be incorporated into RNA. An example of some typical results is shown below:

Unwashed filters:	Washed filters:
t_0 = 50,958 cpm	t_{0w} = 839 cpm
t_{45} = 51,978 cpm	t_{45w} = 39,267 cpm

   Total counts:
   $(t_{45w} - t_{0w})$ x 100 (dilution factor) x 25 (reaction volume in μl)
   i.e., (39,267 - 839) x 100 x 25 = 9.6 x 10^7

   Percent incorporation:
   $(t_{45w} - t_{0w}$/average counts on unwashed filters) x 100%
   i.e., (39,267 – 839/51,468) x 100% = 74.6%

Shearing of RNA

1. Pellet the RNA by spinning at 12,000*g* for 10 minutes. Remove the supernatant. Wash the RNA pellet once with 200–300 μl of 70% ethanol. Allow the last traces of ethanol to evaporate from the tube, then resuspend the RNA in 50 μl of RNase-free (DEPC-treated) H_2O.

2. The RNA must be sheared to an optimum length of approximately 150–200 bp. The approximate time required for optimal RNA shearing can be calculated using the following equation (Cox et al. 1984):

   $$t = (L_o - L_f)/(KL_oL_f)$$

   where L_o is the starting length of RNA in kb (determined by calculating the distance from the transcriptional start site to the cleavage site of the restriction enzyme used to linearize the plasmid); L_f is the final length of RNA in kb; and K is 0.11. The calculated time will be in minutes.

3. Mix the RNA resuspended in 50 μl of H_2O with 30 μl of 0.2 M Na_2CO_3 and 20 μl of 0.2 M $NaHCO_3$. Incubate for the calculated length of time at 60°C.

4. Stop the reaction by adding 3 μl of 3 M sodium acetate (pH 6.0) and 5 μl of 10% glacial acetic acid. (Save an aliquot to run on a formaldehyde gel; 2.5 μl will be adequate in most experiments.)

5. Precipitate the RNA by adding 3 M sodium acetate (pH 6.0) to a final concentration of 0.3 M and two volumes of ice-cold 100% ethanol. Incubate for 1 hour at –20°C.

6. Spin at 12,000*g* for 10 minutes, and wash the pellet with 70% ethanol. Resuspend the RNA in an appropriate volume of 4x hybridization stock solution B. Generally, 25 μl is a suitable volume.

7. Determine the amount of RNA recovered by counting a 1-μl sample.

8. Run a formaldehyde gel to determine the size of the sheared RNA (Sambrook et al. 1989).

Prehybridization Treatment of Tissue

When preparing plant tissue for in situ hybridization, it is essential to protect the RNA from digestion. To minimize the contamination of samples with RNases, always wear gloves, treat solutions with diethylpyrocarbonate (DEPC), and bake all glassware (Sambrook et al. 1989).

MATERIALS

Diethyl pyrocarbonate (DEPC)
DEPC-treated (RNase-free) H_2O (at least 4 liters; use to prepare all aqueous solutions)
Xylene (100%)
Ethanol (100%, 95%, 75%, 50%, and 25% in H_2O)

REAGENTS

Proteinase K solution
Triethanolamine (pH 8.0; 0.1 M)
Acetic anhydride (0.25% in 0.1 M triethanolamine [pH 8.0])
2x SSC
(For recipes, see Preparation of Reagents, pp. 137–139)

SAFETY NOTES

- DEPC is a suspected carcinogen and should be handled with care.
- Xylene should always be handled in a fume hood. It is flammable and may be narcotic at high concentrations.
- Acetic anhydride is extremely destructive to the skin, eyes, mucous membranes, and upper respiratory tract. It is harmful if ingested, inhaled, or absorbed through the skin. It should be used in a chemical fume hood. Gloves and safety glasses should be worn.

PROCEDURE

1. Remove the Paraplast Plus™ from the mounted tissue sections by washing each slide twice in 100% xylene for 10 minutes. Transfer the slides to 100% ethanol and wash them by repeatedly dipping until all streaks disappear.

2. Rehydrate the mounted tissue sections by washing the slides in each of a graded series of dilutions of ethanol in H_2O (95%, 75%, 50%, and 25%) at room temperature. At each stage, wash the slides by repeatedly dipping them into the solution until all streaks disappear. Wash the slides twice in H_2O. Save the ethanol/H_2O solutions for the dehydration procedure later in the day (see step 7).

3. Incubate the slides in proteinase K solution for 30 minutes at 37°C. (100 ml of proteinase K solution should be sufficient for treating 20 slides in a standard slide-staining dish.) Wash the slides twice in H_2O.

   Note: Partial digestion of the tissue sections with proteinase K will facilitate the penetration of the RNA probe.

4. To reduce nonspecific binding of the RNA probe, it is necessary to acetylate any remaining positive charges in the tissue or on the slides.

   Note: This step is extremely harsh. Do one dish at a time. Some sections may fall off the slides. BE GENTLE.

   Equilibrate the slides in approximately 100 ml of 0.1 M triethanolamine (pH 8.0) for 5 minutes at room temperature. Pour off the triethanolamine and *immediately* add approximately 100 ml of freshly prepared 0.25% acetic anhydride in 0.1 M triethanolamine (pH 8.0). Incubate for 10 minutes at room temperature.

5. Wash the slides in 2x SSC by dipping them up and down approximately 15 times.

6. Dehydrate the tissue sections by washing the slides in each of a graded series of dilutions of ethanol in H_2O (25%, 50%, 75%, and 95%) at room temperature. Wash the slides twice in fresh 100% ethanol. At each stage, wash the slides until all streaks disappear.

7. Dry the tissue sections under a vacuum for approximately 1 hour or at ambient pressure and temperature for several hours.

8. Check the integrity of the mounted tissue sections and label the slides with a code that will permit quick identification of the tissue type, hybridization probe, etc. The slides are most easily labeled on the frosted end with a soft lead pencil. Number the slides in a pattern that is convenient for the hybridization.

Hybridization

MATERIALS AND EQUIPMENT

RNA probe (see pp. 121–126)
RNasin (Promega Corporation)
Silanized 22 x 22-mm coverslips (for preparation, see p. 130)
Sigmacote (for silanizing coverslips) (Sigma SL 2)
Diethyl pyrocarbonate (DEPC)
DEPC-treated (RNase-free) H_2O

REAGENTS

Hybridization stock solution B (4x)
Hybridization stock solution A (4x)
Deionized formamide (100%)
(For recipes, see Preparation of Reagents, pp. 137–139)

SAFETY NOTE

- DEPC is a suspected carcinogen and should be handled with care.

PROCEDURE

1. Calculate the amount of hybridization buffer required to treat the slides that have been prepared for hybridization. Add the required amount of RNA probe to the appropriate amount of 4x hybridization stock solution B. Incubate the diluted RNA probe for 5 minutes at 80°C, then quickly cool on ice.

   Note: For the hybridization reaction, each mounted tissue section is incubated in 25 µl of hybridization buffer, which is prepared by combining 4x hybridization stock solution B containing the denatured RNA probe, 4x hybridization stock solution A, and 100% deionized formamide in the proportions 1:1:2. To prepare enough hybridization buffer for at least 75 tissue sections, the following volumes are appropriate:

   0.5 ml of 4x hybridization stock solution B containing the RNA probe
   0.5 ml of 4x hybridization stock solution A
   1.0 ml of 100% deionized formamide

The final concentration of the RNA probe (antisense or sense) in the hybridization buffer should be approximately 10^8 cpm/ml (i.e., ~4 x 10^8 cpm/ml in 4x hybridization stock solution B).

2. Mix an equal volume of 4x hybridization stock solution A with the 4x hybridization stock solution B containing the denatured probe.

3. Add a volume of 100% deionized formamide equal to the combined volume of 4x hybridization stock solutions A and B. Add RNasin to a final concentration of 25 units/ml. Store the hybridization buffer on ice until you are ready to add it to the mounted tissue sections.

   Note: The final concentrations of the components of the hybridization buffer are as follows: 50% formamide; 300 mM NaCl; 10 mM Tris-HCl (pH 7.5); 1 mM EDTA; 1x Denhardt's solution; 10% dextran sulfate; 70 mM dithiothreitol; 25 units/ml RNasin; 500 µg/ml poly A; 150 µg/ml tRNA; 10^8 cpm/ml RNA probe.

4. Check that the slides are labeled in a convenient pattern so that there are no mix-ups in adding the correct hybridization buffer to each slide.

   Note: As mentioned previously, we have found it convenient to mount two adjacent tissue sections on each microscope slide, thereby permitting the use of antisense and sense probes on a single slide. If the sections have been mounted appropriately, there should be no problem with inadvertent mixing of probes or overlap of coverslips.

5. Warm the hybridization buffer for 2 minutes at 37°C. Add 25 µl of warm hybridization buffer to each mounted tissue section and cover with a silanized coverslip. **AVOID BUBBLES!**

   Note: Coverslips are easily silanized by dipping them into a silanizing solution such as Sigmacote and then rinsing them with DEPC-treated (RNase-free) H_2O. Allow the coverslips to air dry completely before use.

6. Place the slides in a dish containing wet paper towels or sponges. Cover to prevent evaporation and incubate overnight at 42°C.

   Note: Some protocols suggest immersing the slides in mineral oil (Cox and Goldberg 1988) or sealing the coverslips in place with rubber cement (Raikhel et al. 1989) to prevent dehydration.

Washing after Hybridization

MATERIALS

Dithiothreitol (DTT)
RNase A (Calbiochem 556746)
Ethanol (100%, 95%, 75%, 50%, and 25% in H_2O)

REAGENTS

4x SSC containing 5 mM DTT (preequilibrate to room temperature and add DTT just before use)
RNase buffer containing 50 μg of RNase A/ml (preequilibrate to 37°C and add RNase A just before use)
RNase buffer containing 5 mM DTT (preequilibrate to 37°C and add DTT just before use)
2x SSC containing 5 mM DTT (preequilibrate to room temperature and add DTT just before use)
0.1x SSC containing 1 mM DTT (preequilibrate to 57°C and add DTT just before use)
(For recipes, see Preparation of Reagents, pp. 137–139)

PROCEDURE

1. After the slides have been incubated overnight in hybridization buffer (see p. 130), transfer them to a slide holder and place them in a slide dish filled with 4x SSC containing 5 mM DTT. When the slides have been immersed in this solution for a few minutes, the coverslips will become loose and gently fall off as you lift up the slides.

2. As the coverslips fall off, transfer the slides to fresh 4x SSC containing 5 mM DTT and wash them for 30 minutes at room temperature. Repeat the wash with fresh solution.

3. Incubate the slides in RNase buffer containing 50 μg of RNase A/ml for 3 minutes at 37°C. (100 ml of RNase buffer should be sufficient for treating 20 slides in a standard slide-staining dish.)

Note: Some researchers have reported that background counts can be lowered by treating with RNase after the high-stringency wash (see step 6) (Angerer and Angerer 1989).

4. Remove the RNase A by washing the slides in approximately 100 ml of RNase buffer containing 5 mM DTT for 20 minutes at 37°C. Repeat this step twice.

5. Perform a low-stringency wash by incubating the slides in 500 ml of 2x SSC containing 5 mM DTT for 30 minutes at room temperature with stirring. Repeat the wash with fresh solution.

6. Perform a high-stringency wash by incubating the slides in 500 ml of 0.1x SSC containing 1 mM DTT for 30 minutes at 57°C with stirring. Repeat the wash with fresh solution.

7. Dehydrate the tissue sections by washing the slides in each of a graded series of dilutions of ethanol in H_2O (25%, 50%, 75%, and 95%) at room temperature. Wash the slides twice in 100% ethanol. Wash the slides at each stage by repeatedly dipping until all streaks disappear.

8. Dry the tissue sections under a vacuum for approximately 1 hour at room temperature.

Autoradiography

MATERIALS

Autoradiography emulsion (Kodak NTB2 165 4433)
X-ray film developer (Kodak D-19 146 4593)
X-ray film fixer (Kodak 197 1738; do not use Rapid Fix)

PROCEDURE

1. Melt the autoradiography emulsion at 45°C in absolute darkness or under a proper safelight (e.g., a 15-watt bulb with a Kodak No. 2, 10 x 12-inch filter [152 1723]). This process takes 1–1.5 hours. When the emulsion has melted, dilute it 1:1 with H_2O (pre-equilibrated to 45°C) to achieve an appropriate consistency for dipping slides.

   Note: At this time, you can transfer aliquots of the diluted autoradiography emulsion into individual vials and store them in the dark at 4°C. The aliquots can be remelted separately as required. Once the diluted emulsion has been remelted, always discard after use. Avoid unnecessary agitation when working with the emulsion as this may lead to increased background levels of silver grains in the emulsion.

2. Dip each slide into the melted, diluted emulsion. To avoid streaks, dip the slides slowly and one time only. Stand the slides on end in an appropriate rack, and dry them in the dark for 1 hour at room temperature.

   Note: When dipping the slides, it is important that the melted emulsion is maintained at 45°C. To facilitate dipping, we constructed a dipping chamber with inside dimensions 6 cm (height) x 3 cm (width) x 1 cm (depth) out of 1/4-inch plexiglass. We equipped the dipping chamber with a large base, which allowed for stabilization in a water bath with a lead donut.

3. Place the dipped slides into dark boxes, and expose them for an appropriate length of time at 4°C.

   Note: Exposure time is determined empirically for each experiment. Always make extra slides for use in determining the correct exposure time.

4. Develop the slides after exposing them for 1–2 days. Place the slides in fresh X-ray film developer at 15–20°C and dip them up and down five times. Let the slides sit in developer for a total of 2.5 minutes. Overdevelopment of the emulsion may increase the background.

5. Remove the slides from the developer and drain them briefly (i.e., for 1–2 seconds).

6. Place the slides in H_2O and dip them up and down for approximately 30 seconds.

7. Remove the slides from the H_2O and drain them briefly.

8. Place the slides into X-ray film fixer at 15–20°C and dip them up and down five times. Let the slides sit in fixer for a total of 5 minutes.

9. Remove the slides from the fixer and drain them briefly.

10. Place the slides in H_2O and rinse them in cool, running H_2O for 30 minutes. Stain the slides immediately after rinsing. Do not allow the slides to dry before you stain them.

Staining

MATERIALS

Toluidine blue (0.05% in H_2O)
Isopropanol (100%, 75%, 50%, and 25% in H_2O)
Xylene (100%)
Permount™ (Fisher Scientific SP15-100)

SAFETY NOTE

- Xylene should always be handled in a fume hood. It is flammable and may be narcotic at high concentrations.

PROCEDURE

1. Place the slides in 0.05% toluidine blue and incubate them for approximately 2 minutes at room temperature. It is wise to stain the first few slides individually because the staining time varies with the type of plant tissue.

2. Remove excess stain by rinsing the slides twice in H_2O for 10–15 seconds.

3. Dehydrate the stained tissue sections *quickly* (dip only 10–15 times in each solution) by washing the slides in each of a graded series of dilutions of isopropanol in H_2O (25%, 50%, and 75%) at room temperature. Wash the slides twice in 100% isopropanol.

4. Dip the slides in 100% xylene until all streaks disappear.

5. Dip the slides several times in fresh 100% xylene. Leave the slides in 100% xylene and move on to step 6.

6. Remove one slide from the xylene and place it onto a stack of 4–5 Kimwipes. Quickly add two drops of Permount™ (~25 µl/22 x 22-mm coverslip or ~60 µl/22 x 50-mm coverslip) to the surface of the slide. Immediately place a coverslip onto the slide. Repeat this process until all slides are treated. Let the slides dry overnight.

7. View the slides under a microscope with darkfield illumination using 5–20x objectives.

   Note: The microscopic image of the tissue sections may be improved by scraping the excess emulsion from the back of the slide. A spare microscope slide works well for this purpose.

• PREPARATION OF REAGENTS

Acetic anhydride

Add 625 µl of acetic anhydride to 250 ml of freshly prepared RNase-free 0.1 M triethanolamine (pH 8.0) (for recipe, see p. 139). Use fresh chemicals. Prepare immediately before use.

Deionized formamide

Fisher Scientific BP 228-100

Deionize the formamide by stirring on a magnetic stirrer with Dowex XG8 mixed-bed resin for 1 hour and filtering twice through Whatman No. 1 paper. Use fresh chemicals. Deionized formamide should be stored in small aliquots under nitrogen at –70°C (Sambrook et al. 1989).

Hybridization stock solution A (4x)

(10 ml)

5 M NaCl	2.4 ml
1 M Tris-HCl (pH 7.5)	0.4 ml
0.5 M EDTA (pH 8.0)	0.1 ml
100x Denhardt's solution	0.4 ml
Dithiothreitol	216 mg
Dextran sulfate	4 g

Use fresh chemicals and RNase-free stock solutions. Adjust the volume to 10 ml with DEPC-treated (RNase-free) H_2O. Divide into aliquots and store at –20°C.

Denhardt's solution (100x)

Ficoll (type 400, Pharmacia Biotech, Inc.)	10 g
Polyvinylpyrrolidone	10 g
Bovine serum albumin (Fraction V, Sigma)	10 g

Use fresh chemicals. Dissolve in DEPC-treated (RNase-free) H_2O and then adjust the volume to 500 ml. Filter sterilize. Divide into aliquots and store at –20°C (Sambrook et al. 1989).

Hybridization stock solution B (4x)

(10 ml)

Dithiothreitol	216 mg
Poly A (Pharmacia Biotech, Inc. 27-7836-02)	20 mg
tRNA	6 mg

Use fresh chemicals. Dissolve in DEPC-treated (RNase-free) H_2O and then adjust the volume to 10 ml. Divide into aliquots and store at –20°C.

Na_2CO_3

0.2 M

Dissolve 2.12 g of Na_2CO_3 in DEPC-treated (RNase-free) H_2O to give a final volume of 100 ml. Store at room temperature.

$NaHCO_3$

0.2 M

Dissolve 1.68 g of $NaHCO_3$ in DEPC-treated (RNase-free) H_2O to give a final volume of 100 ml. Store at room temperature.

Proteinase K solution (For use in the transcription reaction)

Stock solution: 20 mg/ml in H_2O.

Incubate the stock solution for 30 minutes at 37°C to kill RNase activity. Store in aliquots at –20°C. Immediately before use, dilute the stock solution to 1 mg/ml in H_2O.

Proteinase K solution (For use in the prehybridization treatment of tissue)

Dissolve proteinase K in 100 mM Tris-HCl/50 mM EDTA (pH 7.5) that has been prewarmed to 37°C to give a final concentration of 1 µg/ml. Any RNase activity present in the proteinase K preparation will be degraded by the proteinase K during the prehybridization treatment at 37°C. Make fresh.

RNase buffer

0.5 M NaCl
1 mM EDTA
10 mM Tris-HCl

Adjust the pH to 7.5. Store at room temperature.

SSC stock solution (20x)

Dissolve 175.3 g of NaCl and 88.2 g of sodium citrate in 800 ml of H_2O. Adjust the pH to 7.0 with a few drops of a 10-N solution of NaOH. Adjust the volume to 1 liter with H_2O. Autoclave. Store at room temperature (Sambrook et al. 1989).

SSC (0.1x)

Dilute 20x SSC stock solution 1 in 200 with H_2O. Store at room temperature.

SSC (2x)

Dilute 20x SSC stock solution 1 in 10 with H_2O. Store at room temperature.

SSC (4x)

Dilute 20x SSC stock solution 1 in 5 with H_2O. Store at room temperature.

Triethanolamine

0.1 M

Dissolve 15 g of triethanolamine in 1 liter of DEPC-treated (RNase-free) H_2O. Adjust the pH to 8.0 with HCl. Prepare immediately before use.

Tris/EDTA (pH 7.6)
(For resuspending the linearized DNA template)

10 mM Tris-HCl
1 mM EDTA

Adjust the pH to 7.6. Use fresh chemicals and make the stock solutions up in DEPC-treated (RNase-free) H_2O.

• *Section 8*
In Vitro Import of Proteins into Chloroplasts

The import of proteins into the chloroplast is a complex process that includes at least five steps: (i) targeting of the cytoplasmically synthesized precursor protein to the organelle; (ii) binding of the precursor to the envelope; (iii) translocation of the precursor across the outer and inner membranes; (iv) usually, proteolytic processing of the precursor to the mature form by removal of an amino-terminal extension called the transit peptide; and (v) intraorganellar localization of the protein to the correct compartment. Although this basic outline is known, the mechanisms underlying each step are not well understood. The observation that isolated chloroplasts can import precursor polypeptides synthesized in vitro, first demonstrated with the small subunit of ribulose-1,5-bisphosphate carboxylase/oxygenase (Rubisco) (Chua and Schmidt 1978; Highfield and Ellis 1978), allows researchers to use a biochemical approach to explore the requirements for protein import and to identify components that make up the infrastructure of the import machinery.

Recently, much progress has been made in furthering our knowledge of the protein import process, and several studies are worthy of note. In vitro experiments have demonstrated that ATP is required for both the binding and translocation steps (Olsen et al. 1989). The transit peptide appears to mediate precursor recognition by a receptor component of the envelope (Schnell et al. 1990; Waegemann and Soll 1991), and competition studies have indicated that the receptor may be part of a shared import apparatus (Perry et al. 1991; Schnell et al. 1991; Oblong and Lamppa 1992a). A soluble enzyme involved in the proteolytic processing of precursors has been identified (Robinson and Ellis 1984; Lamppa and Abad 1987; Oblong and Lamppa 1992b). Mutational analyses have shown that precursor proteins contain sequence determinants for removal of the transit peptide and these determinants probably influence overall precursor conformation (Wasmann et al. 1988; Clark et al. 1989; Clark and Lamppa 1991). The temporal relationship between precursor import and processing has not been established, but precursors can be translocated across the envelope without cleavage when mutations are introduced at the transit peptide/mature protein junction (Clark et al. 1989; Clark and Lamppa 1991) or when an active processing enzyme is absent (Chitnis et al. 1988). Studies with fusion proteins, in which transit peptides have been exchanged, and modified forms of the precursors for the small subunit of Rubisco, ferredoxin, the light-harvesting chlorophyll binding protein, plastocyanin, and other proteins have helped to define the structural domains of the precursor that are required for localization to the stroma, thylakoid membranes, and thylakoid lumenal space (Smeekens et al. 1986; Lamppa 1988; Hand et al. 1989). Several reviews have summarized early research in this area and give a perspective on problems yet to be examined and resolved (Keegstra 1989; Smeekens et al. 1990; Theg and Scott 1993).

Specific precursor polypeptides can be synthesized in vitro if a cDNA clone (or a genomic clone without introns) is available to insert into a transcription vector. Following transcription, the RNA product is translated in vitro in the presence of a radiolabeled amino acid (Melton et al. 1984). Expression vectors may contain either the bacteriophage SP6 promoter or the T7 promoter, thereby allowing transcription by SP6 or T7 RNA polymerase, respectively. Typically 10 μg of RNA can be synthesized per 1 μg of linearized plasmid DNA. The RNA may have to be capped at the 5′ end for efficient translation, but this depends on the RNA utilized. If capping is necessary, $m^7G(5')ppp(5')G$ can be added directly to the transcription reaction, after adjusting the rGTP concentration (Konarska et al. 1984; Promega Protocols and Applications Guide 1991). Either a reticulocyte lysate or a wheat germ extract, both of which are available commercially, can be used for the in vitro translation reaction.

An alternative method for generating radiolabeled precursors is to synthesize them in *E. coli* using, for example, the system described by Studier (1990). The precursor will probably accumulate in inclusion-like bodies, and will have to be solubilized (and concomitantly denatured) by treatment with 6–8 M urea. However, the sequestering of the precursor in inclusion bodies offers a rapid way to purify the protein. This approach was recently employed to isolate the precursor for the major light-harvesting chlorophyll binding protein of the thylakoids (preLHCP) (Abad et al. 1991) and preFerredoxin (Pilon et al. 1990). Interestingly, recombinant preLHCP was not efficiently imported into the chloroplast, yet it was able to block import of preS via a thermolysin-sensitive factor (Oblong and Lamppa 1992a). The advantage of synthesizing precursor proteins in *E. coli* is that chemical amounts (~1 mg) can be obtained. It is not possible to synthesize such large amounts of protein by any of the in vitro methods thus far available.

In this section, we describe procedures for investigating the import of the precursor for LHCP into chloroplasts. The gene for preLHCP is inserted into an SP6 expression vector, and basic protocols for in vitro transcription using SP6 polymerase (in the absence of capping) and in vitro translation using a reticulocyte lysate are outlined. Intact chloroplasts are isolated from pea seedlings, which readily yield biologically active organelles. The products of the import reaction from both the membrane and soluble fractions are analyzed by SDS-polyacrylamide gel electrophoresis (PAGE) followed by autoradiography. Supplemental methods are provided to quantitate the radiolabeled protein in the membrane and soluble fractions and also to reisolate intact chloroplasts after an import reaction.

REFERENCES

Abad, M., J. Oblong, and G. Lamppa. 1991. Soluble chloroplast enzyme cleaves preLHCP made in *Escherichia coli* to a mature form lacking a basic N-terminal domain. *Plant Physiol.* **96:** 1220–1227.

Bartlett, S., A. Grossman, and N.-H. Chua. 1982. In vitro synthesis and uptake of cytoplasmically synthesized chloroplast proteins. In *Methods in chloroplast molecular biology* (ed. M. Edelman et al.), pp. 1081–1091. Elsevier, Amsterdam.

Chitnis, P., D. Morishige, R. Nechushtai, and P. Thornber. 1988. Assembly of the barley light-harvesting chlorophyll a/b proteins in barley etiochloroplasts involves processing of the precursor on thylakoids. *Plant Mol. Biol.* **11:** 95–107.

Chua, N.-H. and G. Schmidt. 1978. Post-translational transport into intact chloroplasts of a precursor to the small subunit of ribulose-1,5-bisphosphate carboxylase. *Proc. Natl. Acad. Sci.* **75:** 6110–6114.

Clark, S. and G. Lamppa. 1991. Determinants for cleavage of the chlorophyll a/b binding protein precursor: A requirement for a basic residue that is not universal for chloroplast imported proteins. *J. Cell Biol.* **114:** 681–688.

Clark, S., M. Abad, and G. Lamppa. 1989. Mutations at the transit peptide-mature protein junction separate two cleavage events during chloroplast import of the chlorophyll a/b binding protein. *J. Biol. Chem.* **264:** 17544–17550.

Hand, J., L. Szabo, A. Vasconcelos, and A. Cashmore. 1989. The transit peptide of a chloroplast thylakoid membrane protein is functionally equivalent to a stromal targeting sequence. *EMBO J.* **8:** 3195–3206.

Highfield, P. and R. J. Ellis. 1978. Synthesis and transport of the small subunit of chloroplast ribulose bisphosphate carboxylase. *Nature* **217:** 420–424.

Jagendorf, A. 1982. Isolation of chloroplast coupling factor (CF_1) and of its subunits. In *Methods in chloroplast molecular biology* (ed. M. Edelman et al.), pp. 881–898. Elsevier, Amsterdam.

Keegstra, K. 1989. Transport and routing of proteins into chloroplasts. *Cell* **56:** 247–253.

Konarska, M., R. Padgett, and P. Sharp. 1984. Recognition of cap structure in splicing in vitro of mRNA precursors. *Cell* **38:** 731–736.

Lamppa, G. 1988. The chlorophyll a/b binding protein inserts into the thylakoids independent of its cognate transit peptide. *J. Biol. Chem.* **263:** 14996–14999.

Lamppa, G. and M. Abad. 1987. Processing of a wheat light-harvesting chlorophyll a/b binding protein precursor by a soluble enzyme from higher plant chloroplasts. *J. Cell Biol.* **105:** 2641–2648.

Lamppa, G., G. Morelli, and N.-H. Chua. 1985. Structure and developmental regulation of a wheat gene encoding the major chlorophyll a/b binding polypeptide. *Mol. Cell Biol.* **5:** 1370–1378.

Melton, D., P. Krieg, M. Rebagliatie, T. Maniatis, K. Zinn, and M. Green. 1984. Efficient in vitro synthesis of biologically active RNA and RNA hybridization probes from plasmids containing a bacteriophage SP6 promoter. *Nucleic Acids Res.* **12:** 7035–7056.

Oblong, J. and G. Lamppa. 1992a. Precursor for the light-harvesting chlorophyll a/b binding protein (LHCP) synthesized in *E. coli* blocks import of the small subunit of Rubisco. *J. Biol. Chem.* **267:** 14328–14334.

———. 1992b. Identification of two structurally related proteins involved in proteolytic processing of precursors targeted to the chloroplast. *EMBO J.* **11:** 4401–4409.

Olsen, L., S. Theg, B. Selman, and K. Keegstra. 1989. ATP is required for the binding of precursor proteins to chloroplasts. *J. Biol. Chem.* **264:** 6724–6729.

Perry, S., W. Buvinger, J. Bennett, and K. Keegstra. 1991. Synthetic analogues of a transit peptide inhibit binding or translocation of chloroplastic precursor proteins. *J. Biol. Chem.* **266:** 11882–11889.

Pilon, M., A. Douwe de Boer, S. Luc Knols, M. Koppelman, R. van der Graaf, B. de Kruijff, and P. Weisbeek. 1990. Expression of *Escherichia coli* and purification of a translocation-competent precursor of the chloroplast protein ferredoxin. *J. Biol. Chem.* **265:** 3358–3361.

Promega Protocols and Applications Guide. 1991. Promega Corporation, Madison, Wisconsin.

Robinson, C. and R.J. Ellis. 1984. Transport of proteins into chloroplasts: Partial purification of a chloroplast protease involved in the processing of imported precursor polypeptides. *Eur. J. Biochem.* **142:** 337–342.

Schnell, D., G. Blobel, and D. Pain. 1990. The chloroplast import receptor is an integral membrane protein of chloroplast envelope contact sites. *J. Cell Biol.* **111:** 1825–1838.

———. 1991. Signal peptide analogs derived from two chloroplast precursors interact with the signal recognition system of the chloroplast envelope. *J. Biol. Chem.* **266:** 3335–3342.

Smeekens, S., P. Wiesbeek, and C. Robinson. 1990. Protein transport into and within chloroplasts. *Trends Biochem. Sci.* **15:** 73–76.

Smeekens, S., C. Bauerle, J. Hageman, K. Keegstra, and P. Weisbeek. 1986. The role of the transit peptide in the routing of precursors toward different chloroplast compartments. *Cell* **46:** 365–375.

Smith, B.J. 1984. SDS-polyacrylamide gel electrophoresis of proteins. In *Methods in molecular biology* (ed. J.M. Walker), vol. 1, pp. 41–56. Humana Press, Inc., Clifton, New Jersey.

Studier, F.W., A. Rosenber, J. Dunn, and J. Dubendorff. 1990. Use of T7 RNA polymerase to direct expression of cloned genes. *Methods Enzymol.* **185:** 60–89.

Theg, S. and S. Scott. 1993. Protein import into chloroplasts. *Trends Biochem. Sci.* **3:** 186–190.

Waegemann, K. and J. Soll. 1991. Characterization of the protein import apparatus in isolated outer envelopes of chloroplasts. *Plant J.* **1:** 149–158.

Wasmann, C., B. Reiss, and H. Bohnert. 1988. Complete processing of a small subunit of ribulose-1,5-bisphosphate carboxylase/oxygenase from pea requires the amino acid sequence ile-thr-ser. *J. Biol. Chem.* **263:** 617–619.

• *TIMETABLE*
Analysis of Protein Import into Chloroplasts In Vitro

STEPS	*TIME REQUIRED*
In Vitro Transcription	
Transcribe DNA template	1 hour
Extract RNA	1–2 hours
Analyze RNA by electrophoresis	1 hour
In Vitro Translation Using a Rabbit Reticulocyte Lysate	
Translate RNA	1.5 hours
Isolation of Chloroplasts	
Prepare Percoll® gradients	30 minutes
Homogenize plant tissue	15 minutes
Isolate chloroplasts	1 hour
Adjust chlorophyll concentration in chloroplast suspension	15 minutes
Import of Protein into Chloroplasts	
Incubate protein with chloroplast suspension	20 minutes
Digest aliquot of reaction mixture with thermolysin	20 minutes
Lyse thermolysin-treated and untreated chloroplasts	1.5 hours

STEPS	***TIME REQUIRED***
Import of Protein into Chloroplasts (continued)	
Prepare soluble and membrane proteins for SDS-PAGE	2 hours
SDS-PAGE Analysis of the Import Products	
Set up SDS-PAGE apparatus	2 hours
Run gel	8 hours
Processing of SDS-Polyacrylamide Gels for Autoradiography	
Stain and destain gel	1.5 hours
Treat gel with salicylate solution	1 hour
Dry gel	2 hours
Expose gel to X-ray film	1–10 days
Quantitation of Radiolabeled Protein in Membrane and Soluble Fractions	
Excise radiolabeled bands from dried gel and rehydrate gel slices	10 minutes
Incubate gel slices in TS-1 tissue solubilizer	2 hours
Count in scintillation counter	variable (depends on efficiency of import)
Rapid Isolation of Intact Chloroplasts	
Ensure chloroplasts are all intact after import reaction	10 minutes

In Vitro Transcription

When transcribing DNA, it is essential to protect the RNA product from digestion. To minimize the contamination of samples with RNases, always wear gloves, treat solutions with diethyl pyrocarbonate (DEPC), and autoclave all tubes and tips. Employ fresh pipette tips for each manipulation so as not to cross-contaminate the solutions.

MATERIALS

Diethyl pyrocarbonate (DEPC)
In vitro transcription kit (Riboprobe® II Core System, Promega Corporation P1250) (for further information, consult Promega Protocols and Applications Guide 1991), which contains the following:
- Sterile DEPC-treated (RNase-free) H_2O
- Transcription buffer (5x)
- Dithiothreitol (DTT) (100 mM)
- RNasin®
- SP6 RNA polymerase

RQ1 DNase (RNase-free)
DNA template (i.e., a *Hin*dIII-linearized plasmid vector [1 μg/μl] with an SP6 promoter and a wheat gene coding for the LHCP precursor [Lamppa et al. 1985])
Phenol (saturated with H_2O)
Chloroform
Ether
Sodium acetate (2.5 M; pH 5.0)
Ethanol (100% and 70% in H_2O) (ice-cold)
Agarose (ultrapure)
Ethidium bromide

REAGENTS

rNTP mixture (2.5 mM each)
TBE (1x and 10x)
TBE loading buffer for minigels (10x)
(For recipes, see Preparation of Reagents, pp. 167–171)

SAFETY NOTES

- DEPC is suspected to be a carcinogen and should be handled with care.
- Phenol is highly corrosive and can cause severe burns. Wear gloves, protective clothing, and safety glasses when handling it. All manipulations should be carried out in a chemical fume hood. Any areas of skin that come in contact with phenol should be rinsed with a large volume of water or PEG 400 and washed with soap and water; do not use ethanol!
- Chloroform is irritating to the skin, eyes, mucous membranes, and respiratory tract. It should only be used in a chemical fume hood. Gloves and safety glasses should also be worn. Chloroform is a carcinogen and may damage the liver and kidneys.
- Ether is extremely volatile and extremely flammable. It is irritating to the eyes, mucous membranes, and skin. It is also a CNS depressant with anesthetic effects. Ether can exert its effects through inhalation, ingestion, or absorption through the skin. It should only be used in a chemical fume hood. Gloves and safety glasses should be worn. Explosive peroxides can form during storage or on exposure to air or direct sunlight.
- Ethidium bromide is a powerful mutagen and is moderately toxic. Gloves should be worn when working with solutions that contain this dye. After use, decontaminate and discard solutions according to local safety office regulations.

PROCEDURE

Synthesis and Extraction of RNA

1. Set up the following reaction mixture at room temperature:

DEPC-treated H_2O	21 µl
Transcription buffer (5x)	10 µl
DTT (100 mM)	5 µl
RNasin (RNase inhibitor)	2 µl
rNTP mixture (2.5 mM each)	10 µl
DNA template	1 µl
SP6 RNA polymerase	1 µl (10 units)
Total volume:	50 µl

2. Incubate for 60 minutes at 37–40°C.

3. Add 1 unit of RQ1 DNase (RNase-free) per μg of DNA template. Incubate for 15 minutes at 37°C.

   Note: If RNase-free DNase is not available, it is often alright to omit this step provided you have used a DNA template prepared by a CsCl-ethidium bromide gradient. If your translations are not very good, you must ensure that the DNA template is removed.

4. To extract the RNA from the reaction mixture and inactivate the enzymes, add 12 μl of phenol saturated with H_2O and 12 μl of chloroform. Mix by tapping the tube.

5. Spin in a microfuge at maximum speed (14,000*g*) for 4 minutes.

6. Transfer the upper phase to a fresh tube, and add 800 μl of ether. Mix by inverting the tube several times.

7. Spin in microfuge at maximum speed for 1 minute.

8. Carefully aspirate off the upper phase. If a little ether remains, let it evaporate by leaving the tube open in a fume hood for about 5 minutes. Save the lower phase.

9. Ethanol precipitate the RNA in the lower phase by adding 6 μl of 2.5 M sodium acetate (pH 5.0) and 140 μl of ice-cold 100% ethanol. Mix by inverting the tube several times.

10. Place the mixture on dry ice for 20 minutes. It will turn opaque and solid.

11. Spin in a microfuge at maximum speed for 20 minutes.

12. Aspirate off the supernatant, taking care not to disturb the pellet.

13. Add 1 ml of ice-cold 70% ethanol to the pellet and mix well.

14. Spin in a microfuge at maximum speed for 15 minutes, and aspirate off the supernatant.

15. Dry the pellet in a speed-vacuum system (e.g., Savant Instruments, Inc.) for 15 minutes. The pellet should be "fluffy" white, rather than sticky. If it is not, repeat the ethanol precipitation (steps 9–15) because there is probably a little phenol still present that will "kill" the translation reaction.

16. Resuspend the RNA pellet in 20 μl of sterile DEPC-treated H_2O. Store at –20° C.

Analysis of RNA

This step is optional and is usually performed when a batch of DNA template is used for the first time.

1. While the transcription reaction is incubating, prepare a 1% mini-agarose gel. Add 1 g of ultrapure agarose to 100 ml of 1x TBE buffer in a 200-ml flask. Cover the flask with plastic wrap and dissolve the agarose by incubating in a microwave oven for 4 minutes at a high setting. Remove the flask from the oven, allow the agarose solution to cool for 10 minutes, and then cast a small gel for analyzing a 10-μl sample.

2. Combine 1 μl of RNA with 1 μl of 10x TBE and 8 μl of sterile H_2O. Incubate for 15 minutes at 60–65°C.

3. Place the sample on ice for a *maximum* of 10 minutes.

4. At room temperature, add 1 μl of 10x TBE loading buffer with dyes and load the gel.

5. Run the gel at 100 volts for 30–60 minutes.

6. Stain the gel with ethidium bromide for 15 minutes. Destain by incubating in 1x TBE for 30 minutes. View on a UV light box to estimate the quantity of RNA present and to attain a rough estimate of quality (i.e., is the RNA intact or degraded).

In Vitro Translation Using a Rabbit Reticulocyte Lysate

When translating RNA, it is essential to protect the RNA template from digestion. To minimize the contamination of samples with RNases, always wear gloves, treat solutions with DEPC, and autoclave all tubes and tips.

MATERIALS

Diethyl pyrocarbonate (DEPC)
Rabbit reticulocyte lysate kit (GIBCO/BRL 18127-019) containing the following:
- Reticulocyte lysate
- Potassium acetate (1 M; pH 7.2)
- Magnesium acetate (20 mM)
- Translation mix (containing tRNAs and amino acids other than methionine)
- DEPC-treated (RNase-free) H_2O

L-[^{35}S]-Methionine (translation grade; 1000 Ci/mmole in stabilized aqueous buffer) (DuPont NEN® NEG-009A)
RNA template (see pp. 147–150)

SAFETY NOTE

- DEPC is suspected to be a carcinogen and should be handled with care.

PROCEDURE

1. Before starting the in vitro translation reaction, dispense all the reticulocyte lysate into aliquots in microfuge tubes that have been prechilled on dry ice. Freeze rapidly and store at –80°C. Aliquots of 22 μl are handy as these provide enough lysate for two reactions.

   Note: When you thaw the aliquots, you must use the reticulocyte lysate. It will not translate RNA efficiently if it is frozen and thawed again. Thaw the lysate by gentle tapping or by rolling the tube in your hand. Do not vortex!

2. Set up the following reaction mixture on ice:

Potassium acetate (1 M; pH 7.2) (see note below)	2.4 μl
Magnesium acetate (20 mM) (see note below)	0.9 μl
^{35}S-Methionine (10 μCi)	2.0 μl
Translation mix (containing tRNAs and amino acids other than methionine)	3.0 μl
RNA template (0.5–1.0 μg)	1.0 μl
DEPC-treated H_2O	10.7 μl
Reticulocyte lysate	10.0 μl
Total volume:	30.0 μl

Mix gently.

Note: The concentration of potassium and magnesium required for the translation reaction depends on the particular RNA used. In this experiment, the concentrations employed are slightly higher than those suggested by the vendor. Please see notes on the handling and use of in vitro translation systems supplied by the vendor.

3. Incubate for 30–60 minutes at 25°C. The sample can be stored on ice until ready for use, or frozen overnight. Be aware that some precursors are not as efficiently imported into chloroplasts after freezing.

Isolation of Chloroplasts

MATERIALS AND EQUIPMENT

Pea seeds (*Pisum sativum*, Laxton's Progress) (L.L. Olds Seeds Co.)
Miracloth (Calbiochem 47855)
Acetone (100% and 80%)

REAGENTS

Gradient mix
PBF Percoll® (Pharmacia Biotech, Inc.)
Grinding buffer
HSM (1x)
(For recipes, see Preparation of Reagents, pp. 167–171)

PROCEDURE

In this experiment, chloroplasts are isolated from pea plants grown for 8–10 days on a 16-hour-light/8-hour-dark cycle. A pea tissue homogenate is spun through a 40%/80% Percoll® step gradient. Intact chloroplasts sediment through the 40% layer, but do not enter the 80% Percoll®. Nuclei and dense debris pellet to the bottom of the tube, whereas broken chloroplasts, mitochondria, and ribosomes, etc. do not enter the 40% Percoll®. The procedure is a modification of that described by Bartlett et al. (1982).

Preparation of Percoll® Gradients

1. Prepare the 40% and 80% Percoll® solutions as follows:

	40% Percoll® (ml)	80% Percoll® (ml)
Gradient mix	4.42	1.38
PBF Percoll®	8.0	5.0
H_2O	7.58	–
Total volume:	20.0	6.38

This is enough Percoll® for two gradients.

2. Add 3 ml of 80% Percoll® to each of two 30-ml Corex tubes. Overlay each 80% Percoll® layer with 9.5 ml of 40% Percoll®. Store the gradients on ice or in a coldroom until you are ready to use them. The gradients should be used within 2 hours.

Homogenization of Plant Tissue

1. Add 1.25 g of bovine serum albumin and 0.5 g of sodium ascorbate to 500 ml of grinding buffer.

2. Harvest approximately 20 g (fresh weight) of pea shoots. Place them in a beaker on ice and cut the tissue into small pieces with scissors.

3. Add 450 ml of ice-cold grinding buffer to the beaker and swirl to wet the tissue pieces. (Save 50 ml of the buffer for isolating chloroplasts [see below].)

4. In a coldroom, homogenize the tissue for 1 minute using a Polytron set at 3.5 (gentle rotation) or a Waring blender set at low speed. The chloroplasts contain starch crystals that cause breakage of the organelles if the tissue is homogenized too vigorously.

5. Stir the homogenate with a spatula to mix the remaining tissue pieces and homogenize again for 1 minute. The length of time required to homogenize the tissue will vary depending on the age of the plants, but grinding too vigorously will yield only broken chloroplasts.

6. Pour the homogenate through four layers of cheesecloth and four layers of Miracloth. Allow the homogenate to flow through without squeezing the cloths.

Isolation of Chloroplasts

1. Distribute the strained homogenate between four sterile 250-ml Nalgene bottles and spin in a SORVALL® GSA rotor at 4000 rpm for 1 minute at 4°C.

2. In one movement, carefully pour off supernatants. Save the pellets.

3. Add approximately 4 ml of grinding buffer to each of two of the tubes. Using a paintbrush, gently resuspend the pellets. Transfer the resuspended pellets (no clumps!) to the remaining two tubes

and resuspend the pellets. You may have to add a little more grinding buffer.

4. Layer the resuspended pellets onto the two 40%/80% Percoll® step gradients. *BALANCE THE TUBES.*

5. Spin in swinging bucket (SORVALL® HB-4) rotor at 7500 rpm for 8 minutes at 4°C.

6. You should observe two bands on the Percoll® gradient. The upper band contains broken chloroplasts and the lower band contains chloroplasts that have maintained an intact envelope. Aspirate off the liquid to the top of the lower band. Remove the lower band with a pasteur pipette and transfer it to a fresh 30-ml Corex tube. Do not remove any of the 80% Percoll® layer.

7. Adjust the volume of the intact chloroplast suspension to approximately 25 ml with grinding buffer. Spin in a SORVALL® HB-4 rotor at 3500 rpm for 5 minutes.

8. Quickly pour off the supernatant in one movement. It is important to remove all of the Percoll®. Add 25 ml of 1x HSM to the pellet, gently resuspend, and spin in a SORVALL® HB-4 rotor at 3500 rpm for 5 minutes at 4°C.

9. Gently resuspend the pellet in approximately 3 ml of 1x HSM.

10. Check that the chloroplasts are intact using a phase microscope. Intact chloroplasts appear bright, not dark and opaque.

Chlorophyll Determination

1. Set up the following mixture:

Chloroplast suspension	20 μl
Acetone (100%)	80 μl
Acetone (80%)	2 ml

   Vortex.

2. Transfer 1 ml of the mixture to a microfuge tube and spin at maximum speed for 1 minute. Save the supernatant.

3. Add 1 ml of 80% acetone to a cuvette. This will serve as the blank. Add 1 ml of the supernatant from the microfuge tube to a second cuvette.

4. Take readings against the blank at 645 nm and 663 nm.

5. Calculate the chlorophyll concentration using the following equation (Jagendorf 1982):

$$[(A_{645} \times 202) + (A_{663} \times 80.2)] \times 10.5 = \mu g \text{ chlorophyll/ml}$$

6. If necessary, adjust the chlorophyll concentration in the chloroplast suspension to approximately 900 µg/ml with 1x HSM.

Import of Protein into Chloroplasts

MATERIALS

Translation reaction mixture (see p. 152)
HEPES-KOH (5 mM; pH 7.5)
Sterile H_2O
ATP (100 mM)
$MgCl_2$ (1 M)
Chloroplast suspension (see pp. 153–156)
EGTA (200 mM in 1x HSM; pH 8.0)
Trichloroacetic acid (TCA) (20%; ice-cold)
Acetone (80%)

REAGENTS

SDS-PAGE loading buffer
HSM (1x and 5x)
Phenylmethylsulfonylfluoride (PMSF) (100 mM in 100% ethanol and 1 mM in H_2O)
Thermolysin solution (2x)
(For recipes, see Preparation of Reagents, pp. 167–171)

SAFETY NOTES

- PMSF is extremely destructive to the mucous membranes of the respiratory tract, the eyes, and the skin. It may be fatal if inhaled, swallowed, or absorbed through the skin. It is a highly toxic cholinesterase inhibitor. It should be used in a chemical fume hood. Gloves and safety glasses should be worn.
- TCA is highly caustic. Always wear gloves when handling this compound.

PROCEDURE

Chloroplasts should be used within one hour of preparation for the best import results. The two criteria typically used to establish that precursor polypeptides have successfully translocated the chloroplast envelope and entered the organelle are: (i) a change in the size of the polypeptide due to the removal of the transit peptide; and (ii) resistance of the polypeptide to exogenously added protease such as thermolysin. Proteins of the outer envelope, however, may not exhibit these properties.

1. Remove 2 μl from the translation reaction mixture, add 75 μl of SDS-PAGE loading buffer, and freeze. This sample will serve as a positive control during SDS-PAGE. Prepare a translation mix by adding 72 μl of 5 mM HEPES-KOH (pH 7.5) to the remaining 28 μl of the translation reaction mixture.

2. Set up the following mixture in a sterile 15-ml Corex tube:

Sterile H_2O	101.5 μl
HSM (5x)	60 μl
ATP (100 mM)	35 μl
$MgCl_2$ (1 M)	3.5 μl
Translation mix (see step 1 above)	100 μl
Total volume:	300 μl

3. Add 50 μl of the chloroplast suspension (containing 900 μg of chlorophyll/ml; i.e., ~10^{10} chloroplasts) to the mixture.

4. Incubate the import reaction mixture for 20 minutes at 25°C with gentle rotation. This reaction can also be performed at room temperature. The tube must be rotated occasionally so that the chloroplasts do not sediment.

   Note: Longer incubation times do not result in substantially higher yields of imported protein.

5. Divide the sample in two by transferring two 175-μl aliquots to microfuge tubes labeled 1 and 2. To tube 1, add 1 ml of 1x HSM and 10 μl of 100 mM PMSF. Store the tube on ice.

   Note: If you are using a mutant precursor construct, you should process this sample immediately to assure that the imported protein is not turned over. With a wild-type precursor, the imported protein is usually quite stable.

6. To tube 2, add 175 μl of 2x thermolysin solution, and gently mix the contents by inverting the tube. Leave the sample on ice for 20 minutes to digest polypeptide precursors that have not crossed the chloroplast envelope. Stop the reaction by adding 1 ml of 1x HSM. Mix, and then add 75 μl of 200 mM EGTA (pH 8.0) in 1x HSM. EGTA chelates the calcium that thermolysin requires for activity.

   Note: Once treated with thermolysin, the chloroplasts are very fragile.

Add 9 μl of 100 mM PMSF to inhibit contaminating proteases (some of which will have leaked from broken chloroplasts).

7. Spin the chloroplasts in tubes 1 and 2 in a microfuge at maximum speed for 30 seconds.

8. Aspirate off the supernatants. Resuspend the pellets in 400 μl of 1 mM PMSF in H_2O.

9. For complete lysis, leave the chloroplasts on ice for 45–60 minutes. Spin the chloroplasts in a microfuge at maximum speed for 10 minutes. Save the supernatants and the pellets.

10. Transfer the supernatants from tubes 1 and 2 to fresh tubes. To precipitate the soluble proteins, add an equal volume of ice-cold 20% TCA to each supernatant. Mix, and leave on ice for at least 30 minutes. Spin the tubes in a microfuge at maximum speed for 20 minutes and save the pellets. Wash the pellets twice with 1 ml of 80% acetone, and then dry them in a speed vacuum. It is important to remove all the TCA before drying the pellets. Add 75 μl of SDS-PAGE loading buffer to each pellet. The pellets may be difficult to dissolve; pipette up and down vigorously, or use a special pestle for microfuge tubes.

11. To prepare the membrane proteins for SDS-PAGE analysis, add 75 μl of SDS-PAGE loading buffer to the pellets from tubes 1 and 2. Solubilize.

SDS-PAGE Analysis of the Import Products

MATERIALS

Ethanol (100%)
Ammonium persulfate (APS) (10%)
N,N,N′,N′,tetramethylethylenediamine (TEMED)
Isobutanol (saturated with H_2O)

REAGENTS

Acrylamide (30%)/bisacrylamide (0.8%) stock solution
Laemmli resolving gel buffer (8x)
Laemmli stacking gel buffer (4x)
Upper reservoir buffer
Lower reservoir buffer
(For recipes, see Preparation of Reagents, pp. 167–171)

SAFETY NOTE

- Acrylamide and bisacrylamide are potent neurotoxins and are absorbed through the skin. Their effects are cumulative. Wear gloves and a mask when weighing acrylamide and bisacrylamide. Polyacrylamide is considered to be nontoxic, but it should be treated with care because it may contain small quantities of unpolymerized material.

PROCEDURE

It is useful to analyze the total products in both the soluble and membrane fractions after the import reaction to compare the distribution of the imported protein in the different compartments. (For a comprehensive review of how to perform SDS-PAGE, see Smith [1984].)

1. Set up the gel apparatus. Clean the gel plates by washing them with soap (e.g., Comet, or similar type) and rinsing them in hot

H_2O, distilled H_2O, and finally 100% ethanol. Air dry. Any dirt on the plates will show up as bubbles in the gel and will result in irregular bands.

2. Prepare a 12% acrylamide resolving gel solution by mixing:

Acrylamide (30%)/bisacrylamide (0.8%) stock solution (bring to room temperature first)	12 ml
Laemmli resolving gel buffer (8x)	3.75 ml
H_2O	14.25 ml

This is sufficient for a gel approximately 15 x 16 x 0.1 cm in size. Degas the solution under vacuum for 15 minutes. Add 125 μl of 10% APS, freshly prepared, and mix well. Add 30 μl of TEMED. Ideally, you will have about 10 minutes before polymerization begins. Overlay the gel with isobutanol saturated with H_2O or insert the flat edge of the comb upside down. Draw a small amount of unused 12% acrylamide resolving gel solution into a pasteur pipette to monitor polymerization.

3. Prepare a 5% acrylamide stacking gel solution by mixing:

Acrylamide (30%)/bisacrylamide (0.8%) stock solution (bring to room temperature first)	1.67 ml
Laemmli stacking gel buffer (4x)	2.5 ml
H_2O	5.84 ml

This is sufficient for a gel approximately 4 x 16 x 0.1 cm in size. Mix well. Degas the solution for 15 minutes. *STOP HERE until the resolving gel is ready.* Add 88 μl of 10% APS, mix, and then add 10 μl of TEMED.

4. When the resolving gel has completely polymerized (this usually takes ~45 minutes), drain off the isobutanol and rinse the well with tap H_2O and then distilled H_2O. Dry the well space with Kimwipes, taking care not to nick the gel. Add the stacking gel solution. Insert the comb. Allow the stacking gel to polymerize. At this point you may store the gel overnight at 4°C. For overnight storage, wrap the gel securely in moist paper towels and plastic wrap.

5. If you have stored the gel overnight at 4°C, bring the gel to room temperature (this takes ~30 minutes). Remove the comb by running warm H_2O over the plates for a few minutes, and then quickly switch to cold H_2O. The comb should slide out without pulling the gel teeth; otherwise, repeat this process. Rinse out the wells with H_2O and then—*this is very important*—wash them out with upper reservoir buffer.

6. Load the samples into the gel wells in the following order:
   a. Positive control for the translation reaction (see step 1, p. 158)
   b. Membrane fraction
   c. Membrane fraction from chloroplasts treated with thermolysin
   d. Soluble fraction
   e. Soluble fraction from chloroplasts treated with thermolysin.

   Run the gel at approximately 15 mA until the bromophenol blue is about 1 inch from the bottom of the gel.

Processing of SDS-Polyacrylamide Gels for Autoradiography

EQUIPMENT

X-ray film (8 x 10 inches) (Eastman Kodak X-OMAT-AR 1651454)

REAGENTS

Coomassie Brilliant Blue R-250 (0.25%)
Destain
Salicylate solution
(For recipes, see Preparation of Reagents, pp. 167–171)

SAFETY NOTE

- Salicylate can elicit allergic reactions and is readily absorbed through the skin. Gloves should be worn while handling gels in salicylate solution.

PROCEDURE

1. Cut off the stacking gel. Stain the resolving gel with Coomassie Brilliant Blue R-250 (0.25%) for 30 minutes with gentle shaking. Use enough dye to float the gel.

2. Drain off the excess dye, and transfer the gel to destain (~150 ml) for 1 hour. If necessary, leave the gel in destain overnight.

   Note: The gel is fragile, so handle it carefully by holding two corners at the same time.

3. Repeat the destain step if you want to visualize the protein bands clearly. This is useful because it allows you to assess whether the chloroplasts remained intact during thermolysin treatment.

4. Transfer the gel to salicylate solution (~150 ml) and shake for 60 minutes.

5. Place the gel on two pieces of Whatman 3MM paper. Smooth out any bubbles, cover the gel and paper with plastic wrap, and place this "sandwich" on a gel dryer. Dry the gel for 2 hours on low heat (60°C) under a vacuum.

   Note: It is very important not to break the vacuum too early. A premature "check" to see if the gel is dry will cause it to shatter.

6. Place the dried gel in a cassette with X-ray film. Make sure that it lies flat and is tightly clamped against the film. Expose the film in a freezer at –80°C.

Quantitation of Radiolabeled Protein in Membrane and Soluble Fractions

MATERIALS

TS-1 tissue solubilizer (Research Products International)
Glacial acetic acid
3a20 scintillation mixture (Research Products International)

SAFETY NOTE

- Glacial acetic acid is volatile and should be used in a chemical fume hood. Concentrated acids should be handled with great care; gloves and a face protector should be worn.

PROCEDURE

1. Using a nonessential film as a template, excise the radiolabeled bands from a dried gel.

2. Rehydrate the gel slices briefly in H_2O—the slices swell immediately.

3. Transfer the gel slices (without paper) to scintillation vials. Add 500 µl of TS-1 tissue solubilizer to each vial.

4. Incubate for 2 hours at 50°C.

5. Cool the vials to room temperature. Add 10 µl of glacial acetic acid and 5 ml of 3a20 scintillation mixture to each vial. Mix.

6. Count the vials in scintillation counter on the ^{35}S-channel.

Rapid Isolation of Intact Chloroplasts

MATERIALS

40% Percoll® (prepared as described on p. 153)

REAGENTS

Phenylmethylsulfonylfluoride (PMSF) (1 mM in H_2O)
(For recipe, see Preparation of Reagents, pp. 167–171)

SAFETY NOTE

- PMSF is extremely destructive to the mucous membranes of the respiratory tract, the eyes, and the skin. It may be fatal if inhaled, swallowed, or absorbed through the skin. It is a highly toxic cholinesterase inhibitor. It should be used in a chemical fume hood. Gloves and safety glasses should be worn.

PROCEDURE

This is an optional procedure that is useful for ensuring that chloroplasts are all intact after an import reaction, or for isolating a small number of chloroplasts.

1. Add 0.4 ml of 40% Percoll® to a microfuge tube.

2. Layer the import reaction or the chloroplast suspension over the 40% Percoll® pad.

3. Spin in the microfuge at 7000 rpm for 5 minutes at 4°C.

4. Collect the pellet and lyse the chloroplasts in 1 mM PMSF in H_2O. Follow steps 8–11 on page 159 to prepare for electrophoresis.

• PREPARATION OF REAGENTS

Acrylamide (30%)/bisacrylamide (0.8%) stock solution
(1 liter)

Acrylamide (Bio-Rad Laboratories)	300 g
Bisacrylamide	8 g

Dissolve the components in distilled H_2O over a low heat. Adjust the volume to 1 liter with H_2O. Add one spoonful of mixed-bed resin (Bio-Rad Laboratories, AG501-X8[D]) and mix with a magnetic stir bar for approximately 30 minutes. Pass the solution through glass wool to remove the resin, and then filter through a 0.45-μm Millipore filter. Store at 4°C.

Coomassie Brilliant Blue R-250 (0.25%)
(1 liter)

Coomassie Brilliant Blue dye	2.5 g
Methanol	500 ml
Glacial acetic acid	70 ml

Adjust the volume to 1 liter with H_2O. Store at room temperature.

SAFETY NOTE

- Glacial acetic acid is volatile and should be used in a chemical fume hood. Concentrated acids should be handled with great care; gloves and a face protector should be worn.

Destain
(1 liter)

Methanol	500 ml
Glacial acetic acid	70 ml

Adjust the volume to 1 liter with H_2O. Store at room temperature.

Gradient mix
(enough for two gradients)

MMHS solution	6.27 ml
0.5 M EDTA	0.116 ml
L-Ascorbic acid	25 mg
Glutathione (reduced form)	10 mg

Mix well.

MMHS solution
(enough for 30 gradients)

1 M $MgCl_2$	0.5 ml
1 M $MnCl_2$	0.5 ml
1 M HEPES-KOH (pH 7.5)	25 ml
2 M Sorbitol	82.5 ml

Store in the refrigerator at 4°C.

Grinding buffer
(1 liter)

		Final concentration
0.2 M EDTA (pH 8.0)	10 ml	2 mM
1 M $MgCl_2$	1 ml	1 mM
1 M $MnCl_2$	1 ml	1 mM
1 M HEPES (pH 7.5)	50 ml	50 mM
Sorbitol	60. 14 g	0.33 M

Adjust the volume to 1 liter with H_2O. Divide the buffer into two 500-ml aliquots and autoclave. Cool and store at 4°C. On the day of the experiment, add 0.5 g of sodium ascorbate and 1.25 g of bovine serum albumin to one 500-ml aliquot.

HSM (1x)
(50 mM HEPES/0.33 M sorbitol/8.4 mM methionine)

Dilute 5x HSM (see below) 1 in 5 with sterile H_2O.

HSM (5x)
(250 mM HEPES/1.65 M sorbitol/42 mM methionine)
(100 ml)

1 M HEPES (adjusted to pH 8.0 with KOH)	25 ml
Sorbitol	30 g
Methionine	0.62 g

Adjust the volume to 100 ml with sterile H_2O. Adjust the pH to 8.0 (this takes 3–5 pellets of KOH). Filter sterilize through a 0.45-µm Millipore filter.

Laemmli resolving gel buffer (8x)
(3 M Tris-HCl [pH 8.8 at 25°C])
(1 liter)

Dissolve 363 g of Trizma base in H_2O. Adjust the pH to 8.8 with HCl, and adjust the volume to 1 liter with H_2O. Autoclave.

Note: Autoclaving is not necessary if all of the buffer is to be used within a few days.

Laemmli stacking gel buffer (4x)
(0.5 M Tris-HCl [pH 6.8])
(1 liter)

Dissolve 60.5 g of Trizma base in H_2O. Adjust the pH to 6.8 with HCl, and adjust the volume to 1 liter with H_2O. Autoclave.

Lower reservoir buffer
(1 liter)

Take 200 ml of 5x SDS-PAGE reservoir buffer stock and adjust the volume to 1 liter with H_2O.

SDS-PAGE reservoir buffer stock (5x)
(1 liter)

Trizma base	15.1 g
Glycine	72 g

Dissolve in 1 liter of H_2O. Do not adjust the pH.

PBF Percoll®
(enough for two gradients)

Polyethylene glycol (M.W. 3350; Sigma P 3640)	450 mg
Bovine serum albumin	150 mg
Ficoll (Type 400, Sigma)	150 mg

Add 15 ml of Percoll® (Sigma P 1644) and mix well. This solution can be prepared the night before it is required and stored at 4°C, but do not leave any longer.

Phenylmethylsulfonylfluoride (PMSF)

Stock solution: 100 mM in 100% ethanol. Store at –20°C.

To prepare 1 mM PMSF in H_2O, dilute the PMSF stock solution 1 in 100 with H_2O. Use immediately.

SAFETY NOTE

- PMSF is extremely destructive to the mucous membranes of the respiratory tract, the eyes, and the skin. It may be fatal if inhaled, swallowed, or absorbed through the skin. It is a highly toxic cholinesterase inhibitor. It should be used in a chemical fume hood. Gloves and safety glasses should be worn.

---

rNTP mixture

Mix together equal amounts of 10 mM ATP, 10 mM CTP, 10 mM GTP, and 10 mM UTP. The final concentration of each rNTP in the mixture is 2.5 mM. Use 10 µl of the rNTP mixture in the in vitro transcription reaction. Freeze at –20°C until ready to use, if you do not use immediately.

Salicylate solution

Prepare a solution of 1 M salicylic acid (Sigma S 3007) containing 20% methanol and 1% glycerol. Store at room temperature.

SDS-PAGE loading buffer

60 mM Tris-HCl (pH 6.8)
5% 2-Mercaptoethanol
2% Sodium dodecyl sulfate (SDS)
10% Glycerol
0.03% Bromophenol blue

SAFETY NOTES

- 2-Mercaptoethanol may be fatal if swallowed and is harmful if inhaled or absorbed through the skin. High concentrations are extremely destructive to the mucous membranes, upper respiratory tract, skin, and eyes. Use only in a chemical fume hood. Gloves and safety glasses should be worn.
- Wear a mask when weighing SDS.

TBE (1x)

Dilute 10x TBE (see below) 1 in 10 with H_2O

TBE (10x)

(1 liter)

		Final concentration
Tris-base	107.8 g	0.89 M
Boric acid	55.6 g	0.89 M
EDTA (Na salt)	9.3 g	0.025 M

Dissolve in 1 liter of H_2O and adjust the pH to 7.8 with concentrated HCl. Store at room temperature.

TBE loading buffer for minigels (10x)

0.89 M Tris-base
0.89 M Boric acid
0.025 M EDTA
0.25% Bromophenol blue
0.25% Xylene cyanol
25% Ficoll (Type 400)

Thermolysin solution (2x)

2 mg/ml Thermolysin (protease Type X, Sigma P 1512) in 50 mM HEPES-KOH (pH 8.0)	200 μl
1 M $CaCl_2$	8 μl
5x HSM	400 μl
H_2O	1.39 ml

Store at –20°C.

Upper reservoir buffer

(1 liter)

Take 200 ml of 5x SDS-PAGE reservoir buffer stock, add 10 ml of sodium dodecyl sulfate, and adjust the volume to 1 liter with H_2O.

SDS-PAGE reservoir buffer stock (5x)

(1 liter)

Trizma base	15.1 g
Glycine	72 g

Dissolve in 1 liter of H_2O. Do not adjust the pH. Store at room temperature.

• *Section 9*

Microinjection and the Study of Tissue Patterning in Plant Apices

In plants, unlike in animals, organ development continues beyond the period of embryogenesis, and the rudimentary body plan is elaborated upon during the entire life of the plant (Sussex 1989). In most dicots, the apical meristem is composed of three layers, designated L_1, L_2, and L_3. The epidermis is generated from L_1, and the other tissues in the developing shoot are derived from L_2 and L_3 (Fig. 1A; Sussex 1989). Only a few cells from within each layer function as the initials, or mother cells, for that layer. Studies in which central cells of the apical meristem were marked via phenotypically identifiable mutations (see, e.g., Stewart and Dermen 1970) established that one to three initials are present in each cell layer. From these initials, all other cells are clonally derived. Furthermore, cell lineage does not necessarily reflect the pattern of physiological organization. Microdissection experiments have established that cells in different lineages of the apical meristem are not irreversibly determined to specific functions, but rather, their fate is dependent upon their relative position; i.e., position overrules lineage (Sussex and Rosenthal 1973; Sussex 1989).

The structure of the apex can also be defined in terms of cellular patterns or zones. A general topological zonation describes the apex as a tunica with one or more peripheral layers that covers a central mass of cells, termed the corpus. The tunica shows anticlinal divisions, whereas the corpus has divisions in all directions. Other patterns in the apex might reflect geometric consequences of space filling, and thus might not represent physiologically distinct populations of cells (Niklas and Mauseth 1980).

In view of both the dynamic, self-organizing nature of the plant apical meristem (Soma 1973; Sussex 1989) and the supracellular concept of plants (Lucas et al. 1993), it is clear that, as with the study of animal embryogenesis (De Laat et al. 1980; Serras and van den Biggelaar 1990), microinjection techniques should prove valuable in establishing the role played by symplasmic domain formation in the progression of tissue development. (Symplasmic domains are formed within the symplasmic continuum of plants by a downregulation of the intercellular cytoplasmic channels [plasmodesmata] between a

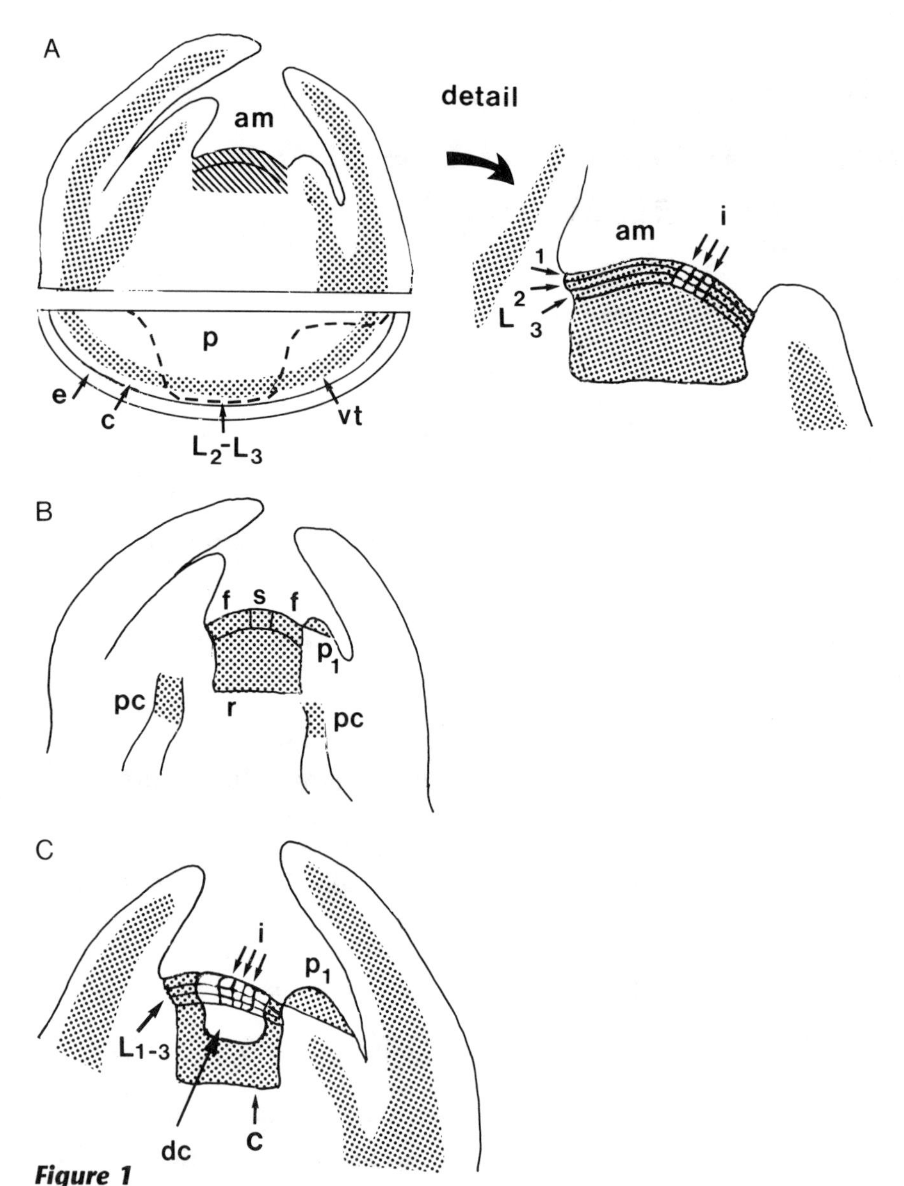

Figure 1

Schematic illustration of the classification of areas present within plant apices. (*A*) Longitudinal section (upper) and partial transverse section (below) of the shoot apical meristem of a dicot illustrating the tissue patterns that develop from the apical meristem. Abbreviations: (am) apical meristem; (c) cortex; (e) epidermis; (i) initials; (L_1–L_3) top layers of the apical meristem (note that these cells would not normally be in the same plane); (---) variable boundary between derivatives of the L_2 and L_3 layers in the apical meristem; (p) pith; (vt) vascular tissue. (Adapted, with permission, from Sussex 1989, copyright by Cell Press.) (*B*) Longitudinal section of the shoot apical meristem of potato. Abbreviations: (f) flanks; (p_1) first primordium; (pc) distal part of procambium; (r) rib meristem; (s) summit. (Adapted, with permission, from Clowes and MacDonald 1987, copyright by Academic Press Limited.) (*C*) Projection of a dye-coupled area in the apical meristem of a dormant potato tuber onto the histocytological scheme of a potato apex. Abbreviations: (c) corpus; (dc) dye-coupled area; (i), initials; (L_1–L_3) top layers of the apical meristem; (p_1) first primordium.

group of cells and the surrounding tissue. Plasmodesmata within such a group remain unaltered.) Support for this concept has been provided by the pioneering microinjection studies of Goodwin and coworkers (Erwee and Goodwin 1985; Goodwin and Erwee 1985; Santiago and Goodwin 1988). In addition to symplasmic domain formation, general changes in symplasmic permeability (the symplasmic pathway) might occur by downregulation of all plasmodesmata throughout the tissue. The mitotic index of long-day-induced *Silene* apices has been shown to correlate closely with down-regulation of the symplasmic pathway in this tissue (Santiago and Goodwin 1988).

VEGETATIVE APICES OF POTATO AS A MODEL SYSTEM

In general, potato shoot apices are highly stratified (Cutter 1978; Stewart et al. 1981). Clowes and MacDonald (1987) discerned five regions within the potato meristem (see Fig. 1B): (i) the summit (also called the central zone), which comprises the central 35% (as seen in sectional view) of the three apical layers; (ii) the flanks (also called the peripheral zone), which are three cells deep and extend down to the insertion of the youngest leaf primordia on both sides of the median sections; (iii) the rib meristem, which is located below the summit and flanks and extends down to the level of the insertion of the second youngest primordium; (iv) the youngest leaf primordium P1; and (v) the distal part of the procambium. The three apical layers (L_1–L_3) in the potato tuber bud are considered independent layers, but L_3 is sometimes considered to have no independent existence (Cutter 1978 and references therein). Leaf shape is principally affected by changes in L_2, whereas changes in L_1, and occasionally in L_3, only have a modifying effect. Adventitious buds in potato appear to arise from one or two cells derived from L_3.

SYMPLASMIC DOMAINS IN POTATO APICAL MERISTEMS

Dissected potato tubers have been used to explore the nature of the symplasm within the vegetative apices of this tissue. Iontophoretic microinjection (see p. 178) of the membrane-impermeant, fluorescent dye, lucifer yellow CH (LYCH), revealed that the symplasm in the apical meristem is organized in such a way that groups of cells are transiently isolated from each other, presumably for developmental purposes (Fig. 2A–F). Plasmodesmata at certain boundaries within the apex (Fig. 2A, B), as well as in the first-order bud (Fig. 2E, F), appear to be transiently shut down below the size exclusion limit of LYCH (457 daltons). The physical boundary demarcated by strongly dye-coupled cells (cells exchanging the dye) was wider and deeper than the group

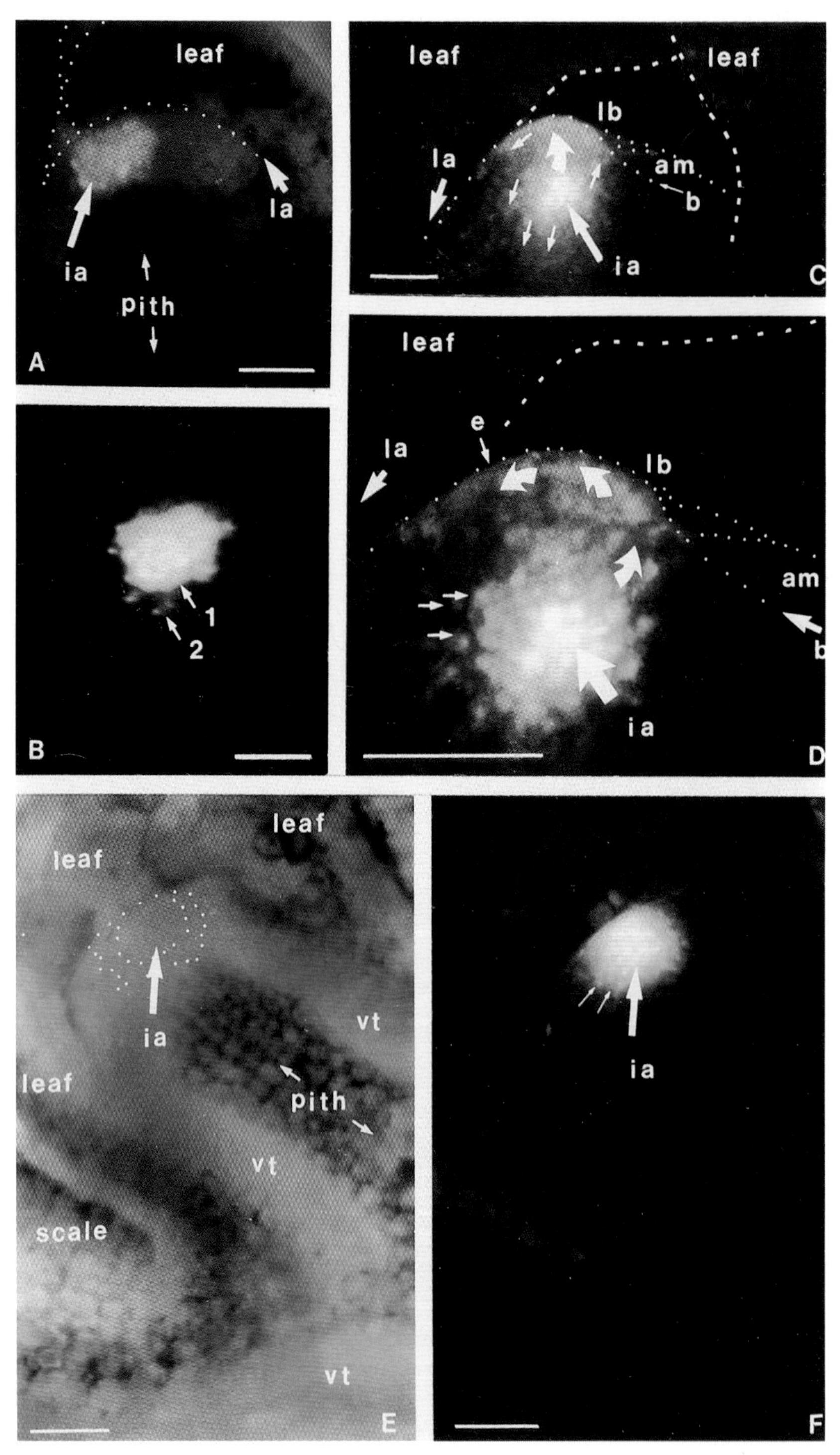

Figure 2 *(See facing page for legend.)*

of initials located within L_1–L_3. This suggests that a symplasmic domain is present that encompasses the group of initial cells in L_1–L_3, their immediate derivatives, and some adjacent cells. In histocytological terms, these dye-coupled cells represent the summit meristem, parts of the peripheral meristem, and the upper part of the corpus (see Fig. 1C). It has been previously established that cell-cycle times of potato initials and their immediate derivatives differ from those of peripheral meristematic cells (Clowes and MacDonald 1987). Thus, the symplasmic domains illustrated in Figure 2 may well reflect the cell initials plus their immediate derivatives.

Further evidence in support of the hypothesis that the formation of symplasmic domains plays a central role in tissue patterning is provided by the presence of a barrier for symplasmic exchange between cells of the leaf buttress and those in the peripheral meristem (Fig. 2C, D). The crucial question that must be addressed in future studies relates to the molecular mechanisms involved in establishing symplasmic boundaries. It will be equally important to unravel the cell-to-cell signals involved in differentiation of plant tissues.

Figure 2

Intercellular spread of microinjected LYCH in apices (*A–D*) and in a first-order bud (*E, F*) of cold-stored (4°C) tubers of *Solanum tuberosum* L., cv Desiree. Sections were covered by the bathing medium. Iontophoretic injection was carried out 30–60 minutes after excision of the tissue slice (bar, 50 μm [*A–C, E, F*] or 25 μm [*D*]).

(*A*) E_m = –38 mV. LYCH moves in a restricted apical area of a longitudinally cut apex. The tissue was viewed under both blue epi-illumination and transmitted white light. (*B*) As in *A*, except that only blue epi-illumination was employed. Note the densely stained central group of cells (1), and the loosely connected peripheral cells (2). (*C*) LYCH moves (unlabeled arrows) from the injected area (E_m = –43 mV) to other cells in the leaf buttress. The tissue was viewed under blue epi-illumination. (*D*) Detail as in *C*. Dye spreads (unlabeled arrows) to other leaf buttress cells and to epidermal cells. Note individual cells containing LYCH (small unlabeled arrows). The dye does not move to the apical meristem. (*E*) Longitudinally sectioned first-order bud, viewed under transmitted white light. The stippled region is the projection of the dye-coupled area illustrated in *F*. (*F*) LYCH spreads within a restricted apical area. Note that the central group of cells is uniformly and densely stained, whereas loose files of cells radiate out to the periphery (small unlabeled arrows). Projection onto the brightfield image (*E*) shows a group of cells located between the dye-coupled cells and the pith. Abbreviations: (am) apical meristem; (b) boundary between apical meristem and leaf buttress; (e) epidermis; (ia) injection area; (la) leaf attachment point; (lb) leaf buttress; (vt) vascular tissue.

PRACTICAL ASPECTS OF MICROINJECTION INTO PLANT CELLS

Microinjection procedures were originally developed for animal cells, and are normally carried out on the stage of an inverted microscope. The advantage of an inverted microscope is that there is ample space for microelectrode movement and surgical manipulation of the cells. A disadvantage is that the impalement procedure is visually monitored from the opposite side of the tissue. With animal cells, this does not impose serious problems since the injected cells are usually growing in a monolayer. However, with multi-layered plant tissues, this approach is simply not feasible as the tip of the microelectrode would be invisible. In addition, the variation in plant cell shape and complicated tissue topography usually necessitates very precise impalement of specific cells. Furthermore, plant cells have turgor, which makes cell impalement more difficult in terms of forming a tight membrane seal around the tip of the microelectrode. In view of the specific characteristics of and complications associated with plant cells, iontophoresis (where a charged molecule like LYCH is introduced by passing a small current into the cell via the injection electrode) is preferable to pressure-mediated delivery (where a transient, controlled increase in pressure within the injection pipette, in excess of the turgor pressure within the impaled cell, forces fluid into the cytoplasm).

Coupling a fluorescence microscope to a basic electrophysiological set-up has the following advantages: (i) small cells are not subjected to significant changes in volume; (ii) iontophoresis is vibration-free, and thus does not destabilize the membrane seal around the tip of the microelectrode; (iii) the physiological status of the impaled cell can be judged by monitoring the membrane potential (E_m); and (iv) the physiological condition of the impaled cell can be checked at any given moment during the dye-coupling experiment. Provided the iontophoretic current is kept within appropriate limits (which depend on such parameters as cell surface area, plasmodesmatal density, etc.), the E_m should return to the resting potential when current injection is terminated.

It should be noted that both current injection (iontophoresis) and pressure injection may induce side effects on the experimental tissue. Current can transiently enhance plasmodesmatal conductance and thus cause a modification to dye coupling. However, as long as dye transfer is not quantitatively assessed, this effect should not introduce a serious problem; rather it has the advantage of enhancing contrast. Pressure injection always causes an abrupt change in pressure (20–40 psi; Warner and Bate 1987), which can influence plasmodesmatal conductance. Depending on the sensitivity of the tissue, such fluctuations in pressure can induce callose formation at the neck region of the plasmodesmata (Robards and Lucas 1990 and references therein).

OUTLINE OF EXPERIMENT

In this section, we describe the steps involved in using microinjection of LYCH to investigate symplasmic domains associated with meristematic tissues within the vegetative buds of the potato tuber. The same general protocol can be applied to address a wide range of questions in plant biology.

REFERENCES

Clowes, F.A.L. and M.M. MacDonald. 1987. Cell cycling and the fate of potato buds. *Ann. Bot.* **59:** 141–148.

Cutter, E.G. 1978. Structure and development of the potato plant. In *The potato crop* (ed. P.M. Harris), pp. 70–152. Chapman and Hall, London.

De Laat, S.W., L.G.J. Tertoolen, A.W.C. Dorrestein, and J.A.M. van den Biggelaar. 1980. Intercellular communication patterns are involved in cell determination in early molluscan development. *Nature* **287:** 546–548.

Erwee, M.G. and P.B. Goodwin. 1985. Symplast domains in extrastellar tissues of *Egeria densa* Planch. *Planta* **163:** 9–19.

Goodwin, P.B. and M.G. Erwee. 1985. Intercellular transport studied by microinjection. In *Botanical microscopy* (ed. A.W. Robards), pp. 335–358. Oxford University Press, Oxford, England.

Halliwell, J.V. and M.J. Whitaker. 1987. Using microelectrodes. In *Microelectrode techniques* (ed. N.B. Standen et al.), pp. 1–12. Company of Biologists, Cambridge, England.

Kater, S.B., C. Nicholson, and W.J. Davis. 1973. A guide to intracellular staining techniques. In *Intracellular staining in neurobiology* (ed. S.B. Kater and C. Nicholson), pp. 432–450. Springer Verlag, Berlin.

Lucas, W.J., B. Ding, and C. van der Schoot. 1993. Plasmodesmata and the supracellular nature of plants. *New Phytologist* **125:** 435–476.

Mertz, S.M. and N. Higinbotham. 1976. Transmembrane electropotential in barley roots as related to cell type, cell location, and cutting and aging effects. *Plant Physiol.* **57:** 123–128.

Niklas, K.J. and J.D. Mauseth. 1980. Simulations of cell dimensions in shoot apical meristems: Implications concerning zonate apices. *Am. J. Bot.* **67:** 715–732.

Pierce, W.S. and D.L. Hendrix. 1981. Electrochemical aging responses in *Pisum*: Cellular adaptions or recovery from injury? *Plant Physiol.* **67:** 864–868.

Robards, A.W. and W.J. Lucas. 1990. Plasmodesmata. *Annu. Rev. Plant Physiol. Plant Mol. Biol.* **41:** 369–419.

Santiago, J.F. and P.B. Goodwin. 1988. Restricted cell-cell communication in the shoot of *Silene coeli* Rosa during the transition to flowering is associated with a high mitotic index rather than with evocation. *Protoplasma* **146:** 52–60.

Serras, F. and J.A.M. van den Biggelaar. 1990. Progressive restrictions in gap junctional communication during development. In *Parallels in cell to cell junctions in plants and animals* (ed. A.W. Robards et al.), pp. 129–143. Springer Verlag, Berlin.

Soma, K. 1973. Experimental studies on the morphogenesis in the vegetative shoot apex. Induced mutation and chimera in woody plants. *Gamma Field Symp.* **12:** 83–94.

Stewart, F.C., U. Moreno, and W.M. Roca. 1981. Growth, form and composition of potato plants as affected by environment. *Ann. Bot.* (suppl. 2) **48:** 1–45.

Stewart, R.N. and H. Dermen. 1970. Determination of number and mitotic activity of shoot apical initial cells by analysis of mericlinal chimeras. *Am. J. Bot.* **57:** 816–826.

Sussex, I.M. 1989. Developmental programming of the shoot meristem. *Cell* **56:** 225–229.

Sussex, I.M. and D. Rosenthal. 1973. Differential ^{3}H-thymidine labelling of nuclei in the shoot apical meristem of *Nicotiana*. *Bot. Gaz.* **134:** 295–301.

Van der Schoot, C. and A.J.E. van Bel. 1989. Glass microelectrode measurements of sieve tube membrane potentials in internode discs and petiole strips of tomato (*Solanum lycopersicum* L.). *Protoplasma* **149:** 144–154.

Warner, A.E. and C.M. Bate. 1987. Techniques for dye injection and cell labelling. In *Microelectrode Techniques* (ed. N.B. Standen et al.), pp. 169–185. Company of Biologists, Cambridge, England.

• *TIMETABLE*

Microinjection and the Study of Tissue Patterning in Plant Apices

STEPS	*TIME REQUIRED*
Preparation of Tissue	
Dissect tissue	15–30 minutes
Mount apical sections	1 minute
Allow tissue to recover	30–60 minutes
Transfer mounted tissue to microscope stage	5 minutes
Microinjection and Dye Coupling Experiments	
Fabricate, load, and position microelectrode	2–4 minutes
Impale potato section with microelectrode and inject dye	5–15 minutes
Record and analyze image	10 minutes

Preparation of Tissue

MATERIALS AND EQUIPMENT

Potatoes
Stereo microscope (e.g., model SM2–4 zoom; Nikon, Inc.)

REAGENTS

Bathing medium (pre-equilibrated to room temperature)
(For recipe, see Preparation of Reagents, p. 189)

PROCEDURE

1. Select a potato with a well-proportioned apical eye, showing an upright, centrally located apex and well-developed side buds. Using a razor blade, transversely dissect off the upper portion (~1 cm) of the tuber that contains the apical eye (Fig. 3A). Under a stereo microscope (low magnification), cut this portion into halves (Fig. 3B): the cut should be made through the apical eye, just behind the central bud. Take a new razor blade and slice vertically (Fig. 3C) through the middle of the apex (Fig. 3D). The section should be 0.5–1 mm thick. Alternatively, perform the same dissection on a lateral bud. Examine the section immediately under the stereo microscope, and discard it if you observe significant cell damage.

   Note: (i) In principle, all buds on a tuber are suitable for microinjection. However, the buds are in different stages of rest/development, which can influence the size/shape of the symplasmic domains. To minimize such variation, it is preferable to use identical buds in all experiments. We recommend using the central bud of the most apical eye, or either of its two side buds. In this way, several good sections can be obtained per tuber. In practice, often 3–5 tubers are needed to obtain two usable sections. (ii) It takes practice to be able to cut uniformly thin (0.5–1 mm) sections. The apex, including the overarching leaves, is usually 0.3–0.5 mm in diameter, and it lacks support when the potato section is fixed onto a microscope slide. Therefore, very elongate apices should be avoided as they usually present problems during the experiment because they tend to bend downwards out of the focal plane.

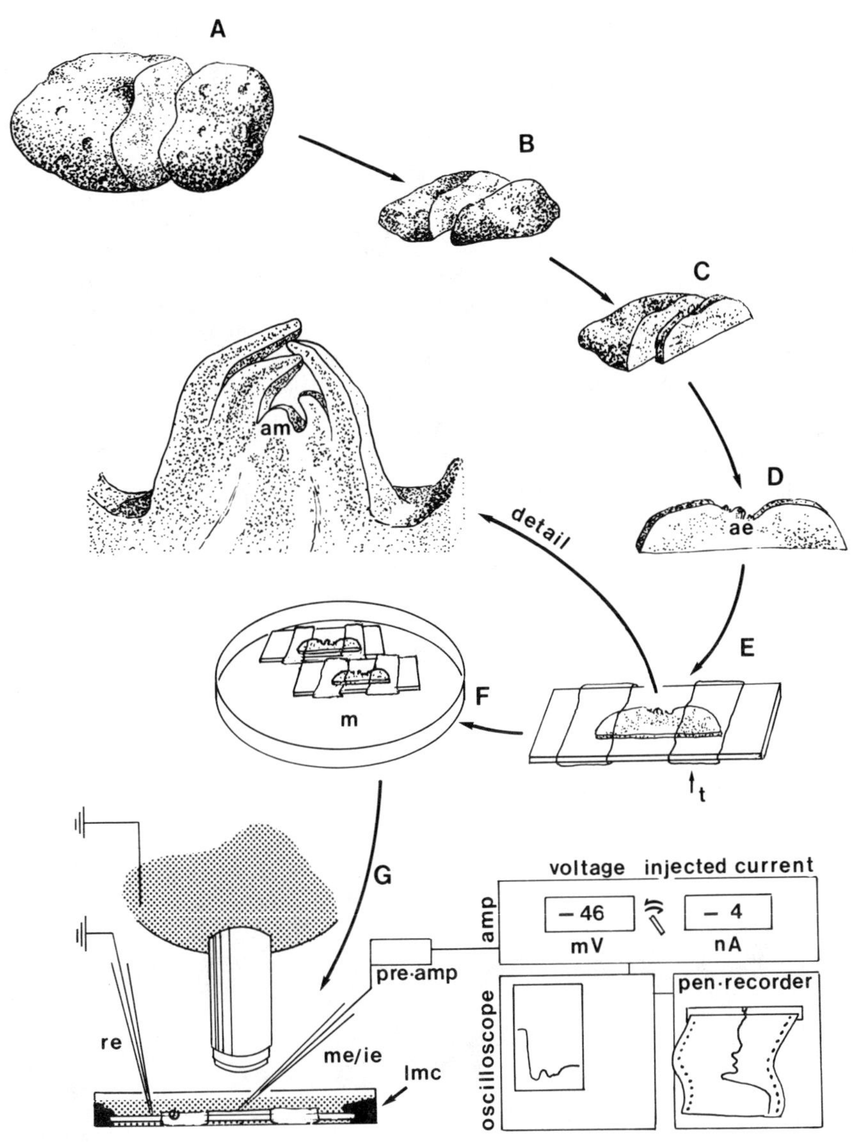

Figure 3
Scheme of experimental procedures used to prepare potato tuber apical meristems for microinjection and dye coupling studies. Abbreviations: (ae) apical eye; (am) apical meristem; (ie) injection electrode; (lmc) laboratory modeling clay; (m) medium; (me) measuring electrode; (re) reference electrode; (t) transparent tape.

2. Transfer each potato section to a dry, clean microscope slide, and fasten the section to the slide with regular (one-sided) transparent tape (Fig. 3E). When appropriately applied, the tape will keep the section in place. If the potato section is too wet, blot it briefly with a Kleenex tissue.

3. Immediately transfer the slides to a petri dish containing bathing medium (pre-equilibrated to room temperature) and allow the potato tissue to recover for 30–60 minutes at room temperature (Fig. 3F).

   Note: It is necessary to allow the tissue to recover because cutting (wounding) has been shown to change the E_m (see, e.g., Mertz and Higinbotham 1976; Pierce and Hendrix 1981). The initial low E_m values (ranging from -30 to -60 mV) are likely reflect a wound response, whereas the subsequent increase in the negative value of the potential (to values ranging from -60 to -100 mV) may reflect cell recovery. However, as highly meristematic tissues of a dormant apical meristem may not be as sensitive to wounding as mature or rapidly expanding tissues, the rise in potential may reflect the onset of regeneration processes.

4. Using the stereo microscope, select a potato section that shows the desired section plane (see Fig. 3E, detail) through the apical (or lateral) bud. It is important to ensure that the surface of the potato section remains covered with the bathing medium. Use a pasteur pipette to place some drops of the medium along the edge of the section.

Microinjection and Dye Coupling Experiments

MATERIALS AND EQUIPMENT

Upright, epi-illumination light microscope (e.g., Optiphot microscope with substage halogen lamp and epifluorescence attachment [Nikon, Inc.]) equipped with either a standard camera system (e.g., Photolux model UFX-II/FX-35WA; Nikon, Inc.) or a video camera (e.g., model C2400; Hamamatsu Photonic Systems) and image processor (e.g., ARGUS-10; Hamamatsu Photonic Systems)
Borosilicate glass tubes with an inner filament (World Precision Instruments, Inc. 1B1/00F-4)
Microelectrode puller (model P-87; Sutter Instrument Co.)
Microelectrode holders with an Ag/AgCl bridge (model MEH1S; World Precision Instruments, Inc.)
Pipette tips, beveled (used to manufacture plastic "needles" required for backfilling micropipettes, see p. 186) (Out Patient Services, Inc. 1033 G)
KCl solution (0.1 M and 3 M)
Amplifier (WPI DUAL 773; World Precision Instruments, Inc.)
Micromanipulator with an MN-2 3D course control, fastened to a mounting bar fixed on the stand in parallel to the microscope stage (model MO-203; Narishige USA, Inc.)
Oscilloscope (model cs-1021; Kenwood)
Chart recorder (model BD5; Kipp & Zonen)

REAGENTS

Bathing medium (preequilibrated to room temperature)
Iontophoresis solution
(For recipes, see Preparation of Reagents, p. 189)

PROCEDURE

1. Transfer the selected potato section (see p. 184) to a transparent bathing chamber.

   Note: A small, rectangular Plexiglas chamber with a thin base is all that is required. The base must be transparent to allow passage of light from the microscope lamp to the tissue slice.

Fix the slide to the bottom of the chamber with laboratory modeling clay (Fig. 3G). Use a pasteur pipette to supply bathing medium to the tissue via the wall of the bathing chamber. It is important to ensure that the surface of the potato section remains covered with bathing medium. Place the bathing chamber on the stage of the light microscope, and then connect the bathing medium to ground, via a reference electrode, by inserting the electrode into the medium at the edge of the section.

Note: An Ag/AgCl reference electrode is made by depositing a thin layer of AgCl onto a silver wire. Alternatively, the electrical circuit can be completed via a salt bridge (see Halliwell and Whitaker 1987). Note that in some cases it may be necessary to ground the microscope to avoid the influence of spurious electrical fields (see Fig. 3G)

2. Identify the appropriate injection area, then lower the microscope stage so that the microelectrode can be brought into position.

3. Pull a microelectrode using a standard microelectrode puller. For iontophoretic injection into small cells, microelectrodes with outer tip diameters of <1 μm and resistances in the range of 20–30 x 10^6 Ω are required. Throughout a particular experimental series, it is advisable to use microelectrodes with the same characteristics.

Note: (i) Microelectrodes for microinjection are made from borosilicate glass tubing using a conventional puller. Although a range of tubing is available from various manufacturers, the most essential features are wall thickness and the presence of an inner filament that is attached to the wall of the tube. When the microelectrode is pulled, this filament extends down into the tip. Injection solution introduced into the microelectrode is pulled right down to the tip by surface tension, thus allowing for quick and easy loading. (ii) Prior to sectioning the potato tissue, run a test series to find a suitable program on the puller. Note that the WPI DUAL 773 amplifier has a built-in circuit for determining electrode resistance. (iii) E_m measurements are intended to check for entry of injected solution into the target cell, the formation of a tight membrane seal, and the continued viability of the microinjected cell. However, when determined appropriately (see step 6), E_m measurements can give valuable additional information. For example, the finding that dye coupling within the phloem tissue of tomato stems was restricted to the sieve tube-companion cell complex was confirmed by membrane potential data demonstrating that this cell complex was electrically isolated from the surrounding phloem cell types (Van der Schoot and Van Bel 1989).

4. Load the tip of the microelectrode with iontophoresis solution (which contains LYCH), and then backfill the microelectrode with 0.1 M KCl (which serves as an electrolyte). Fill the microelectrode holder with 3 M KCl and then insert the microelectrode into the holder. Finally, connect the microelectrode holder to the preamplifier (clamped to the micromanipulator), and the preamplifier to the 10^{12} Ω input of the WPI DUAL 773 amplifier.

Note: (i) Hollow plastic "needles" can be fabricated from plastic pipette tips by pulling them after heating in a low flame. These needles can be fixed onto plastic syringes (filled with electrolyte) for use in backfilling LYCH-loaded microelectrodes. Alternatively, a very fine gauge hypodermic needle can be used to introduce electrolyte into the shank of the LYCH-loaded microelectrode. (ii) The WPI microelectrode holder contains an Ag/AgCl disc that forms the corresponding half-cell to the grounded reference electrode.

5. Use the Narishige micromanipulator to locate the microelectrode tip within the center of the field of view and then adjust the position of the tissue until the selected target cell is just below the focal plane of the microelectrode. As soon as the microelectrode is in the bathing medium, the electrical circuit is complete and a potential should be detected. A potential difference of less than –10 mV can be eliminated using the off-set on the WPI DUAL 773 amplifier.

Note: The characteristics of the microelectrode tip and the composition of the filling solution generate specific junction and tip potentials. The junction potentials are due to diffusional forces at the tip of the microelectrode, resulting from different anionic and cationic mobilities and from differences in solute concentrations. Tip potentials result from restriction of anionic mobility by the surface chemistry of the glass; additionally, they can arise from differences between inside and outside solutions, and are proportional to the microelectrode resistance. Thus, it is important to select tips that have resistances within the 20–30 x 10^6 Ω range (Halliwell and Whitaker 1987).

6. Impale the target cell by carefully advancing the microelectrode toward the cell while observing the output signal from the amplifier on the oscilloscope. Entry of the microelectrode into the cell is indicated by an abrupt, negative jump in the potential (see Fig. 3G). Once the E_m has stabilized, which usually takes a few minutes, switch the amplifier from voltage-recording to current-injection mode.

Note: In these meristematic cells, the membrane potential normally ranges from -30 to -60 mV (C. van der Schoot, unpubl.).

7. Increase the negative current gradually in a stepwise manner until dye is observed to flow out of the microelectrode into the target cell. Take care to ensure that the hyperpolarization of the E_m, resulting from the current injection, is kept within physiological limits (i.e., more positive than -300 mV). If dye does not flow into the impaled cell, the tip of the microelectrode may be blocked. If this is the case, the microelectrode resistance will have increased.

Note: Current flow causes local heating of the cytoplasm and if too much current is passed, coagulation of protoplasmic components can occur, which causes obstruction of the tip (Kater et al. 1973). Blockage can also result from precipitation of LYCH in the very tip of the microelectrode. At times, clogging can be reversed by applying a brief pulse of alternating current to the pipette tip, or by reversing the direction of the current flowing through the tip.

8. Monitor the E_m value intermittently to check cell viability.

9. Visualize cell-to-cell movement of LYCH by epifluorescence microscopy. Symplasmic domains are visible as groups of strongly dye-coupled cells at specific locations within the apical meristem. These domains can be identified irrespective of the actual cell that receives the dye.

   Note: Light from either a mercury- or quartz-halogen lamp is passed through a combination of violet excitation (BP 436/437 nm) and barrier (LP 460 nm) filters. The fluorescence emitted from the excited LYCH molecules can be visually observed, but it is more convenient to employ a video image intensifying camera for enhanced observation and recording. Similarly, current injection, E_m measurement, and data recording can best be performed simultaneously by controlling the experimental system using a computer interface.

• *PREPARATION OF REAGENTS*

Bathing Medium

125 mM Mannitol
10 mM NaOH/2-(*N*-morpholino)ethanesulfonic acid (MES) buffer
0.5 mM KCl
0.5 mM $MgCl_2$
0.5 mM $CaCl_2$

Adjust the pH to 5.7 by adding 100 mM NaOH. The bathing medium is required to maintain the physiological state of the potato sections (C. van der Schoot, unpubl.). When applied to the tissue the bathing medium should be at room temperature. Keep the stock solution at 4°C.

Iontophoresis Solution

Prepare a 1 mM solution of lucifer yellow CH (LYCH) in 0.1 M KCl. Pass the solution through a 0.2 μm filter (Micron Separations, Inc. DDN 0200300) and store it in a microfuge tube on ice until it is required for the iontophoresis experiments.

• *Section 10*

Chloroplast Run-on Transcription: Determination of the Transcriptional Activity of Chloroplast Genes

Cellular RNA levels can be controlled by changes in the transcriptional activity of genes (transcriptional regulation), by the stability of the RNA transcripts themselves (post-transcriptional regulation), or by a combination of both regulatory mechanisms. One approach to distinguish between these modes of regulation for chloroplast genes in higher plants is to examine transcription in isolated organelles. In isolated nuclei from both animal (Tsai et al. 1978; Derman et al. 1981; McKnight and Palmiter 1979) and plant (Gallagher and Ellis 1982) tissues, it is generally assumed that transcriptional activity represents RNA elongation by preinitiated RNA polymerases. This transcriptional activity is termed nuclear run-on (or run-off) transcription. Recently, chloroplast run-on transcription systems have been developed and used to measure the activity of the preinitiated transcription elongation complexes and the relative transcriptional activities of chloroplast genes (Deng et al. 1987; Deng and Gruissem 1987, 1988; Mullet and Klein 1987). Using the chloroplast run-on transcription system, it has been demonstrated that post-transcriptional processes are responsible for the differential accumulation of certain plastid mRNAs during chloroplast development (Deng and Gruissem 1987; Mullet and Klein 1987).

Chloroplast run-on transcription studies involve three steps: (i) isolation of intact chloroplasts in such a way that transcription elongation complexes are preserved; (ii) incubation of the run-on transcription reaction and purification of the run-on transcription products; (iii) quantitation of the run-on transcripts from each gene by hybridization to chloroplast gene probes. In this section, we describe a procedure for determining the transcriptional activity of chloroplast genes from spinach. The protocol can be adapted for experiments with plastids isolated from organs other than leaves, and it has been used successfully with fruit chromoplasts and root plastids (Gruissem et al. 1987; Deng and Gruissem 1988).

REFERENCES

Deng, X.W., D.B. Stern, J.C. Tonkyn, and W. Gruissem. 1987. Plastid run-on transcription. Application to determine the transcriptional regulation of plastid genes. *J. Biol. Chem.* **262:** 9641–9648.

Deng, X.W. and W. Gruissem. 1987. Control of plastid gene expression during development: The limited role of transcriptional regulation. *Cell* **49:** 379–387.

———. 1988. Constitutive transcription and regulation of gene expression in nonphotosynthetic plastids of higher plants. *EMBO J.* **7:** 3301–3308.

Derman, E., K. Krauter, L. Walling, C. Weinberger, M. Ray, and J.E. Darnell, Jr. 1981. Transcriptional control in the production of liver-specific mRNAs. *Cell* **23:** 731–739.

Gallagher, T.F. and R.J. Ellis. 1982. Light-stimulated transcription of genes for two chloroplast polypeptides in isolated pea leaf nuclei. *EMBO J.* **1:** 1493–1498.

Gruissem, W., K. Callan, J. Lynch, T. Manzara, M. Meighan, J. Marita, B. Picchulla, M. Sugita, M. Thelander, and L. Wanner. 1987. Plastid and nuclear gene expression during tomato fruit formation. In *Tomato biotechnology* (ed. D.J. Nevins and R.A. Jones), pp. 239–250. A.R. Liss, New York.

Kannangara, C.G., S.P. Gough, B. Hansen, J.N. Rasmussen, and D.J. Simpson. 1977. A homogenizer with replaceable razor blades for bulk isolation of active barley plastids. *Carlsberg Res. Commun.* **42:** 431–439.

Klein, R.R. and J.E. Mullet. 1990. Light-induced transcription of chloroplast genes. *psbA* transcription is differentially enhanced in illuminated barley. *J. Biol. Chem.* **265:** 1895–1902.

McKnight, G.S. and R.D. Palmiter. 1979. Transcriptional regulation of the ovalbumin and conalbumin genes by steroid hormones in chick oviduct. *J. Biol. Chem.* **254:** 9050–9058.

Mullet, J.E. and R.R. Klein. 1987. Transcription and RNA stability are important determinants of higher plant chloroplast RNA levels. *EMBO J.* **6:** 1571–1579.

Sambrook, J., E.F. Fritsch, and T. Maniatis. 1989. *Molecular cloning: A laboratory manual.* Cold Spring Harbor Laboratory Press, Cold Spring Harbor, New York.

Tsai, S.Y., D.R. Roop, M.J. Tsai, J.P. Stein, A.R. Means, and B.W. O'Malley. 1978. Effect of estrogen on gene expression in the chick oviduct. Regulation of the ovomucoid gene. *Biochemistry* **17:** 5773–5780.

• *TIMETABLE*
Determination of the Transcriptional Activity of Chloroplast Genes

STEPS	*TIME REQUIRED*
Isolation of Intact Chloroplasts from Spinach	
Prepare Percoll® gradients	1 hour
Homogenize spinach tissue	1 hour
Isolate chloroplasts	1.5 hours
Determine chlorophyll concentration	0.5 hour
Run-on Transcription	
Incubate run-on transcription reaction	8 minutes
Extract labeled RNA	2 hours
Quantitation of Run-on Transcript Levels by Hybridization	
Apply chloroplast gene probes to nylon membrane and prehybridize	2–4 hours
Hybridize labeled RNA to DNA probes	18–24 hours
Wash nylon membrane	2 hours
Expose nylon membrane to X-ray film	24 hours
Quantitate relative transcription activity of each chloroplast gene	3 hours

Isolation of Intact Chloroplasts from Spinach

MATERIALS AND EQUIPMENT

Glutathione
Distilled H_2O (autoclaved)
Spinach plants
Miracloth (Calbiochem 47855)
Acetone (80%)

REAGENTS

GM mix (5x)
PCBF
GM mix (1x)
IC mix
(For recipes, see Preparation of Reagents, pp. 204–207)

PROCEDURE

To reduce the transcriptional activity in the chloroplasts, it is critical to maintain cold temperatures during the chloroplast isolation procedure. Therefore, all the steps described below should be carried out on ice in a cold room. At low temperatures, the RNA polymerase complex has reduced transcription elongation activity, which effectively "freezes" the transcription complexes on the DNA template.

Preparation of Percoll® Gradients

1. Prepare 10% and 80% Percoll® solutions as follows:

	10% Percoll®	80% Percoll®
GM mix (5x) (with freshly added 50 mM dithiothreitol)	20 ml	20 ml
Glutathione	6 mg	6 mg
PCBF	10 ml	80 ml
Distilled H_2O (autoclaved)	70 ml	–

2. Using a gradient maker, prepare a linear gradient (40 ml total volume) in a 50-ml SORVALL® centrifuge tube (or equivalent) with 80% Percoll® at the bottom and 10% Percoll® at the top.

Homogenization of Plant Tissue

1. Harvest approximately 50 g (wet weight) of fresh young spinach leaves. Wash the leaves with distilled H_2O and blot them dry. Wearing gloves, remove the leaf veins by hand and tear the leaves into pieces approximately 1 cm^2.

   Note: It is important to minimize the time between harvest and chloroplast isolation. Keep the leaves in a moist chamber during transport to the laboratory. In our experience, it is best to use spinach leaves harvested directly from the greenhouse (or from a field). Spinach purchased at produce stores is not good material for the experiment described here.

2. Add 150 ml of ice-cold 1x GM mix to the leaf pieces. Homogenize the tissue in a Waring blender equipped with four additional razor blades mounted on an extended shaft (Kannangara et al. 1977) at low speed for a few seconds (until the leaf pieces are covered in the buffer) and then at high speed for 4–5 seconds.

   Note: The razor blade equipment for the Waring blender is not commercially available. However, the cited reference provides sufficient detail for the apparatus to be easily built by a professional mechanic in a shop. The razor blade device significantly reduces the time required to homogenize the leaf material and results in a higher proportion of intact chloroplasts. It has also been used successfully and with good efficiency for leaves that are difficult to homogenize, such as corn, rye, and barley.

3. Pour the homogenate through four layers of sterile Miracloth that has been prewetted and autoclaved. To increase the chloroplast yield, squeeze the debris until it is almost dry.

   Note: The Miracloth must be squeezed gently because it breaks easily when wet.

Isolation of Chloroplasts

1. Distribute the filtrate between four sterile 50-ml centrifuge tubes and spin in a SORVALL® SS-34 rotor at 6000 rpm for 30 seconds at 2°C.

2. Pour off the slightly green supernatant and carefully remove the upper layer of the chloroplast pellet, which contains mostly broken chloroplasts.

3. Using a small paintbrush that has been autoclaved, resuspend the remainder of the chloroplast pellet in 1x GM mix until no more clumps are apparent. The final volume of the resuspended pellet should be approximately 2 ml.

   Note: The paintbrush facilitates resuspension of the pellet with minimal breakage of chloroplasts.

4. Using a wide-bore pasteur pipette, layer the 2 ml of chloroplast suspension onto the 10–80% linear Percoll® gradient (volume 40 ml). Spin in a SORVALL® HB-4 (swinging bucket) rotor at 7000 rpm for 20 minutes at 2°C.

   Note: Avoid overloading the gradient. Overloaded gradients result in poor separation of broken and intact chloroplasts. It may be necessary to run pilot gradients to establish the capacity of the gradient for experiments other than the one described here.

5. You should observe two discrete bands on the Percoll® gradient. The upper band contains broken chloroplasts and the lower band contains intact chloroplasts. Aspirate the liquid to the top of the lower band. Do this *slowly* to avoid contaminating the lower portion of the gradient with broken chloroplasts. Using a wide-bore pasteur pipette, transfer the lower band of intact chloroplasts to a clean centrifuge tube.

6. Dilute the chloroplast-Percoll® suspension with two volumes of IC mix and pellet the chloroplasts by spinning in a SORVALL® HB-4 rotor at 7000 rpm for 1 minute at 2°C.

7. Using a small paintbrush, resuspend the chloroplast pellet in approximately 10 ml of IC mix. Spin again at 7000 rpm for 1 minute at 2°C.

   Note: Steps 6 and 7 are critical to remove the Percoll®, which may interfere with subsequent steps in the protocol.

8. Using a small paintbrush, resuspend the chloroplasts in a very small volume (50–100 μl) of IC mix. Transfer the chloroplast suspension to a microfuge tube on ice. Cover the tube with aluminum foil to avoid exposure to light.

Chlorophyll Determination

1. Add 1 μl of the chloroplast suspension to 1 ml of 80% acetone and vortex. Pellet the debris by spinning in a microfuge for 5 minutes. Remove the supernatant and measure the absorbance (OD) at 663 nm and 645 nm. If the OD value at 663 nm is >0.8, dilute the su-

pernatant with 80% acetone, read the OD again, and remember to take the dilution factor into account in the calculation.

$$\text{Chlorophyll concentration (mg/ml)} = [20.2 \times OD_{645nm} + 8.02 \times OD_{663nm}] \times \text{the dilution factor}$$

Note: Alternatively, the number of chloroplasts per unit volume can be determined by counting an aliquot of the chloroplast suspension using a hemocytometer. A known number of chloroplasts can then be used in the run-on transcription reaction (Mullet and Klein 1987).

2. Adjust the chlorophyll concentration in the chloroplast suspension to 5 mg/ml by adding an appropriate volume of IC mix. Store the chloroplast suspension on ice before use, but not for longer than 30 minutes.

Note: Aliquots of the chloroplast suspension should be reserved for DNA, RNA, and protein concentration determinations (Sambrook et al. 1989).

Run-on Transcription

MATERIALS AND EQUIPMENT

Diethyl pyrocarbonate (DEPC)
DEPC-treated (RNase-free) H_2O
[α-^{32}P]UTP (2 mCi; ~410 Ci/mmol)
Chloroplast suspension (see p. 197)
Phenol/chloroform/isoamyl alcohol (25:25:1; stored at 4°C)
Ammonium acetate (5 M; autoclaved)
Ethanol (100% precooled to –20°C and 70% at room temperature)
Sephadex™ G-50 spin column (prepacked) or Stratagene push column (Stratagene 400701)

REAGENTS

R mix
Nucleotide mix (10x)
Lysis buffer
(For recipes, see Preparation of Reagents, pp. 204–207)

SAFETY NOTES

- DEPC is suspected to be a carcinogen and should be handled with care.
- Wear gloves when handling radioactive substances. Consult the local safety office for further guidance in the appropriate use of radioactive materials.
- Phenol is highly corrosive and can cause severe burns. Wear gloves, protective clothing, and safety glasses when handling it. All manipulations should be carried out in a chemical fume hood. Any areas of skin that come in contact with phenol should be rinsed with a large volume of water or PEG 400 and washed with soap and water; do not use ethanol!
- Chloroform is irritating to the skin, eyes, mucous membranes, and respiratory tract. It should only be used in a chemical fume hood. Gloves and safety glasses should also be worn. Chloroform is a carcinogen and may damage the liver and kidneys.

PROCEDURE

1. Add the following reagents (in the order listed) to a microfuge tube on ice:

R mix	60 μl
Nucleotide mix (10x)	10 μl
RNase-free H_2O	10 – Y μl
[α-^{32}P]UTP (100 μCi)	Y μl
Total volume:	80 μl

Note: Some published protocols (e.g., Klein and Mullet 1990) suggest including heparin in the run-on transcription reaction to prevent rapid degradation of the newly synthesized RNA. However, in our laboratory, we have not found any significant degradation of run-on RNA either in the absence or presence of heparin (Deng et al. 1987). If you decide to include heparin, the final concentration of the polyanion in the run-on transcription reaction mixture should be 20 μg/ml (Deng et al. 1987).

2. To start the reaction, add 20 μl of the chloroplast suspension to the reaction mixture using a P200 Pipetman®. Pipette the mixture up and down six times. Transfer the tube to a water bath and incubate for 8 minutes at 25°C.

Note: (i) The pipetting will help to break the chloroplast envelopes, which act as a barrier to the uptake of all nucleotide triphosphates except ATP. (ii) This is a good time to start the prehybridization, see p. 201.

3. To stop the reaction, add 10 μl of lysis buffer, followed immediately by 100 μl of phenol/chloroform/isoamyl alcohol (25:25:1). Cap the tube tightly and vortex thoroughly.

Note: At this point, the sample (or samples, if a time course has been performed) can be stored at room temperature. The combination of lysis buffer and phenol will immediately stop the reaction and denature nucleases and proteases.

4. Spin the sample in a microfuge for 5 minutes at room temperature.

5. Transfer the upper (aqueous) phase to a clean microfuge tube. Add 100 μl of phenol/chloroform/isoamyl alcohol (25:25:1), cap tightly, and vortex.

Note: Avoid collecting material from the interface. With chloroplasts isolated from leaves harvested the morning of the experiment, the interface should be very small, but with starch-containing chloroplasts or other plastid types, there may a substantial interface.

6. Repeat step 4.

7. The upper (aqueous) phase should be almost clear and free of green or yellow color. Transfer the upper phase to a clean microfuge tube, add 1/10 volume of 5 M ammonium acetate, 300 µl of 100% ethanol (precooled to –20°C), and vortex.

   Note: It is generally not necessary to add exogenous carrier DNA or RNA because enough carrier nucleic acid is usually extracted from the chloroplasts.

8. Leave on ice for 20 minutes. Spin in a microfuge for 30 minutes at 4°C.

   Note: This is a good time to set up the Sephadex™ G-50 spin column.

9. Carefully discard the supernatant. The pellet contains total nucleic acid, including the labeled run-on transcripts.

10. Wash the pellet by carefully adding 100 µl of 70% ethanol (at room temperature). The pellet will remain intact. Spin in a microfuge for 5 minutes at room temperature.

11. Carefully remove all the liquid and let the tube sit uncapped for 2 or 3 minutes at room temperature.

    Note: The pellet can also be dried briefly in a vacuum centrifuge.

12. Resuspend the pellet in 100 µl of RNase-free H_2O.

    Note: At this point, you can include an optional step to remove the chloroplast DNA by digesting the resuspension with RNase-free DNase I and isolating the RNA by phenol:chloroform extraction and ethanol precipitation. However, we have found that if the RNA is not heated before it is added to the hybridization solution (see p. 202), there is no detectable effect of contaminating chloroplast DNA on hybridization.

13. Load the resuspended pellet onto a Sephadex™ G-50 spin column. Spin in clinical bench-top centrifuge at speed 7 (i.e., ~1640*g* at the tube tip) for 2 minutes at room temperature. The labeled RNA will elute from the column and the unincorporated free nucleotides will remain in the column.

    Note: (i) DNA elutes from the Sephadex™ G-50 spin column with labeled RNA, but it does not interfere with the hybridization reaction (see note to step 12 above). (ii) If you use the Stratagene push column device, remove unincorporated nucleotides according to the manufacturer's instructions.

14. Collect the eluate from the spin column. Remove 1 µl and count in a scintillation counter to determine the incorporation of radioactive UTP into the run-on transcripts. Store the remainder of the sample at –20°C until you are ready to use it as a hybridization probe.

Quantitation of Run-on Transcript Levels by Hybridization

MATERIALS AND EQUIPMENT

Nylon membrane (e.g., HYBOND™ N, Amersham Life Science, Inc.)
Chloroplast gene probes
Chloroplast run-on transcripts (see p. 200)

REAGENTS

Hybridization solution
SSC (2x)/SDS (0.1%) (prewarmed to 65°C)
SSC (1x)/SDS (0.1%) (prewarmed to 65°C)
SSC (0.5x)/SDS (0.1%) (prewarmed to 65°C)
(For recipes, see Preparation of Reagents, pp. 204–207)

PROCEDURE

Prehybridization

1. Apply equimolar amounts of chloroplast gene probes to a piece of nylon membrane. Ensure that the blot contains a molar excess of each DNA probe relative to the labeled run-on RNA for which it is specific. This is important to drive the hybridization reaction and to minimize the competition between unlabeled endogenous transcripts and run-on transcripts.

   Note: If you intend to quantitate the run-on transcripts (see pp. 202–203), it is important to include controls to estimate the efficiency of hybridization to the DNA probes. Such control experiments can be carried out by synthesizing precise concentrations of labeled RNA in vitro (for details, see Section 11), and then hybridizing the labeled RNA to the DNA probes.

2. Insert the DNA blot into a plastic bag, add 10 ml of hybridization solution, and heat-seal the bag.

   Note: These hybridizations can also be carried out in bottles in a hybridization oven.

3. Prehybridize with shaking for 2–4 hours at 65°C.

   Note: Provided that step 1 is performed ahead of time, steps 2 and 3 can be carried out in parallel with the extraction of labeled RNA (see p. 199).

Hybridization

1. Carefully remove as much hybridization solution from the bag as possible. Add 5 ml of fresh hybridization solution (prewarmed to 65°C) to the bag and then add the chloroplast run-on transcripts. Heat-seal the bag, making sure that you do not trap any air bubbles.

   Note: It is not necessary to heat the RNA sample before adding it to the hybridization solution.

2. Hybridize with shaking overnight at 65°C.

Washing after Hybridization

1. Remove the blot from the plastic bag and transfer it to a plastic dish containing 100 ml of 2x SSC/0.1% SDS prewarmed to 65°C. Wash the blot with shaking for 30 minutes at 65°C.

2. Decant the washing solution and replace it with 100 ml of 1x SSC/0.1% SDS prewarmed to 65°C. Wash the blot with shaking for 30 minutes at 65°C.

3. Repeat step 2 with 100 ml of 0.5x SSC/0.1% SDS prewarmed to 65°C.

   Note: At this point, the nylon membrane can be treated with RNase to digest any nonhybridized RNA, which may interfere with quantitation of the hybridized RNA. However, in our experience, the nonhybridized RNA is degraded sufficiently during the hybridization procedure such that RNase treatment does not remove significantly more labeled RNA from the membrane (Deng et al. 1987). We therefore recommend this only as an optional step.

Autoradiography

1. Dry the nylon membrane on a paper towel and wrap it in Saran® Wrap.

2. Mark the corners of the nylon membrane with radioactive ink and expose it to X-ray film overnight.

   Note: Marking of the membrane is necessary if you intend to quantitate the run-on transcripts by excising the bands from the membrane and counting them in a scintillation counter.

3. To determine the relative transcription activity of each chloroplast gene, scan the X-ray film in a densitometer, use a phosphorimager, or excise and count the regions of the nylon membrane that cor-

respond to bands on the autoradiograph. Use the measurements from the control experiments (see p. 201) to correct the estimated levels of run-on transcripts for hybridization efficiency. Use the method described by Deng and Gruissem (1987) to correct for the hybridization of any unlabeled RNA present in the run-on RNA samples. Because the DNA probes you have used in the hybridization reaction are likely to be different lengths, you should then normalize the estimated level of run-on transcription for each gene to a standard length, e.g., 1 kb of DNA.

• *PREPARATION OF REAGENTS*

GM mix (1x)
(500 ml)

5x GM mix (see below)	100 ml
Dithiothreitol	700 mg
Distilled H_2O (autoclaved)	400 ml

Prepare fresh before each use and chill to 4°C.

GM mix (5x)
(1 liter)

$Na_4P_2O_7 \cdot 10H_2O$	2.225 g
HEPES	59.58 g
Sorbitol	300.6 g
EDTA	3.72 g
$MgCl_2$	1.016 g
$MnCl_2$	0.99 g

Dissolve the $Na_4P_2O_7 \cdot 10H_2O$ in approximately 40 ml of boiling autoclaved distilled H_2O. Dissolve the remaining components in approximately 700 ml of autoclaved distilled H_2O. Mix the solutions together. Adjust the pH to 6.8 with NaOH, and adjust the volume to 1 liter with autoclaved distilled H_2O. Store in aliquots at –20°C, and thaw completely before use.

Hybridization solution
(25 ml)

20x SSC	7.5 ml
100x Denhardt's solution	1.25 ml
20% Sodium dodecyl sulfate (SDS)	0.625 ml
10 mg/ml Yeast tRNA	0.125 ml

Add the 20x SSC to 10 ml of autoclaved distilled H_2O and mix together. Add the 100x Denhardt's solution, 20% SDS, and yeast tRNA. Adjust the volume to 25 ml with autoclaved distilled H_2O. SDS will precipitate if added directly to 20x SSC.

SSC (20x)
(1000 ml)

NaCl	175.3 g
Sodium citrate	88.2 g

Dissolve in 800 ml of distilled H_2O. Adjust the pH to 7.0 with NaOH and adjust the volume to 1 liter with distilled H_2O. Autoclave. Store at room temperature.

Denhardt's solution (100x)

Ficoll (type 400, Pharmacia)	10 g
Polyvinylpyrrolidone	10 g
Bovine serum albumin (Fraction V, Sigma)	10 g

Dissolve in H_2O and then adjust the volume to 500 ml with H_2O. Filter sterilize (Sambrook et al. 1989). Divide into aliquots and store at -20°C.

SDS (20%)

Add 2 g of SDS to 10 ml of autoclaved distilled H_2O. Mix gently by inversion to avoid foaming. Alternatively (to avoid weighing SDS), add 40 ml of autoclaved distilled H_2O to a new 10-g bottle of SDS. Mix gently until dissolved (e.g., on a magnetic stir plate), then adjust the volume to 50 ml with autoclaved distilled H_2O.

SAFETY NOTE

- Wear a mask when weighing SDS.

IC mix
(500 ml)

$Na_4P_2O_7 \cdot 10H_2O$	0.223 g
HEPES	5.96 g
Sorbitol	30.1 g

Dissolve the $Na_4P_2O_7 \cdot 10H_2O$ in approximately 20 ml of boiling distilled H_2O. Dissolve the remaining components in approximately 300 ml of distilled H_2O. Mix the solutions together. Adjust the pH to 7.9 with NaOH, and adjust the volume to 500 ml with distilled H_2O. Autoclave. Store in aliquots at -20°C, and thaw completely before use.

Lysis buffer
(100 ml)

Sodium sarcosinate	5 g
1 M Tris-HCl (pH 8.0) (autoclaved)	5 ml
200 mM EDTA (pH 8.0) (autoclaved)	12.5 ml

Adjust the volume to 100 ml with autoclaved distilled H_2O. Store at room temperature.

Nucleotide mix (10x)
(100 μl)

Mix 3 μl of 10 mM UTP with 5 μl each of 100 mM ATP, 100 mM CTP, and 100 mM GTP. Adjust the volume to 100 μl with H_2O. Store in aliquots at –20°C.

PCBF
(100 ml)

Percoll®	100 ml
Polyethylene glycol (M.W. 6000)	3.0 g
Bovine serum albumin	1.0 g
Ficoll	1.0 g

Prepare fresh before each use and chill to 4°C.

R mix
(10 ml)

		Final concentration
1 M $MgCl_2$	170 μl	17 mM
1 M KCl	660 μl	66 mM
25 mM HEPES (pH 7.9)	9.17 ml	23 mM

Each solution should be autoclaved before preparing R mix. Store at room temperature.

SSC (2x)/SDS (0.1%)
(100 ml)

20x SSC	10 ml
Distilled H_2O (autoclaved)	89.5 ml
20% SDS	0.5 ml

Add the 20x SSC to the autoclaved distilled H_2O and mix together, then add the 20% SDS. Prepare fresh before each use and preequilibrate to the wash temperature.

SSC (20x)
(1000 ml)

NaCl	175.3 g
Sodium citrate	88.2 g

Dissolve in 800 ml of distilled H_2O. Adjust the pH to 7.0 with NaOH and adjust the volume to 1 liter with distilled H_2O. Autoclave. Store at room temperature.

SDS (20%)

Add 2 g of SDS to 10 ml of autoclaved distilled H_2O. Mix gently by inversion to avoid foaming. Alternatively (to avoid weighing SDS), add 40 ml of autoclaved distilled H_2O to a new 10 g bottle of SDS. Mix gently until dissolved (e.g., on a magnetic stir plate), then adjust the volume to 50 ml with autoclaved distilled H_2O.

SSC (1x)/SDS (0.1%)
(100 ml)

20x SSC	5 ml
Distilled H_2O (autoclaved)	94.5 ml
20% SDS	0.5 ml

Add the 20x SSC to the autoclaved distilled H_2O and mix together, then add the 20% SDS. Prepare fresh before each use and preequilibrate to the wash temperature.

SSC (0.5x)/SDS (0.1%)
(100 ml)

20x SSC	2.5 ml
Distilled H_2O (autoclaved)	97 ml
20% SDS	0.5 ml

Add the 20x SSC to the autoclaved distilled H_2O and mix together, then add the 20% SDS. Prepare fresh before each use and preequilibrate to the wash temperature.

• *Section 11*

In Vitro Processing of Chloroplast RNA and Analysis of RNA-Protein Interactions by Ultraviolet Cross-linking

Following transcription, translatable mRNA is formed from precursor RNA by a series of modifications that may include 5′ and 3′ end processing, splicing, capping, polyadenylation, and editing. During these posttranscriptional processing reactions, the RNA interacts with specific nucleic acids and proteins (Dahlberg and Abelson 1989, 1990). Posttranscriptional processing events and the interactions between RNA and proteins are now widely regarded as important control steps in the regulation of mRNA accumulation and gene expression (Bandziulis et al. 1989; Mattaj 1989; Frankel et al. 1991; Kenan et al. 1991; Gruissem and Schuster 1993).

POSTTRANSCRIPTIONAL PROCESSING OF CHLOROPLAST RNA

In higher plants, most of the protein-coding regions in chloroplast DNA are flanked at their 3′ ends by inverted repeat sequences (IRs) that can fold into stem-loop structures. S1 nuclease protection experiments have shown that the 3′ ends of the IRs coincide with the 3′ ends of the mature mRNAs (Stern and Gruissem 1987). This finding suggests that the 3′ ends of chloroplast mRNAs are produced either by termination of transcription at the respective 3′ IRs or by processing of longer precursors in which the 3′ IRs function as specific recognition signals for endonuclease and/or exonuclease activities. In vitro transcription experiments using chloroplast protein extracts have demonstrated that chloroplast RNA polymerase does not recognize the 3′ IRs as efficient transcription termination signals. However, the 3′ ends of mature chloroplast mRNAs can be generated by a rapid processing event that can be reproduced in vitro with a chloroplast protein extract or with a preparation of isolated, electroporated chloroplasts into which mRNA precursor 3′ ends have been introduced. These results indicate that the IRs are efficient RNA processing elements (Stern and Gruissem 1987). During the posttranscriptional processing of the 3′ ends, the precursor RNAs interact with specific proteins (Stern et al. 1989; Schuster and Gruissem 1991). Interactions between the 3′ IR of the mature mRNA and specific proteins may be important in the regulation of chloroplast mRNA accumulation and in the control of chloroplast gene expression (Gruissem 1989a).

Crude or partially fractionated protein extracts have proven to be powerful tools in the analysis of transcriptional regulatory factors and RNA processing pathways for both nuclear and organelle genes in plant and animal systems. In plants, partially fractionated chloroplast protein extracts were first used to determine the structure of chloroplast promoter regions and to study the processing of precursor tRNAs (Gruissem 1989b). More recently, chloroplast protein extracts have also been used to analyze the processing of chloroplast mRNAs. In a typical in vitro RNA processing experiment, synthetic, radioactively labeled chloroplast mRNA precursors (which contain the 3′ coding sequences, the IR in the 3′ untranslated region of the mRNA, and additional 3′ sequences that are not part of the mature mRNAs) are incubated with a partially purified chloroplast protein extract (Gruissem et al. 1986). Following incubation, the RNAs are purified and the processing of the precursor into the mature mRNA product is analyzed by polyacrylamide gel electrophoresis (PAGE) and autoradiography (see Fig. 1) (Stern and Gruissem 1987).

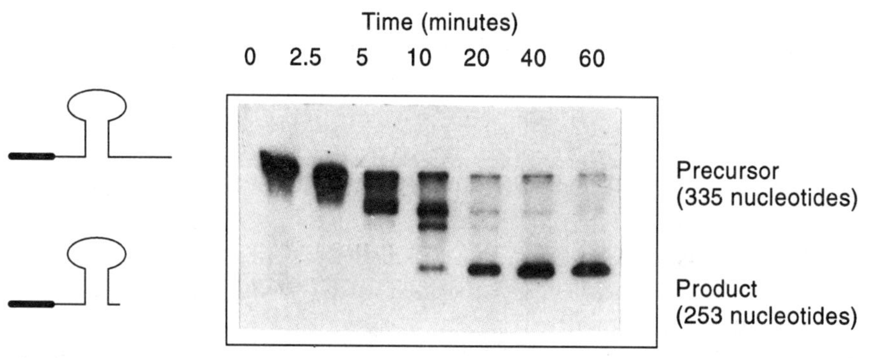

Figure 1
In vitro 3′end processing of spinach chloroplast *psbA* mRNA. The *psbA* gene, which encodes the D1 thylakoid protein of the photosystem II reaction center, was subcloned into the pBluescript® vector and transcribed using T7 RNA polymerase in the presence of α-[^{32}P]UTP. Full-length radiolabeled transcripts were purified by PAGE. The in vitro processing experiment was performed as described in the protocol on pp. 218–221. The potential secondary structure is shown schematically on the left. The coding regions are symbolized by the filled box, and the stem-loop structure symbol indicates the position of the inverted repeat sequence.

ANALYSIS OF RNA-PROTEIN INTERACTIONS BY ULTRAVIOLET CROSS-LINKING

Several techniques can be employed to detect interactions between proteins and nucleic acids. Cross-linking of proteins to nucleic acids with ultraviolet (UV) light is a relatively simple method (as compared to gel retardation assays or DNase I footprinting analysis) that can be used to obtain the approximate molecular weight of the nucleic acid binding proteins in a crude protein extract. In a UV cross-linking assay, proteins are labeled indirectly by transfer of radioactive label from the nucleic acid to the protein. Irradiation of a nucleic acid/protein mixture with UV light produces purine and pyrimidine radicals. A covalent bond can then be formed between these radicals and nearby amino acids (i.e., those at nucleic acid-protein contact sites) such as cysteine, serine, methionine, lysine, arginine, histidine, tryptophan, phenylalanine, and tyrosine. Following UV irradiation, the nucleic acid is completely digested by nucleases (DNase I for DNA and RNase A/RNase T1 for RNA) and the proteins are analyzed by SDS-PAGE and autoradiography. The specificity of the protein binding to nucleic acids can be determined in competition experiments using an excess of nonspecific competitor such as heterologous DNA or poly(dI-dC) (in DNA-protein binding assays) and unlabeled RNAs, tRNAs or poly(U) (in RNA-protein binding assays). UV cross-linking of DNA to proteins was introduced in 1972 (Markowitz 1972) and since then has become a common tool for analysis of DNA- and RNA-binding proteins. In many studies of cross-linking between DNA and proteins, the bromodeoxyuridine (BrdU) analog is incorporated into the DNA molecules to enhance sensitivity.

The interaction of proteins with the 3′ ends of precursor and mature chloroplast mRNA can be examined directly by UV cross-linking experiments using mRNA radiolabeled at the 3′ end. Such studies are an important preliminary to the isolation of proteins that have a direct role in the regulation of mRNA processing and accumulation during plant development (Stern et al. 1989; Chen and Stern 1991; Nickelsen and Link 1991; Schuster and Gruissem 1991).

OUTLINE OF EXPERIMENT

This section describes protocols for analyzing chloroplast mRNA 3′ end processing events and for studying RNA-protein interactions by UV cross-linking (Stern and Gruissem 1987; Stern et. al. 1989; Stern and Gruissem 1989). The RNA used in these experiments is transcribed in vitro from the spinach chloroplast *psbA* gene, which encodes the D1 thylakoid protein of the photosystem II reaction center.

However, the protocols can be adapted for the study of other RNA molecules.

REFERENCES

Bandziulis, R.J., M.S. Swanson, and G. Dreyfuss. 1989. RNA-binding proteins as developmental regulators. *Genes Devel.* **3:** 431–437.

Chen, H.C. and D.B. Stern. 1991. Specific binding of chloroplast proteins in vitro to the 3′ untranslated region of spinach chloroplast *petD* mRNA. *Mol. Cell. Biol.* **11:** 4380–4388.

Dahlberg, J.E. and J.N. Abelson, eds. 1989. RNA processing. Part A: General methods. *Methods Enzymol.* **180**.

———. 1990. RNA processing. Part B: Specific methods. *Methods Enzymol.* **181**.

Frankel, A.D., I.W. Mattaj, and D.C. Rio. 1991. RNA-protein interactions. *Cell* **67:** 1041–1046.

Gruissem, W. 1989a. Chloroplast gene expression: How plants turn their plastids on. *Cell* **56:** 161–170.

———. 1989b. Chloroplast RNA: Transcription and processing. In *Biochemistry of plants: A comprehensive treatise* (ed. A. Marcus), vol. 15, pp. 151–191. Academic Press, New York.

Gruissem, W. and G. Schuster. 1993. Control of mRNA degradation in organelles. In *Control of messenger RNA stability* (ed. G. Brawerman and J. Belasco), pp. 329–365. Academic Press, New York.

Gruissem, W., B.M. Greenberg, G. Zurawski, and R.B. Hallick. 1986. Chloroplast gene expression and promoter identification in chloroplast extracts. *Methods Enzymol.* **118:** 253–270.

Kenan, D.J., C.C. Query, and J.D. Keene. 1991. RNA recognition: Towards identifying determinants of specificity. *Trends Biochem. Sci.* **16:** 214–220.

Markowitz, A. 1972. Ultraviolet light-induced stable complexes of DNA and DNA polymerase. *Biochim. Biophys. Acta* **281:** 522–534.

Mattaj, I.W. 1989. A binding consensus: RNA-protein interactions in splicing, snRNPs, and sex. *Cell* **57:** 1–3.

Merril, C.R., D. Goldman, and M.L. Van Keuren. 1984. Gel protein stain: Silver stain. *Methods. Enzymol.* **104:** 441–447.

Nickelsen, J. and G. Link. 1991. RNA-protein interactions at transcript 3′ ends and evidence for *trnK-psbA* cotranscription in mustard chloroplasts. *Mol. Gen. Genet.* **228:** 89–96.

Promega Protocols and Applications Guide. 1991. Promega, Madison, Wisconsin.

Sambrook, J., E.F. Fritsch, and T. Maniatis. 1989. *Molecular cloning: A laboratory manual*, 2nd edition. Cold Spring Harbor Laboratory Press, Cold Spring Harbor, New York.

Schuster, G. and W. Gruissem. 1991. Chloroplast mRNA 3′ end processing requires a nuclear-encoded RNA-binding protein. *EMBO J.* **10:** 1493–1502.

Stern, B.D. and W. Gruissem. 1987. Control of plastid gene expression: 3′ inverted repeats act as mRNA processing and stabilizing elements, but do not terminate transcription. *Cell* **51:** 1145–1157.

———. 1989. Plastid mRNA processing: Ion requirements and protein-mediated RNA stabilization. *Plant Mol. Biol.* **13:** 615–625.

Stern, B.D., H. Jones, and W. Gruissem. 1989. Function of plastid mRNA 3′ inverted repeats: RNA stabilization and gene-specific protein binding. *J. Biol. Chem.* **264:** 18742–18750.

• *TIMETABLE*
In Vitro Processing of RNA and UV Cross-linking of RNA to Proteins

STEPS	*TIME REQUIRED*
Synthesis of RNA	
Linearize DNA template	1 hour
Transcribe DNA template to yield radiolabeled H-RNA and X-RNA	1 day
In Vitro Processing of Chloroplast mRNA 3′ Ends	
Prepare chloroplast protein extract using protocol outlined in Gruissem et al. 1986	2 days
Incubate RNA processing reaction	1 hour
Extract RNA	1 hour
Analyze RNA by PAGE	2–3 hours
Expose gel to X-ray film	3 hours–overnight
Cross-linking of Proteins to RNA	
Perform UV cross-linking experiment	1–2 hours
Analyze proteins by SDS-PAGE	4 hours (or 2 hours for minigel)
Silver stain gel	2 hours
Expose gel to X-ray film	overnight

Synthesis of RNA

When transcribing DNA, it is essential to protect the RNA product from digestion. To minimize the contamination of samples with RNases, always wear gloves, treat solutions with diethyl pyrocarbonate (DEPC), and bake all glassware (Sambrook et al. 1989).

The *psbA* gene is subcloned into the pBluescript® vector and transcribed using T7 RNA polymerase in the presence of α-[^{32}P]UTP. For the experiments described in this section, it is necessary to synthesize two RNA species—H-RNA (for the in vitro processing assays) and X-RNA (for the UV cross-linking assays). These RNA species are transcribed from the same DNA template and are identical in structure, but a higher specific activity of α-[^{32}P]UTP is used for the synthesis of X-RNA than is used for the synthesis of H-RNA. Full-length radiolabeled transcripts are purified by PAGE. The synthetic RNA is 335 nucleotides long; it includes 41 nucleotides of the coding region and 82 nucleotides 3′ to the inverted repeat, which is 58 nucleotides long.

MATERIALS

DNA template (e.g., spinach chloroplast *psbA* gene subcloned into a transcription vector such as pBluescript® [Stratagene])
Restriction enzyme (appropriate for the DNA template used)
Diethyl pyrocarbonate (DEPC)
DEPC-treated (RNase-free) H_2O
Dithiothreitol (DTT) (40 mM)
RNase inhibitor (RNasin or equivalent)
α-[^{32}P]UTP (20 μCi/μl; 800 Ci/mmol)
SP6, T3, or T7 RNA polymerase
Ammonium acetate (5 M)
Ethanol (100%)
DEPC-treated sodium acetate (3 M; pH 5.5)
DEPC-treated EDTA (0.25 M)
Sodium dodecyl sulfate (SDS) (20%)

REAGENTS

Reaction buffer (5x) (use the recipe appropriate for the RNA polymerase you are using—SP6 or T3/T7)
NTP mixture (use the recipe appropriate for the RNA species you are preparing—H-RNA or X-RNA)
Formamide dye solution
(For recipes, see Preparation of Reagents, pp. 227–231)

SAFETY NOTES

- DEPC is suspected to be a carcinogen and should be handled with care.
- Wear gloves when handling radioactive substances. Consult the local safety office for further guidance in the appropriate use of radioactive materials.
- Wear a mask when weighing SDS.

PROCEDURE

Linearization of DNA Template

1. Digest the plasmid containing the DNA template (Schuster and Gruissem 1991) with an appropriate restriction enzyme according to the supplier's instructions.

   Note: (i) RNA synthesis is usually more reliable from the T7 promoter than from the T3 or SP6 promoters. (ii) Promega recommends avoiding the use of enzymes that produce 3′ overhangs (e.g., *Pst*I, *Sac*I, *Sac*II, and *Kpn*I). It has been their observation that extraneous RNA is produced when these enzymes are used to linearize the template (Promega Protocols and Applications Guide 1991).

2. Check that the digestion is complete by analyzing the linearized plasmid on a DNA agarose gel (Sambrook et al. 1989).

3. Precipitate the DNA with ethanol (Sambrook et al. 1989) and resuspend the DNA pellet in H_2O. The final concentration of the linearized DNA should be 200 μg/ml.

Transcription of DNA Template

1. Set up the following reaction mixture in a microfuge tube:

DNA template	2 μl (0.4 μg)
Reaction buffer (5x)	4 μl
DTT (40 mM)	0.5 μl
NTP mixture (for H-RNA or X-RNA)	2 μl
RNase inhibitor	0.5 μl (~15 units)
H_2O	8.5 μl for H-RNA or 2.5 μl for X-RNA

To avoid the possibility of spermidine precipitation of the DNA, add each reagent at room temperature or 37°C.

Note: The reaction can be scaled up as necessary.

2. Add 2 μl (for H-RNA) or 8 μl (for X-RNA) of α-[^{32}P]UTP to the reaction mixture.

3. Add 0.5–1 μl (i.e., 10–20 units) of SP6, T3 or T7 polymerase to give a final reaction volume of 20–20.5 μl.

4. Incubate for 45–60 minutes at 37°C.

5. Precipitate the RNA by adding 15 μl of 5 M ammonium acetate and 100 μl of 100% ethanol.

6. Incubate for 10 minutes at –20°C, and spin in a microfuge for 10 minutes at 4°C.

7. Remove the supernatant, dry the RNA pellet in air or in a vacuum centrifuge, and resuspend the RNA in 7 μl of formamide dye solution.

8. Boil the resuspended RNA in a boiling water bath for 3 minutes and quench on ice. Load the sample onto a denaturing 6% acrylamide gel (see p. 221).

9. Cover the gel with Saran® Wrap and expose it to X-ray film for approximately 20 seconds.

   Note: You must ensure that after developing the autoradiograph the X-ray film can be reoriented precisely on the gel.

10. Cut out the gel slice containing the RNA and place it in a 1.5-ml microfuge tube.

11. To the tube, add:

DEPC-treated H_2O	180 µl
DEPC-treated sodium acetate (3 M; pH 5.5)	20 µl
DEPC-treated EDTA (0.25 M)	1 µl
SDS (20%)	1 µl

12. Incubate for 1–1.5 hours at 65°C or overnight at 4°C to allow the RNA to diffuse out of the gel.

13. Transfer the RNA solution to new microfuge tube and add 500 µl of 100% ethanol. Incubate for 10 minutes at –20°C. Count the Cerenkov radiation.

14. Pellet the precipitated RNA by spinning in a microfuge for 10 minutes. Remove the supernatant, dry the pellet under vacuum, and resuspend the RNA in an appropriate buffer (e.g., 1 mM Tris-HCl/0.1 mM EDTA [pH 8.0]) before using it for further manipulations.

In Vitro Processing of Chloroplast mRNA 3′ Ends

When studying in vitro processing reactions, it is essential to protect the RNA from digestion. To minimize the contamination of samples with RNases, always wear gloves, treat solutions with diethyl pyrocarbonate (DEPC), and bake all glassware (Sambrook et al. 1989).

The protocol described here can be adapted to perform a number of different experiments. Some options are:

- To measure the kinetics of mRNA 3′ end processing for one or more precursor RNA species. For this experiment, the reaction mixture must be scaled up appropriately to allow removal of 5 μl samples of the reaction mixture at every time point.
- To assay the effect on mRNA 3′ end processing of adding increasing concentrations of nonspecific competitor molecules such as poly(U) and tRNAs. We suggest starting with a 1:1 ratio of competitor to 3′ end RNA. 1 pmol of mRNA 3′ ends is equivalent to approximately 0.1 μg.
- To assay the effect on mRNA 3′ end processing of adding different concentrations of chloroplast protein extract.
- To assay the temperature dependence of mRNA 3′ end processing.
- To investigate any other parameters that may affect mRNA 3′ end processing.

MATERIALS

Chloroplast protein extract (prepared as outlined in Gruissem et al. [1986])
Diethyl pyrocarbonate (DEPC)
DEPC-treated (RNase-free) H_2O
H-RNA (see p. 214–217)
Phenol/chloroform/isoamyl alcohol (25:25:1)
Sodium acetate (3 M; pH 5.5)
tRNA (10 mg/ml)
Ethanol (100%; ice cold)
Ammonium persulfate (APS) (20%)
N,N,N′,N′,tetramethylethylenediamine (TEMED)

REAGENTS

Buffer IVT (20x)
Buffer E
Stop solution
Formamide dye solution
Solution A
Solution B
TBE (0.5x)
(For recipes, see Preparation of Reagents, pp. 227–231)

SAFETY NOTES

- DEPC is suspected to be a carcinogen and should be handled with care.
- Wear gloves when handling radioactive substances. Consult the local safety office for further guidance in the appropriate use of radioactive materials.
- Phenol is highly corrosive and can cause severe burns. Wear gloves, protective clothing, and safety glasses when handling it. All manipulations should be carried out in a chemical fume hood. Any areas of skin that come in contact with phenol should be rinsed with a large volume of water or PEG 400 and washed with soap and water; do not use ethanol!
- Chloroform is irritating to the skin, eyes, mucous membranes, and respiratory tract. It should only be used in a chemical fume hood. Gloves and safety glasses should also be worn. Chloroform is a carcinogen and may damage the liver and kidneys.
- Acrylamide and bisacrylamide, which are components of solution A, are potent neurotoxins and are absorbed through the skin. Their effects are cumulative. Wear gloves and a mask when weighing acrylamide and bisacrylamide. Polyacrylamide is considered to be nontoxic, but it should be treated with care because it may contain small quantities of unpolymerized material.

PROCEDURE

RNA Processing Reaction

1. Set up the following reaction mixture in a 1.5-ml microfuge tube:

Buffer IVT (20x)	0.5 μl
Chloroplast protein extract (20 μg)	L μl
Buffer E	M μl
(L + M = 5 μl)	
DEPC-treated H_2O	X μl
H-RNA (~0.25 fmol; 2000 cpm)	Y μl
(X + Y = 4.5 μl)	
Total volume:	10 μl

Start the reaction by adding the H-RNA, vortex gently, and spin briefly in a microfuge to bring all the reaction mixture down from the sides of the tube.

Note: (i) The reaction can be scaled up depending on the number of samples (5 μl/sample) to be assayed, but do not use more then 30 μl total volume and 40 μg of chloroplast protein extract. (ii) The final concentrations of the buffer components are 10 mM HEPES (pH 7.9), 40 mM KCl, 10 mM $MgCl_2$, 0.05 mM EDTA, 3 mM DTT, and 8.5% glycerol. (iii) 2000 cpm of H-RNA should result in a nice-looking autoradiograph after overnight exposure. However, you can use 10,000 or 20,000 cpm and get similar results in a shorter time.

2. Incubate for 60 minutes (or 0–60 minutes for the kinetics experiment) at 25°C.

   Note: Good time points for the kinetics experiment are 2.5, 5, 10, 20, 40, and 60 minutes (see Fig. 1). Always take a time zero sample and/or include a control sample without proteins.

3. Stop the reaction by transferring 5 μl of the reaction mixture into 45 μl of stop solution. Vortex, and keep on ice.

4. Add 50 μl of phenol/chloroform/isoamyl alcohol (25:25:1), vortex, and spin for 2 minutes at room temperature. Transfer the upper aqueous (red) phase to a new microfuge tube, and discard the phenol-containing phase.

5. Add 1 μl of 3 M sodium acetate (pH 5.5), 1 μl of 10 mg/ml tRNA, and 125 μl of ice-cold 100% ethanol. Incubate for 10 minutes at -20°C and spin in a microfuge for 10 minutes at 4°C.

6. Carefully remove the supernatant and dry the RNA pellet in a vacuum centrifuge for approximately 5 minutes. Resuspend the RNA in 7 μl of formamide dye solution.

7. Boil the resuspended RNA in a boiling water bath for 3 minutes and quench on ice.

Analysis of RNA Product by PAGE

1. Set up the gel apparatus. Wash the glass plates (18 x 18 x 0.08 cm) with soap, rinse them with H_2O, distilled H_2O and ethanol, and then air dry.

2. Prepare a 6% acrylamide gel solution by mixing:

Solution A	4.8 ml
Solution B	7.2 ml
APS (20%)	55 µl
TEMED	8 µl

   Pour the gel and insert the comb. Polymerization of the acrylamide mixture will be complete within 30 minutes.

3. Using 0.5x TBE as the running buffer, pre-run the gel at 30 mA until the temperature is 50°C or the voltage is 1000 V (this should take ~20 minutes).

4. Rinse the wells of the gel with running buffer to remove the urea that has diffused from the gel during the pre-run. Load the RNA sample(s), and run the gel at 20 mA until the bromophenol blue reaches the bottom of the gel (20–30 minutes).

5. Carefully separate the glass plates. Transfer the gel to Whatman 3MM paper, cover with Saran® Wrap, and place an additional sheet of Whatman 3MM paper over the covered gel. Dry the gel on a gel dryer.

6. Expose the dried gel to X-ray film for 6–15 hours at -80°C using an intensifying screen.

Cross-linking of Proteins to RNA

When studying cross-linking of proteins to RNA, it is essential to protect the RNA from digestion. To minimize the contamination of samples with RNases, always wear gloves, treat solutions with diethyl pyrocarbonate (DEPC), and bake all glassware (Sambrook et al. 1989).

The protocol described here can be adapted to perform a number of different experiments. Some options are:

- To assay the binding of chloroplast proteins to different mRNA 3′ ends.
- To determine the specificity of RNA-protein interactions by using different concentrations of nonspecific competitors such as poly(U) or tRNAs. The standard concentration of poly(U) in the UV cross-linking reaction is 6.6 μg/ml (i.e., 0.1 μg/15 μl). At this concentration, 3′ end processing is inhibited.
- To assay the effect of temperature and/or salt conditions on RNA-protein interactions.
- To determine the optimal UV light dose for maximum cross-linking of proteins.

For all experiments, it is important to remember to include the appropriate controls, e.g., no UV light, no proteins, etc.

MATERIALS AND EQUIPMENT

Chloroplast protein extract (prepared as outlined in Gruissem et al. [1986])
Diethyl pyrocarbonate (DEPC)
DEPC-treated (RNase-free) H_2O
Poly(U) (0.1 mg/ml)
X-RNA (see p. 214–217)
UV cross-linker apparatus (Stratagene)
RNase A (0.5 mg/ml)
Ammonium persulfate (APS) (20%)
N,N,N′,N′,tetramethylethylenediamine (TEMED)

X-ray film
Methanol (50%)/glacial acetic acid (12%) in H_2O
Methanol (10%)/glacial acetic acid (5%) in H_2O
Acetic acid (3%)
Glycerol (5%)

REAGENTS

Buffer IVT (20x)
Buffer E
Laemmli sample buffer (5x)
Acrylamide (30%)/bisacrylamide (0.8%)
Separation gel buffer (8x)
Stacking gel buffer (4x)
Running gel buffer (10x)
Potassium dichromate (3.4 mM)/nitric acid (3.2 mM)
Silver nitrate (12 mM)
Sodium carbonate (0.28 M) containing 0.5 ml of 37% formaldehyde solution/liter
(For recipes, see Preparation of Reagents, pp. 227–231)

SAFETY NOTES

- DEPC is suspected to be a carcinogen and should be handled with care.
- Wear gloves when handling radioactive substances. Consult the local safety office for further guidance in the appropriate use of radioactive materials.
- UV light is mutagenic, carcinogenic, and damaging to the eyes. *Never look directly at the UV lamp!* Always wear safety glasses when using UV. Protective gloves should be worn when holding materials under the UV lamp.
- Acrylamide and bisacrylamide are potent neurotoxins and are absorbed through the skin. Their effects are cumulative. Wear gloves and a mask when weighing acrylamide and bisacrylamide. Polyacrylamide is considered to be nontoxic, but it should be treated with care because it may contain small quantities of unpolymerized material.
- Methanol is poisonous and can cause blindness if ingested in sufficient quantities. Adequate ventilation is necessary to limit exposure to vapors.
- Glacial acetic acid and nitric acid are volatile and should be used in

a chemical fume hood. Concentrated acids should be handled with great care; gloves and a face protector should be worn.
- Potassium dichromate is caustic and considered hazardous material. Consult your institutional safety officer for specific handling and disposal procedures.
- Silver nitrate is a strong oxidizing agent and should be handled with care. It is also toxic and caustic. Consult your institutional safety officer for specific handling and disposal procedures.
- Formaldehyde is toxic and a carcinogen. It is readily absorbed through the skin and is destructive to the skin, eyes, mucous membranes, and upper respiratory tract. Formaldehyde should only be used in a chemical fume hood. Gloves and safety glasses should be worn.

PROCEDURE

UV Cross-linking Reaction

1. Add the following components (*except the X-RNA*) to a 1.5-ml microfuge tube:

Buffer IVT (20x)	0.75 μl
Chloroplast protein extract (20 μg)	L μl
Buffer E	M μl
(L + M = 7.5 μl)	
Poly(U) (0.1 mg/ml)	1 μl
DEPC-treated H_2O	X μl
X-RNA (~1.25 fmol; 100,000 cpm)	Y μl
(X + Y = 6.5 μl)	
Total volume:	15.75μl

   Incubate for 5 minutes at room temperature. Add the the X-RNA, vortex gently, and spin briefly in a microfuge to bring all the reaction mixture down from the sides of the tube.

   Note: Poly(U) is a nonspecific competitor of RNA-protein interactions. You can try using tRNA as an alternative to poly(U) or you can perform experiments in the absence of nonspecific competitor.

2. Incubate for 10 minutes at room temperature.

3. Carefully open the tube and place it in a UV cross-linker. Operate the UV cross-linker in energy mode at 1.8 J (i.e., 2 x 9000 on the scale). If there is no UV cross-linker available, use a germicidal lamp at a distance of 6 cm for 30 minutes (Stern et al. 1989).

4. Add 1 μl of a 0.5 mg/ml solution of RNase A and incubate for 20 minutes at 37°C.

   Note: You can use a higher concentration of RNase A, but we try to use the minimum amount of RNase to obtain a nice silver-stained protein profile.

5. Add 4 μl of 5x Laemmli sample buffer and heat the sample for 1 minute at 80–100°C.

Analysis of Proteins by SDS-PAGE

1. Set up the gel apparatus, and clean the glass plates. You can use 18 x 18 x 0.08-cm plates or 8 x 10-cm (minigel) plates.

2. Prepare a 13% acrylamide separation gel solution by mixing:

Acrylamide (30%)/bisacrylamide (0.8%)	8.7 ml
H_2O	8.8 ml
Separation gel buffer (8x)	2.5 ml
APS (20%)	75 μl
TEMED	30 μl

3. Prepare a stacking gel solution by mixing:

Acrylamide (30%)/bisacrylamide (0.8%)	1.3 ml
H_2O	3.9 ml
Stacking gel buffer (4x)	1.8 ml
APS (20%)	45 μl
TEMED	10 μl

4. Run the gel at 20 mA constant current until the bromophenol blue reaches the bottom of the gel (2–3 hours).

5. Carefully separate the glass plates. Transfer the gel to a glass dish and silver stain the proteins using the procedure outlined below.

   Note: The molecular weight of RNase A is approximately 14 kD. When you are working with a very low concentration of chloroplast proteins and you are performing repeated silver staining, an additional band corresponding to RNase A can be detected at approximately 36 kD.

Positive Image Silver Staining

This procedure is adapted from Merril et al. 1984. All the steps described below should be performed at room temperature.

1. Fix the gel by incubating it in 400 ml of 50% methanol/12% glacial acetic acid in H_2O for 10 minutes. Repeat with fresh solution.

2. Wash the gel by incubating it in 400 ml of 10% methanol/5% glacial acetic acid in H_2O for 10 minutes. Repeat with fresh solution.

3. Soak the gel in 200 ml of 3.4 mM potassium dichromate/3.2 mM nitric acid for 15 minutes.

   Note: After performing this step, you can reduce background by rinsing the gel with H_2O for 10 minutes before proceeding to step 4.

4. Soak the gel in 150 ml of 12 mM silver nitrate for 15 minutes.

5. Rinse the gel three times with H_2O.

6. Rapidly wash the gel twice in 150 ml of 0.28 M sodium carbonate containing 0.5 ml of 37% formaldehyde solution/liter. Add 150 ml of fresh solution to the gel and allow development to occur for 3–15 minutes.

7. Discard the developing solution and stop development by incubating in 200 ml of 3% acetic acid for 5 minutes.

8. Wash the developed gel in 200 ml of 5% glycerol for 10 minutes.

9. Dry the gel on a gel dryer at approximately 80°C. Expose to X-ray film overnight at –80°C using an intensifying screen.

• *PREPARATION OF REAGENTS*

Acrylamide (30%)/bisacrylamide (0.8%)
(1 liter)

Acrylamide (Bio-Rad Laboratories)	300 g
Bisacrylamide	8 g

Dissolve the components in distilled H_2O over a low heat. Adjust the volume to 1 liter with H_2O. Add one spoonful of mixed-bed resin (Bio-Rad Laboratories, AG-501-x8D) and mix with a magnetic stir bar for approximately 30 minutes. Pass the solution through glass wool to remove the resin, and then filter through a 0.45-μm Millipore filter. Store at 4°C.

SAFETY NOTE

- Acrylamide and bisacrylamide are potent neurotoxins and are absorbed through the skin. Their effects are cumulative. Wear gloves and a mask when weighing acrylamide and bisacrylamide.

Buffer E

20 mM HEPES (pH 7.9)
60 mM KCl
12.5 mM $MgCl_2$
0.1 mM EDTA
2 mM Dithiothreitol
17% Glycerol

Treat with 0.1% diethyl pyrocarbonate (DEPC) and autoclave.

SAFETY NOTE

- DEPC is suspected to be a carcinogen and should be handled with care.

Buffer IVT (20x)

200 mM KCl
75 mM $MgCl_2$
40 mM Dithiothreitol

Treat with DEPC and autoclave.

Formamide dye solution

Deionized formamide
0.1% Xylene cyanol
0.1% Bromophenol blue

Laemmli sample buffer (5x)

10% Sodium dodecyl sulfate (SDS)
20% Glycerol
4.3 M 2-Mercaptoethanol
0.05% Bromophenol blue

SAFETY NOTES

- Wear a mask when weighing SDS.
- 2-Mercaptoethanol may be fatal if swallowed and is harmful if inhaled or absorbed through the skin. High concentrations are extremely destructive to the mucous membranes, upper respiratory tract, skin, and eyes. Use only in a chemical fume hood. Gloves and safety glasses should be worn.

NTP mixture (for H-RNA)

5 mM GTP
5 mM CTP
5 mM ATP
0.25 mM UTP

NTP mixture (for X-RNA)

5 mM GTP
5 mM CTP
5 mM ATP

Potassium dichromate (3.4 mM)/nitric acid (3.2 mM)

(1 liter)

Potassium dichromate	10 g
Nitric acid	2.05 ml

Add potassium dichromate to 800 ml of distilled H_2O and then add nitric acid slowly, just until the potassium dichromate dissolves (excess nitric acid will decrease the sensitivity of the silver stain). Adjust the volume to 1 liter with distilled H_2O. Dilute this solution 1 in 10 with distilled H_2O before use. (You can also make up a solution that is 50x.)

SAFETY NOTES

- Potassium dichromate is caustic and considered hazardous material. Consult your institutional safety officer for specific handling and disposal procedures.
- Nitric acid is volatile and should be used in a chemical fume hood. Concentrated acids should be handled with great care; gloves and a face protector should be worn.

Reaction buffer (5x) (for SP6 polymerase)

200 mM Tris-HCl (pH 7.9)
30 mM $MgCl_2$
10 mM Spermidine-$(HCl)_3$

5x reaction buffer is usually provided by the supplier of the enzyme.

Reaction buffer (5x) (for T3/T7 polymerases)

200 mM Tris-HCl (pH 7.9)
30 mM $MgCl_2$
10 mM Spermidine-$(HCl)_3$
125 mM NaCl

5x reaction buffer is usually provided by the supplier of the enzymes.

Running gel buffer (10x)
(pH ~8.3)

0.25 M Tris-HCl
1.92 M Glycine
1% Sodium dodecyl sulfate (SDS)

Separation gel buffer (8x)

3 M Tris-HCl (pH 8.8)

Silver nitrate (12 mM)
(150 ml)

Dissolve 0.3 g of silver nitrate in 150 ml of distilled H_2O. Make fresh before use.

SAFETY NOTE

- Silver nitrate is a strong oxidizing agent and should be handled with care. It is also toxic and caustic. Consult your institutional safety officer for specific handling and disposal procedures.

Sodium carbonate (0.28 M) containing 0.5 ml of 37% formaldehyde solution/liter
(500 ml)

Sodium carbonate	14.8 g
37% Formaldehyde solution	0.25 ml

Dissolve in 500 ml of distilled H_2O.

SAFETY NOTE

- Formaldehyde is toxic and a carcinogen. It is readily absorbed through the skin and is destructive to the skin, eyes, mucous membranes, and upper respiratory tract. Formaldehyde should only be used in a chemical fume hood. Gloves and safety glasses should be worn.

Solution A

0.5x TBE (use 10x TBE stock solution)
7 M Urea
15% Acrylamide
0.75% Bisacrylamide

Solution B

0.5x TBE (use 10x TBE stock solution)
7 M Urea

Stacking gel buffer (4x)

0.5 M Tris-HCl (pH 6.8)

Stop solution

6 M Urea
1% Sodium dodecyl sulfate (SDS)
4.5 mM Aurintricarboxylic acid

Treat with DEPC and autoclave.

TBE stock solution (10x)

0.9 M Tris-borate (pH 8.0)
0.02 M EDTA

• *Section 12*

Identification of Promoter Sequences That Interact with DNA-binding Proteins

To understand how gene expression is regulated at the level of transcription, it is critical to be able to identify specific promoter sequences that interact with DNA-binding proteins. Mobility-shift assays and DNase I footprinting are two methods that are frequently employed to characterize DNA-protein interactions. Mobility-shift assays are generally used to identify promoter fragments that interact with DNA-binding proteins, to titrate binding activity, to determine the specificity of binding, and to determine whether a given DNA-binding activity can interact with related DNA sequences. When one or more specific binding activities have been identified, DNase I footprinting can be used to determine the nucleotide sequence of the binding site(s). In both methods, a protein extract derived from whole cells or isolated nuclei is incubated with a radiolabeled DNA fragment of interest in the presence of a nonspecific competitor of binding, usually double-stranded poly(dI-dC).

MOBILITY-SHIFT ASSAYS

In mobility-shift assays, the interaction between the labeled DNA fragment and one or more DNA-binding proteins is detected by running the reaction mixture on a nondenaturing acrylamide gel and visualizing the labeled bands by autoradiography. If the DNA fragment is bound by protein, the electrophoretic mobility of the fragment will be reduced with respect to the mobility of the unbound fragment. The percentage of the DNA fragment that is bound will depend on the activity and concentration of the DNA-binding protein. In general, relatively low protein concentrations are used for mobility-shift assays, so that bands corresponding to both bound and free DNA are visible. In competition mobility-shift assays, an excess of unlabeled competitor DNA is included in the reaction, and successful competition results in the disappearance of the bound band. For mobility-shift assays, probe DNA can be labeled by phosphorylation (provided the extract used does not contain phosphatase activity), by nick translation, or by fill-in of recessed 3′ ends with the Klenow fragment of DNA polymerase I (see Sambrook et al. 1989). The optimization and applications of mobility-shift reactions have been discussed recently by Carey (1991).

DNASE I FOOTPRINTING

The DNase I footprinting technique was first used by Galas and Schmitz (1978) to confirm the binding site of the *E. coli lac* repressor on the *lac* operator. The basis of the technique is that DNA-binding proteins protect DNA from digestion with DNase I and that DNase I cleaves DNA more or less randomly. The binding reaction is similar to that of the mobility-shift assay, but the amount of protein in the reaction is increased so that 100% of the DNA fragment is bound. After binding is complete, the reaction mixture is treated with DNase I under conditions that produce a ladder of bands and the digest is run on a denaturing acrylamide gel. A site at which protein binds to DNA is characterized by an area of reduced band intensity when compared to the corresponding band in the protein-free control lane. For this assay, the probe DNA must be radiolabeled on one end only. Typically, a Maxam and Gilbert sequencing reaction (see Sambrook et al. 1989) of the labeled fragment is run alongside the DNase I ladder so that the sequence of the protected region can be determined directly. Al-

Figure 1
Comparison of DNase I protection patterns produced by nuclear extracts from various tomato organs. The radiolabeled DNA fragment used here consists of the sequence extending from –503 to –137 (with respect to the transcription start site) of the *RbcS3A* gene. The fragment is shown diagrammatically above the autoradiogram, where the grey-tone boxes indicate conserved DNA sequence motifs (see Manzara et al. 1991). All lanes contain 0.75 fmol of the probe, except the Maxam and Gilbert G reaction lane (G rxn). The nuclear extract used is indicated at the top of each lane, where control designates lanes containing no extract. The amount of extract in each lane is as follows: leaf, 4.5 mg; young fruit, 4.5 mg; mature fruit, 10.1 mg; cotyledon, 11.2 mg. A similar protection to that shown for mature fruit was obtained using 8.1 mg of root extract (not shown). The final concentration of DNase I in mg/ml is indicated at the bottom of each lane. The protected regions defined by each extract are subdivided based on comparisons between lanes and have been given letter designations. The pattern of protein binding with respect to the conserved sequence motifs for each nuclear extract is shown below the autoradiogram, where ellipses represent DNA-binding proteins. Note that because the gel runs on a log scale, there is not a linear relationship between the protected regions shown on the autoradiogram and those shown diagrammatically below the gel. (Reprinted, with permission, from Manzara et al. 1991, copyright by American Society of Plant Physiologists.)

though DNase I is the most frequently used cleavage reagent for in vitro footprinting, some sequences are not efficiently cleaved by this enzyme and may require the use of chemical reagents such as 1,10 phenanthroline-copper ion (Kuwabara and Sigman 1987; Sigman et al. 1991) or methidiumpropyl-EDTA-iron (II) (Landolfi et al. 1989), which cleave DNA more uniformly than DNase I. A general review of footprinting techniques and applications, including a discussion of commonly used footprinting reagents, has been provided by Tullius (1989).

DNA-PROTEIN INTERACTIONS IN PLANT GENES

Sites of DNA-protein interaction have been identified in the 5′-upstream regions of a number of plant genes, including several light-regulated genes (for reviews, see Gilmartin et al. 1990; Thompson and White 1991). In our laboratory, the in vivo expression of the five genes encoding ribulose-1,5-bisphosphate carboxylase (RbcS) in tomato has been characterized in detail. Steady-state mRNA levels fluctuate on a diurnal cycle in leaves and fruit, and developmental, light-induced, and organ-specific patterns of gene regulation are apparent (Piechulla et al. 1986; Piechulla and Gruissem 1987; Sugita and Gruis-

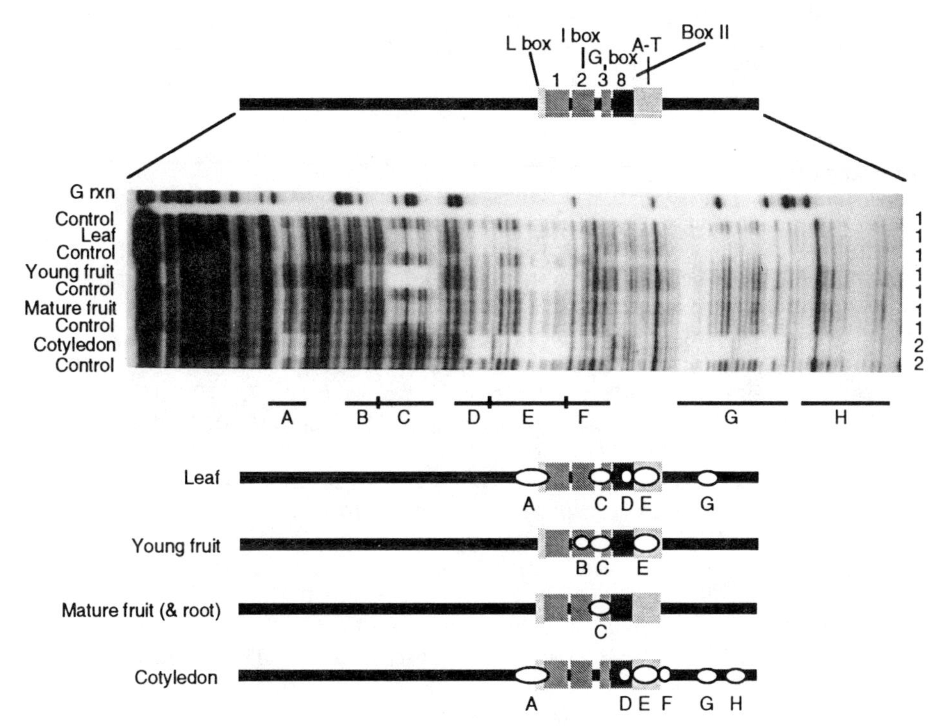

Figure 1 *(See facing page for legend.)*

sem 1987; Piechulla 1988). Furthermore, these genes are regulated, at least in part, at the level of transcription (Wanner and Gruissem 1991), and conserved sequences have been identified in the promoter regions of each gene (Manzara and Gruissem 1988). Using the techniques described above, we have characterized a number of sequences in the tomato *RbcS* promoters that interact with DNA-binding proteins. We have found that although transcription of *RbcS* is enhanced by light, overall patterns of DNA-protein interaction do not change in response to light. In contrast, the pattern of protein binding changes markedly with organ-specific and developmentally regulated changes in *RbcS* transcription (Manzara et al. 1991, 1993; Carrasco et al. 1993). Figure 1 shows the changes in protein binding to a portion of the *RbcS3A* promoter that occur during fruit development and organ differentiation in tomato.

OUTLINE OF EXPERIMENT

This section describes procedures for the isolation of nuclei and the preparation of nuclear extracts from tomato cotyledons, a method for single-end labeling of DNA, and protocols for mobility-shift and DNase I footprinting assays. The procedure for the isolation of nuclear proteins has been used for tomato leaves, roots, and fruits at various stages of development (Manzara et al. 1991). A similar procedure has been described for the isolation of pea nuclear proteins (Green et al. 1989).

REFERENCES

Carrasco, P., T. Manzara, and W. Gruissem. 1993. Developmental and organ-specific changes in DNA-protein interactions in the tomato *rbcS3B* and *rbcS3C* promoter regions. *Plant Mol. Biol.* **21:** 1–15.

Carey, J. 1991. Gel retardation. *Methods Enzymol.* **208:** 103–117.

Galas, D.J. and A. Schmitz. 1978. DNase footprinting: A simple method for the detection of protein-DNA binding specificity. *Nucleic Acids Res.* **5:** 3157–3170.

Gilmartin, P.M., L. Sarokin, J. Memelink, and N.-H. Chua. 1990. Molecular light switches for plant genes. *Plant Cell* **2:** 369–378.

Green, P.J., S.K. Kay, and N.-H. Chua. 1987. Sequence-specific interactions of a pea nuclear factor with light-responsive elements upstream of the *rbcS-3A* gene. *EMBO J.* **6:** 2543–2549.

Green, P.J., S.A. Kay, E. Lam, and N.-H. Chua. 1989. In vitro DNA footprinting. In *Plant molecular biology manual* (ed. S.B. Gelvin and R.A. Schilperoort), section B11, pp. 1–22. Kluwer Academic Publishers, Dordrecht, The Netherlands.

Kannangara, C.G., S.P. Gough, B. Hansen, J.N. Rasmussen, and D.J. Simpson. 1977. A homogenizer with replaceable razor blades for bulk isolation of active barley plastids. *Carlsberg Res. Commun.* **42:** 431–439.

Kuwabara, M.D. and D.S. Sigman. 1987. Footprinting DNA-protein complexes in situ following gel retardation assays using 1,10 phenanthroline-copper ion. *E. coli* RNA polymerase-*lac* promoter complexes. *Biochemistry* **26:** 7234–7238.

Landolfi, N.F., X.-M. Yin, J.D. Capra, and P.W. Tucker. 1989. Protection analysis (or "footprinting") of specific protein-DNA complexes in crude nuclear extracts using methidiumpropyl-EDTA-iron (II). *Biotechniques* **7:** 500–504.

Manzara, T. and W. Gruissem. 1988. Organization and expression of the genes encoding ribulose-1,5-bisphosphate carboxylase in higher plants. *Photosynth. Res.* **16:** 117–139.

Manzara, T., P. Carrasco, and W. Gruissem. 1991. Developmental and organ-specific changes in promoter DNA-protein interactions in the tomato *rbcS* gene family. *Plant Cell* **3:** 1305–1316.

———. 1993. Developmental and organ-specific changes in DNA-protein interactions in the tomato *rbcS1*, *rbcS2* and *rbcS3A* promoter regions. *Plant Mol. Biol.* **21:** 69–88.

Piechulla, B. 1988. Differential expression of nuclear- and organelle-encoded genes during tomato fruit development. *Planta* **174:** 505–512.

Piechulla, B. and W. Gruissem. 1987. Diurnal mRNA fluctuations of nuclear and plastid genes in developing tomato fruits. *EMBO J.* **6:** 3593–3599.

Piechulla, B., E. Pichersky, A.R. Cashmore, and W. Gruissem. 1986. Expression of nuclear and plastid genes for photosynthesis-specific proteins during tomato fruit development and ripening. *Plant Mol. Biol.* **7:** 367–376.

Sambrook, J., E.F. Fritsch, and T. Maniatis. 1989. *Molecular cloning: A laboratory manual*, 2nd edition. Cold Spring Harbor Laboratory Press, Cold Spring Harbor, New York.

Sigman, D.S., M.D. Kuwabara, C.-H.B. Chen, and T.W. Bruice. 1991. Nuclease activity of 1,10 phenanthroline-copper in study of protein-DNA interactions. *Methods Enzymol.* **208:** 414–433.

Sugita, M. and W. Gruissem. 1987. Developmental, organ-specific, and light-dependent expression of the tomato ribulose-1,5-bisphosphate carboxylase small subunit gene family. *Proc. Natl. Acad. Sci.* **84:** 7104–7108.

Svaren, J., S. Inagami, E. Lovegren, and R. Chalkley. 1987. DNA denatures upon drying after ethanol precipitation. *Nucleic Acids Res.* **15:** 8739–8754.

Thompson, W.F. and M.J. White. 1991. Physiological and molecular studies of light-regulated nuclear genes in higher plants. *Annu. Rev. Plant Physiol. Plant Mol. Biol.* **42:** 423–466.

Tullius, T.D. 1989. Physical studies of protein-DNA complexes by footprinting. *Annu. Rev. Biophys. Biophys. Chem.* **18:** 213–237.

Wanner, L. and W. Gruissem. 1991. Expression dynamics of the tomato *rbcS* gene family during development. *Plant Cell* **3:** 1289–1303.

• *TIMETABLE*
Detection of DNA-binding Proteins

STEPS	***TIME REQUIRED***
Isolation of Nuclei from Cotyledons of Tomato	
Grow tomato seedlings	7 days
Isolate nuclei	4–6 hours[1]
Preparation of Nuclear Extract	
Lyse nuclei and prepare protein extract	6 hours
Labeling of DNA Fragment	
Pour gel	15 minutes
Set up and incubate labeling reaction	45 minutes
Determine amount of radioactivity incorporated	40 minutes[2]
Extract total labeled DNA	1 hour
Digest labeled DNA with *Pst*I	1 hour
Purify labeled DNA fragment by PAGE	3–4 hours
Elute labeled DNA fragment from gel	overnight
Recover labeled DNA fragment	2 hours
Mobility-shift Assay	
Pour gel	15 minutes
Set up and incubate binding reaction	1.5 hours

STEPS	*TIME REQUIRED*
Mobility-shift Assay (continued)	
Load and run gel	2–3 hours
Dry gel	0.5–1 hour
Expose gel to X-ray film	overnight
DNase I Footprinting	
Pour gel	30 minutes
Set up and incubate binding reaction	1 hour
Digest samples with DNase I and extract DNA	1.5 hours
Load and run gel	1.5 hours
Fix and dry gel	1.5 hours
Expose gel to X-ray film	overnight

[1]The time required for this procedure varies depending on the amount of material processed and the amount of starch in the plant organs used.
[2]This step can be performed concurrently with the extraction of total labeled DNA.

Isolation of Nuclei from Cotyledons of Tomato

MATERIALS AND EQUIPMENT

Tomato seeds
Nitex nylon mesh (52 μm, 100 μm, and 300 μm) (Tetko, Inc.)
Liquid nitrogen

REAGENTS

Homogenization buffer
Nuclei resuspension buffer
(For recipes, see Preparation of Reagents, pp. 257–260)

SAFETY NOTE

- The temperature of liquid nitrogen is –185°C. Handle with great care; always wear gloves and a face protector.

PROCEDURE

Steps 2–7 should be performed at 4°C using sterile (autoclaved) equipment.

1. Grow tomato seedlings in foil-covered pans on sterile, water-saturated filter paper for 7 days. Harvest the cotyledons using a razor blade.

2. Homogenize 40–50 g of fresh tissue in 300 ml of homogenization buffer in a Waring blender (preferably fitted with an attachment for four additional razor blades, see Kannangara et al. 1977) for 1 x 4 seconds and then 6 x 1 seconds.

   Note: Depending on the tomato organ used, 50–250 g of tissue may be processed per preparation. If you are processing more than 50 g of tissue, divide the tissue into 50-g batches and repeat steps 2–4 as necessary.

3. Pour the homogenate through four layers of cheesecloth placed over three layers of Nitex nylon mesh of pore sizes 300 μm, 100 μm, and 52 μm arranged in descending order (52 μm on the bottom) in a funnel supported by a large beaker. After each batch of homogenate, gather the cheesecloth and squeeze the filtrate into the Nitex, which remains in the funnel. Do not squeeze out the Nitex—the flow must occur by gravity only.

   Note: For preparations involving more than 50 g of tissue, it is convenient to set up two beaker/funnel/Nitex assemblies, so that one can drain while the next batch of tissue is being processed.

4. Pour the filtered homogenate from each 50-g batch of tissue into a 500-ml centrifuge bottle and spin in a SORVALL® GS-3 rotor at 5000 rpm (4225*g*) for 20 minutes at 4°C.

5. Aspirate the supernatants and *gently* resuspend each crude nuclear pellet in homogenization buffer by pipetting up and down repeatedly with a wide-bore plastic pasteur pipette. (DO NOT scrape the pellets off the side of the bottles with the pipette, or you will destroy the nuclei.) Transfer each resuspended pellet to a 50-ml centrifuge tube, and adjust the volume to approximately 40 ml with homogenization buffer.

6. Pellet the nuclei again by spinning in a SORVALL® SS-34 rotor at 4000 rpm (1912*g*) for 10 minutes at 4°C. Repeat steps 5 and 6 three times with spins at 3500 rpm for 10 minutes, 8 minutes, and 6 minutes, respectively. After the final spin, aspirate as much of the homogenization buffer as possible.

7. Resuspend the nuclei from each 50-g batch of tissue in approximately 0.5 ml of nuclei resuspension buffer (or more, if required), freeze in liquid nitrogen, and store at –80°C. If the nuclear extract is to be prepared immediately, it is not necessary to freeze the nuclei.

Preparation of Nuclear Extract

MATERIALS AND EQUIPMENT

Nuclei isolated from tomato cotyledons (see pp. 240–241)
Liquid nitrogen
CENTRICON™ 10 microconcentrator (Amicon, Inc.)
Bradford-based protein assay kit (Bio-Rad Laboratories)

REAGENTS

Nuclei lysing buffer
Dialysis buffer
(For recipes, see Preparation of Reagents, pp. 257–260)

SAFETY NOTE

- The temperature of liquid nitrogen is –185°C. Handle with great care; always wear gloves and a face protector.

PROCEDURE

1. Thaw the nuclei on ice and determine the exact volume of the nuclear suspension.

2. Add nuclei lysing buffer to give a final NaCl concentration of 0.47 M (i.e., 182 μl of nuclei lysing buffer per 1 ml of nuclear suspension).

3. Place the nuclear suspension on a rocker and rock gently for 30 minutes at 4°C.

4. Pellet the chromatin by spinning the lysed nuclei in a microfuge (or in a SORVALL® SS-34 rotor at 10,000 rpm [12,000*g*] if you are handling a large volume) for 20 minutes at 4°C.

5. Carefully remove the supernatant without disturbing the gelatinous, DNA-containing pellet—you want to avoid including any of this material in the extract.

6. Dialyze the supernatant against several changes (e.g., 4 x 1 liter) of dialysis buffer for 3–4 hours at 4°C.

7. To precipitate additional protein that may clog the microconcentrator during the concentration step, freeze the dialyzed extract in liquid nitrogen and thaw it on ice. Spin in a microfuge or in a SORVALL® SS-34 rotor at 10,000 rpm (12,000*g*) for 15 minutes at 4°C. Retain the supernatant and discard the pellet.

8. Concentrate the supernatant using a CENTRICON™ 10 microconcentrator so that the final concentration of protein in the extract is 1–3 mg/ml. We use the Bio-Rad Bradford-based protein assay kit with lysozyme as the standard. To achieve this concentration, it will be necessary to concentrate the extract at least 2–3 times.

9. Set aside a portion of the concentrated nuclear extract for the mobility-shift assay (see p. 250–252) and freeze the remaining extract in liquid nitrogen or on dry ice. Store the extract at –80°C until it is required.

Labeling of DNA Fragment

MATERIALS AND EQUIPMENT

Ammonium persulfate (APS) (10%)
N,N,N′,N′,tetramethylethylenediamine (TEMED)
DNA fragment of interest (subcloned into a plasmid, which is then cut with *Eco*RI or another restriction enzyme that leaves a 3′ overhanging end but does not cleave within the fragment)
Sequenase® DNA sequencing kit (Amersham Life Science, Inc.) (Klenow may be substituted)
[α-^{32}P]dATP (6000 Ci/mmol) (Amersham Life Science, Inc.)
Dithiothreitol (0.1 M)
Sodium acetate (0.35 M)
DE81 filters
Sodium phosphate (0.5 M; pH 7.0)
Ethanol (95% and 80%)
Phenol/chloroform/isoamyl alcohol (25:24:1)
Crushed dry ice
*Pst*I (10 units/μl)
X-ray film

REAGENTS

Acrylamide/bisacrylamide (29:1)
TBE (1x and 10x)
dNTP mix
High-EDTA 10x gel dyes
Elution buffer
Fragment resuspension buffer
(For recipes, see Preparation of Reagents, pp. 257–260)

SAFETY NOTES

- Acrylamide and bisacrylamide are potent neurotoxins and are absorbed through the skin. Their effects are cumulative. Wear gloves and a mask when weighing acrylamide and bisacrylamide. Polyacrylamide is considered to be nontoxic, but it should be treated with care because it may contain small quantities of unpolymerized material.

- Wear gloves when handling radioactive substances. Consult the local safety office for further guidance in the appropriate use of radioactive materials.
- Phenol is highly corrosive and can cause severe burns. Wear gloves, protective clothing, and safety glasses when handling it. All manipulations should be carried out in a chemical fume hood. Any areas of skin that come in contact with phenol should be rinsed with a large volume of water or PEG 400 and washed with soap and water; do not use ethanol!
- Chloroform is irritating to the skin, eyes, mucous membranes, and respiratory tract. It should only be used in a chemical fume hood. Gloves and safety glasses should also be worn. Chloroform is a carcinogen and may damage the liver and kidneys.

PROCEDURE

Preparation of Polyacrylamide Gel

1. Set up the gel apparatus and clean the glass plates.

2. Prepare a 4% acrylamide gel solution by mixing:

Acrylamide/bisacrylamide (29:1)	4 ml
TBE (10x)	3 ml
APS (10%)	200 µl
TEMED	40 µl

   Pour a gel 1 mm thick and ensure the wells are wide enough to accommodate 25-µl samples. The gel will take approximately 1 hour to polymerize. It can be stored for 1 day at room temperature before using.

   Note: The volumes required will vary depending on the size and thickness of the gel.

Labeling of Probe

1. Add the following reagents (in the order listed) to a 1.5-ml microfuge tube:

*Eco*RI-cut plasmid DNA (1 μg or ~800 fmol ends)	2.5 μl
Sequenase® buffer (5x)	2.0 μl
[α-^{32}P]dATP (6000 Ci/mmol)	3.0 μl
Dithiothreitol (0.1 M)	1.0 μl
Double-distilled H_2O	0.5 μl
Sequenase® enzyme (diluted 4x with Sequenase® dilution buffer)	1.0 μl
Total volume:	10.0 μl

Mix by pipetting up and down after each addition.

Note: (i) In this experiment, 1 μg of plasmid DNA is equivalent to approximately 400 fmol. (Note that the relationship of μg to fmol varies with the plasmid size.) When the plasmid is cut, two ends are generated per molecule, that is, 2 x 400 fmol = 800 fmol ends. This is important because the amount of [α-^{32}P]dATP added is determined by the number of fmol that are required to be incorporated into each end. Therefore, the number of ends is more important than the μg of DNA. [α-^{32}P]dATP from Amersham Life Science, Inc. is approximately 1500 fmol/μl. The amount added to the labeling reaction should be approximately three times the number of fmol that can be incorporated. If you wanted to cut with an enzyme other than *Eco*RI, it would be necessary to adjust the amount and type of nucleotides used (see note after step 11, p. 249). (ii) After the radiolabeled nucleotide is added, all procedures should be performed behind a Plexiglas shield. In addition, gloves and protective clothing should be worn, and radioactive waste should be discarded in the appropriate manner.

2. Incubate for 15 minutes at room temperature.
3. Add 0.5 μl of dNTP mix, and mix well.
4. Incubate for an additional 10 minutes at room temperature.
5. To stop the reaction, add 90 μl of 0.35 M sodium acetate and mix by pipetting up and down.

Determination of Amount of Radioactivity Incorporated

This protocol is adapted from Sambrook et al. 1989. It is convenient to perform this procedure simultaneously with step 4, p. 247 (bottom).

1. Transfer 5 μl of the labeling reaction mixture to a new tube. Spot 2 μl of this mixture onto a DE81 filter and drop the filter into a beaker containing approximately 20 ml of 0.5 M sodium phosphate (pH 7.0). Repeat with a second filter.
2. Place the beaker on a platform shaker and allow the solution to swirl gently for 5 minutes. Pour off the buffer into the radioactive waste.
3. Repeat this washing procedure twice with fresh buffer.

4. Wash the filters in double-distilled H_2O for 5 minutes. Repeat twice with fresh H_2O.

5. Wash the filters in 95% ethanol for 5 minutes.

6. Allow the filters to dry completely. Transfer each filter to a counting vial, add scintillation cocktail, and count.

7. Calculate the percent of theoretical maximum incorporation. The theoretical maximum incorporation is calculated as follows: (i) 400 fmol of plasmid DNA are labeled per reaction. (ii) Label is incorporated at both ends of the *Eco*RI-cut plasmid DNA, and two deoxyadenosine phosphate residues can be incorporated at each end. (iii) Therefore, 1600 fmol of labeled nucleotide can be incorporated per 400 fmol of plasmid DNA. (iv) The specific activity of the labeled nucleotide is 6000 Ci/mmol, i.e., 13,000 cpm/fmol. (v) Therefore, the maximum theoretical incorporation is 13,000 cpm x 1600 fmol = 2.08×10^7 cpm per 400 fmol of plasmid DNA or 1.04×10^7 cpm per 400 fmol of fragment. Note that only half of the incorporated counts will be contained in the fragment of interest. Subsequent enzyme digestion will result in liberation of the fragment of interest (labeled at one end) and the vector DNA (also labeled at one end).

Isolation of Labeled Probe

1. Add 95 μl of phenol/chloroform/isoamyl alcohol (25:24:1) to the remaining 95 μl of the labeling reaction mixture. Mix the phases thoroughly.

   Note: It is extremely important that no leakage occurs during the phenol/chloroform/isoamyl alcohol extraction and ethanol precipitation steps. Therefore, it is advisable to use screw-cap microfuge tubes with gaskets for these procedures.

2. Separate the phenol and aqueous layers by spinning in a microfuge for 3–5 minutes at room temperature.

3. Carefully transfer 85 μl of the aqueous (upper) phase to a new tube, and discard the phenol-containing tube in the radioactive waste. Add 200 μl of 95% ethanol to the aqueous phase, make certain the lid of the tube is securely closed, and vortex.

4. Place the tube in crushed dry ice for 15–20 minutes, then spin in a microfuge for 20 minutes. (This is a convenient time to determine the amount of labeled nucleotide incorporated into the DNA fragment, see p. 246.)

5. Using a Pipetman®, carefully remove the ethanol-containing supernatant and discard it in the radioactive waste. Wash the pellet by adding 600 µl of 80% ethanol to the tube. Spin in a microfuge for 1–2 minutes, and remove the ethanol with a Pipetman®.

   Note: You must be extremely careful that you do not discard the pellet at this step.

6. Dry the pellet in a SpeedVac® for 2–3 minutes.

7. Resuspend the pellet in 20 µl of 1x Sequenase® buffer and then add 0.5 µl of 10 units/µl *Pst*I to cleave the labeled DNA fragment of interest away from the vector DNA (which is also labeled). Incubate for 30–60 minutes at 37°C.

8. Add 2.5 µl of high-EDTA 10x gel dyes to the digestion mixture. Load the sample onto the preparative 4% acrylamide gel (see p. 245) and run the gel in 1x TBE at approximately 10 V/cm for approximately 2.5 hours. The time will depend on the size of the fragment you are labeling—the idea is to get a good separation of the fragment from the plasmid, but not to run it so long that the fragment runs off the gel. (An insert of approximately 400 bp runs close to the xylene cyanol marker on this percentage gel.)

9. After electrophoresis, wrap the gel carefully in Saran® Wrap and expose it to X-ray film for 5–10 minutes. Use a Sharpie® or other waterproof marker to mark the film and the borders of the gel so that it will be possible to re-align the film with the gel after the film has been developed.

10. Excise the region of the gel corresponding to the labeled DNA fragment of interest and place it into a 1.5-ml microfuge tube. Elute the DNA by incubating the gel slice in 300 µl of elution buffer overnight at 4°C. Remove the elution buffer and transfer it to a clean microfuge tube. Add 250 µl of fresh elution buffer to the gel slice. Freeze at –80°C and thaw one time, then remove the elution buffer and combine it with the elution buffer collected previously. Spin the combined elution buffer fractions in a microfuge for approximately 10 minutes to pellet any small fragments of polyacrylamide. Transfer the supernatant to a clean tube and precipitate the DNA by adding two volumes of 95% ethanol, placing on crushed dry ice for 20 minutes, and spinning in a microfuge for 20 minutes. Discard the supernatant, and wash the pellet with approximately 500 µl of 80% ethanol. Dry the pellet in a SpeedVac® for 2–3 minutes.

    Note: It is important not to over-dry the pellet, as this will result in the formation of single-stranded DNA fragments (Svaren et al. 1987).

11. Resuspend the purified DNA fragment in 20 μl of fragment resuspension buffer. Count a small aliquot to determine how many fmol of the fragment were recovered. We typically recover approximately 50% of the original amount labeled. The efficiency of elution from the gel is approximately 80%, but it varies depending on the size of the DNA fragment. The purified DNA fragment can be stored at 4°C for a maximum of 2 weeks.

    Note: The DNA fragments we use for footprinting are all 200–500 bp in length and they are cloned into the multiple cloning site of plasmid pUC118/119 between the *Hinc*II and *Sma*I sites. We always label the DNA either at the *Eco*RI or the *Hind*III site of pUC. To label at the *Hind*III end, follow the same labeling protocol as described above except (i) use 0.5 μl of cold dGTP at a concentration of 10,000 fmol/μl instead of 0.5 μl of double-distilled H_2O; (ii) use only half the amount of [α-^{32}P]dATP and replace the rest with an equivalent amount of [α-^{32}P]dCTP. To release the insert from the plasmid DNA after labeling at the *Hind*III site, use *Sst*I rather than *Pst*I.

Mobility-shift Assay

MATERIALS AND EQUIPMENT

Ammonium persulfate (APS) (10%)
N,N,N′,N′,tetramethylethylenediamine (TEMED)
Poly(dI-dC)·poly(dI-dC) (10 mg/ml in sterile, double-distilled H_2O)
(Pharmacia Biotech, Inc.)
EDTA (32 mM; pH 8.0)
^{32}P-labeled DNA fragment (see pp. 244–249)
Nuclear extract (see pp. 242–243)
Xylene cyanol
Bromophenol blue
X-ray film

REAGENTS

Acrylamide/bisacrylamide (29:1)
TBE (1x and 10x)
Fragment resuspension buffer
Dialysis buffer
(For recipes, see Preparation of Reagents, pp. 257–260

SAFETY NOTES

- Acrylamide and bisacrylamide are potent neurotoxins and are absorbed through the skin. Their effects are cumulative. Wear gloves and a mask when weighing acrylamide and bisacrylamide. Polyacrylamide is considered to be nontoxic, but it should be treated with care because it may contain small quantities of unpolymerized material.
- Wear gloves when handling radioactive substances. Consult the local safety office for further guidance in the appropriate use of radioactive materials.

PROCEDURE

Preparation of Polyacrylamide Gel

1. Set up the gel apparatus and clean the glass plates.

2. Prepare a 4% nondenaturing acrylamide gel solution by mixing:

Acrylamide/bisacrylamide (29:1)	4 ml
TBE (10x)	3 ml
APS (10%)	200 μl
TEMED	40 μl

Pour a gel 1–1.5 mm thick. The gel will take approximately 1 hour to polymerize. It can be stored for 1 day at room temperature before using.

Note: The volumes required will vary depending on the size and thickness of the gel.

Binding Reaction

1. Dilute the 10 mg/ml poly(dI-dC)·poly(dI-dC) stock solution with sterile, double-distilled H_2O to give a concentration of 6 μg/μl. Prepare the poly(dI-dC)·poly(dI-dC) + EDTA working solution by combining equal volumes of 6 μg/μl poly(dI-dC)·poly(dI-dC) and 32 mM EDTA (pH 8.0). Dilute the ^{32}P-labeled DNA fragment to a concentration of 2 fmol/μl in fragment resuspension buffer.

2. For the standard binding reaction, add the following reagents (in the order listed) to a 1.5-ml microfuge tube:

^{32}P-labeled DNA fragment (2 fmol/μl)	0.5 μl
Poly(dI-dC)·poly(dI-dC) + EDTA working solution (i.e., 3 μg/μl poly[dI-dC]·poly[dI-dC] in 16 mM EDTA)	0.5 μl
Nuclear extract and/or dialysis buffer	9 μl
Total volume:	10.0 μl

Mix by pipetting up and down several times.

Note: (i) To monitor the mobility of the free DNA fragment, you should include a control tube prepared with with 9 μl of dialysis buffer instead of nuclear extract. (ii) To titrate the binding activity of the nuclear extract, you should include several tubes in which increasing volumes (up to 9 μl) of extract are added to a given amount of labeled DNA fragment. This experiment will also allow you to determine the amount of nuclear extract required for DNase I footprinting. To observe a footprint using the method described on pp. 253–256, 100% of the labeled DNA fragment must be bound by protein. If 100% of the fragment is not bound using the maximum volume of extract (i.e., 9 μl), it may be necessary to concentrate the nuclear extract further. (iii) The mobility-shift assay can also be used to determine whether binding to a particular DNA fragment is specific. Specificity of binding can be tested by adding an excess of competitor unlabeled DNA fragment to the reaction; we typically add a 100-fold and 500-fold molar excess of each competitor. All competitors are used at both concentrations. Use one lane for the 100-fold excess, and a separate lane for the 500-fold excess. If binding is specific, the homologous fragment should compete for binding, whereas an unrelated fragment should not compete. To obtain specific binding, it may also be necessary to vary the amount of poly(dI-dC)·poly(dI-dC) competitor used in the reaction or to vary the type of nonspecific competitor used (e.g., it may be preferable to use sheared salmon sperm DNA).

3. Incubate for 45 minutes at room temperature. During the incubation, pre-run the gel at 100 V for 10–15 minutes.

4. At the end of the incubation period, load the samples directly onto the gel without adding dye. Monitor the progress of the electrophoretic separation by adding 10 μl of dialysis buffer containing 0.05% xylene cyanol and 0.05% bromophenol blue to one well (i.e., a well that does not contain sample).

5. Run the gel in 1x TBE at approximately 10 V/cm for approximately 2 hours, depending on the size of the fragment you are using (see Sambrook et al. 1989).

   Note: In the literature, you may have noticed that this type of gel is often run in recirculating Tris/EDTA or glycine buffer and that it may be loaded under tension. In our experience, the sharpness of the bands and the percentage of the DNA fragment bound do not seem to vary significantly with the buffer/loading conditions.

6. Carefully transfer the gel to Whatman 3MM paper and cover it with Saran® Wrap. Dry the gel on a gel dryer for approximately 1 hour at 80°C.

7. Expose the dried gel to X-ray film overnight at room temperature.

DNase I Footprinting

MATERIALS AND EQUIPMENT

Urea (dry)
Ammonium persulfate (APS) (10%)
N,N,N′,N′,tetramethylethylenediamine (TEMED)
Poly(dI-dC)·poly(dI-dC) (10 mg/ml) (Pharmacia Biotech, Inc.)
EDTA (32 mM; pH 8.0)
^{32}P-labeled DNA fragment (see pp. 244–249)
Nuclear extract (see pp. 242–243)
DNase I (RNase-free) (Life Technologies, Inc.)
Phenol/chloroform/isoamyl alcohol (25:24:1)
Ethanol (95% and 80%)
Crushed dry ice
Sequenase® DNA sequencing kit (Amersham Life Science, Inc.)
Methanol/glacial acetic acid/H_2O (5:5:90)
X-ray film (35 x 43 cm)

REAGENTS

Acrylamide/bisacrylamide (38:2)
Sequencing TBE (1x and 20x)
Fragment resuspension buffer
Dialysis buffer
DNase I dilution buffer
DNase I stop buffer
Formamide dye solution (from Sequenase® kit)
(For recipes, see Preparation of Reagents, pp. 257–260)

SAFETY NOTES

- Acrylamide and bisacrylamide are potent neurotoxins and are absorbed through the skin. Their effects are cumulative. Wear gloves and a mask when weighing acrylamide and bisacrylamide. Polyacrylamide is considered to be nontoxic, but it should be treated with care because it may contain small quantities of unpolymerized material.
- Wear gloves when handling radioactive substances. Consult the local safety office for further guidance in the appropriate use of radioactive materials.

- Phenol is highly corrosive and can cause severe burns. Wear gloves, protective clothing, and safety glasses when handling it. All manipulations should be carried out in a chemical fume hood. Any areas of skin that come in contact with phenol should be rinsed with a large volume of water or PEG 400 and washed with soap and water; do not use ethanol!
- Chloroform is irritating to the skin, eyes, mucous membranes, and respiratory tract. It should only be used in a chemical fume hood. Gloves and safety glasses should also be worn. Chloroform is a carcinogen and may damage the liver and kidneys.
- Methanol is poisonous and can cause blindness if ingested in sufficient quantities. Adequate ventilation is necessary to limit exposure to vapors.
- Glacial acetic acid is volatile and should be used in a chemical fume hood. Concentrated acids should be handled with great care; gloves and a face protector should be worn.

PROCEDURE

Preparation of Polyacrylamide Gel

1. Set up the apparatus for a sequencing gel and clean the glass plates.

2. Prepare a 6% acrylamide gel solution by mixing:

Acrylamide/bisacrylamide (38:2)	11.3 ml
Urea	31.5 g
Sequencing TBE (20x)	3.75 ml
APS (10%)	300 μl
TEMED	30 μl

Pour the gel. The gel will take approximately 2 hours to polymerize. It can be stored for 1 day at room temperature before using.

Note: The volumes required will vary depending on the size and thickness of the gel.

Binding Reaction

1. Dilute the 10 mg/ml poly(dI-dC)·poly(dI-dC) stock solution with H_2O to give a concentration of 6 μg/μl. Prepare the poly(dI-dC)·poly(dI-dC) + EDTA working solution by combining equal volumes of 6 μg/μl poly(dI-dC)·poly(dI-dC) and 32 mM EDTA (pH 8.0). Dilute the ^{32}P-labeled DNA fragment to a concentration of 2 fmol/μl in fragment resuspension buffer.

2. For the standard binding reaction, add the following reagents (in the order listed) to a 1.5-ml microfuge tube:

^{32}P-labeled DNA fragment (2 fmol; usually ~20,000 cpm)	1 μl
Poly(dI-dC)·poly(dI-dC) + EDTA working solution (i.e., 3 μg/μl poly[dI-dC]·poly[dI-dC] in 16 mM EDTA)	1 μl
Nuclear extract and/or dialysis buffer (20–50 μg of protein, depending on the activity of the DNA-binding protein in your extract)	18 μl
Total volume:	20.0 μl

Mix by pipetting up and down several times.

Note: (i) Before adding the nuclear extract to the reaction mixture, keep it on ice at all times. (ii) The total number of samples will vary according to your experimental design. For a sample protocol for a preliminary experiment, see Table 1. You will need to stagger the incubations by at least 1 minute to allow sufficient time to perform the DNase I digestions at the end of the binding reaction.

3. Incubate for 30 minutes at room temperature. During this incubation, dilute the DNase I in DNase I dilution buffer. (For our standard binding reaction, the concentration of the diluted DNase I solution is 20 μg/ml.)

4. Add 2 μl of the diluted DNase solution to each binding reaction mixture and mix by pipetting up and down several times. (In our standard reaction, the final concentration of DNase I is ~2 μg/ml.) Incubate for 10 seconds at room temperature—start the timing period with the first push of the Pipetman® into the binding reaction mixture.

Note: We have minimized the time of the DNase I incubation because the magnesium in the DNase I buffer may disrupt the binding of some proteins.

Table 1 **Sample Protocol for DNase I Footprinting**

Lane/ Tube No.	Labeled Fragment	poly(dI-dC)· poly(dI-dC) + EDTA	Nuclear Extract	Dialysis Buffer	DNase I Final Concentration (μg/ml)
1	+	+	–	+	0
2	+	+	–	+	0.5
3	+	+	–	+	1.0
4	+	+	–	+	2.0
5	+	+	–	+	4.0
6	+	+	+	as required	0

This experiment serves to determine the appropriate DNase I concentration required to produce an even ladder of bands, and to determine whether endogenous nuclease activity is contained in the nuclear extract. In subsequent experiments, the optimal amount of protein to produce a footprint and the optimal DNase I concentration to be used with the extract should be determined.

5. Stop the digestions by adding 80 μl of DNase I stop solution (Green et al. 1987) followed by 100 μl of phenol/chloroform/isoamyl alcohol (25:24:1).

6. Vortex each sample for several seconds, taking care that the microfuge tubes are securely closed. Spin in a microfuge for 3–5 minutes at room temperature.

7. Transfer 85 μl of each aqueous (upper) phase to a clean tube, and discard the phenol-containing tubes in the radioactive waste. Add 200 μl of 95% ethanol to each aqueous phase. Vortex briefly and place the tubes in crushed dry ice for 15–20 minutes.

8. Spin the tubes in a microfuge for 20 minutes. Carefully remove the supernatants and discard them. Add 500 μl of 85% ethanol to each pellet. Spin the tubes in a microfuge for 2–3 minutes. Remove each supernatant as before, leaving the pellets as dry as possible.

9. Dry the pellets in a SpeedVac® for approximately 5 minutes (make sure each pellet is completely dry). Add 3 μl of formamide dye solution to each tube, and mix on an Eppendorf shaker or by vortexing.

10. Boil the samples for 3 minutes, and immediately cool them on ice. Briefly spin the tubes in a microfuge to bring all the sample down from the sides of the tube, and then return them to the ice. Load each sample on the 6% sequencing gel and run it in 1x sequencing TBE at 3000–4000 V at 45°–50°C until the bromophenol blue reaches the bottom of the gel. The voltage may need to be increased during run to maintain the gel temperature at 45°–50°C.

    Note: (i) Ideally, you should run Maxam and Gilbert sequencing reactions (Green et al. 1989; Sambrook et al. 1989) on each labeled fragment so that the location of the protection site can be easily determined. However, an M13 sequencing ladder (from the Sequenase® kit) labeled with ^{32}P-nucleotide can be substituted if necessary. (ii) The low-ionic-strength TBE used here allows the gel to be run in approximately 1 hour.

11. Fix the gel in methanol/acetic acid/H_2O (5:5:90) for 15 minutes. Transfer the gel to Whatman 3MM paper and dry it on a gel dryer.

12. Expose the dried gel to X-ray film overnight at –80°C, using an intensifying screen.

• *PREPARATION OF REAGENTS*

Acrylamide/bisacrylamide (29:1)
(100 ml)

Acrylamide (Bio-Rad Laboratories)	29 g
Bisacrylamide	1 g

Dissolve the acrylamide and bisacrylamide in approximately 50 ml of H_2O. Adjust the volume to 100 ml with H_2O. Store at 4°C.

Acrylamide/bisacrylamide (38:2)
(100 ml)

Acrylamide (Bio-Rad Laboratories)	38 g
Bisacrylamide	2 g

Dissolve the acrylamide and bisacrylamide in approximately 50 ml of H_2O. Adjust the volume to 100 ml with H_2O. Store at 4°C.

SAFETY NOTE

- Acrylamide and bisacrylamide are potent neurotoxins and are absorbed through the skin. Their effects are cumulative. Wear gloves and a mask when weighing acrylamide and bisacrylamide. Polyacrylamide is considered to be nontoxic, but it should be treated with care because it may contain small quantities of unpolymerized material.

Dialysis buffer

20 mM HEPES (pH 7.6)
40 mM NaCl
0.2 mM EDTA
20% Glycerol
1 mM Dithiothreitol (DTT)

Combine the appropriate volumes of H_2O, buffer, salt, EDTA, and glycerol stock solutions and autoclave this solution. Prepare the DTT stock solution in sterile double-distilled H_2O and add DTT to the dialysis buffer immediately before use.

DNase I dilution buffer

25 mM $MgCl_2$
25 mM $CaCl_2$

Autoclave.

DNase I stop buffer

6.25 mM EDTA (pH 8.0)
0.125% Sodium dodecyl sulfate (SDS)
0.375 M Sodium acetate
62.5 μg/ml Yeast tRNA

SAFETY NOTE

•Wear a mask when weighing SDS.

---

dNTP mix

2.5 mM dATP
2.5 mM dTTP
2.5 mM dCTP
2.5 mM dGTP

Elution buffer

50 mM Tris-HCl (pH 8.0)
500 mM NaCl
20 mM EDTA

Autoclave.

Formamide dye solution

95% Deionized formamide
20 mM EDTA (pH 8.0)
0.05% Bromophenol blue
0.05% Xylene cyanol FF

This solution is included in the Sequenase® DNA sequencing kit (Amersham Life Science, Inc.).

Fragment resuspension buffer

10 mM Tris-HCl (pH 8.0)
20 mM NaCl
1 mM EDTA (pH 8.0)

Autoclave.

High-EDTA 10x gel dyes

30% Ficoll
0.2% Bromophenol blue
0.2% Xylene cyanol
100 mM EDTA (pH 8.0)

Dissolve in 10x TBE (for recipe, see p. 260).

Homogenization buffer

25 mM PIPES (pH 7.0)
10 mM NaCl
5 mM EDTA (pH 8.0)
250 mM Sucrose
0.15 mM Spermine
0.5 mM Spermidine
20 mM 2-Mercaptoethanol
0.1% Nonidet P-40 (Sigma)
0.2 mM Phenylmethylsulfonyl fluoride (PMSF)

Combine the appropriate volumes of H_2O, buffer, salt, EDTA, and sucrose stock solutions and autoclave this solution. Add 2-mercaptoethanol, Nonidet P-40, and PMSF (from a stock solution in 100% ethanol) to the homogenization buffer immediately before use.

SAFETY NOTES

- 2-Mercaptoethanol may be fatal if swallowed and is harmful if inhaled or absorbed through the skin. High concentrations are extremely destructive to the mucous membranes, upper respiratory tract, skin, and eyes. Use only in a chemical fume hood. Gloves and safety glasses should be worn.
- PMSF is extremely destructive to the mucous membranes of the respiratory tract, the eyes, and the skin. It may be fatal if inhaled, swallowed, or absorbed through the skin. It is a highly toxic cholinesterase inhibitor. It should be used in a chemical fume hood. Gloves and safety glasses should be worn.

Nuclei lysing buffer

50 mM HEPES (pH 7.6)
2.5 M NaCl
5 mM $MgCl_2$
10 mM KCl
20% Glycerol
1 mM Dithiothreitol (DTT)
0.5 μg/ml Leupeptin
50 μg/ml Antipain (Sigma)

Combine the appropriate volumes of H_2O, buffer, salt, and glycerol stock solutions and autoclave this solution. Prepare the DTT, leupeptin, and antipain stock solutions in sterile double-distilled H_2O and add them to the nuclei lysing buffer immediately before use.

Nuclei resuspension buffer

50 mM HEPES (pH 7.6)
100 mM NaCl
5 mM $MgCl_2$
10 mM KCl
50% Glycerol
1 mM Dithiothreitol (DTT)
0.5 μg/ml Leupeptin
50 μg/ml Antipain (Sigma)

Combine the appropriate volumes of H_2O, buffer, salt, and glycerol stock solutions and autoclave this solution. Prepare the DTT, leupeptin, and antipain stock solutions in sterile double-distilled H_2O and add them to the nuclei resuspension buffer immediately before use.

Sequencing TBE (20x)

0.89 M Tris base
0.89 M Boric acid
0.002 M EDTA

TBE (10x)

0.89 M Tris base
0.89 M Boric acid
0.02 M EDTA

• *Section 13*
Electrophoretic Mobility-shift Assay to Characterize Protein/ DNA-binding Sites

Elucidation of the mechanisms by which specific genes are expressed in a temporal and tissue-specific manner or expressed in response to extracellular signals has become one of the central issues in molecular biology. Initiation of transcription is one major level at which gene expression is regulated in eukaryotes. Transcriptional control requires the RNA polymerase transcription initiation complex and results from an interplay between regulatory DNA sequences and sequence-specific DNA-binding proteins. Hence, much effort has been devoted to the development of technologies to detect and characterize DNA-binding proteins.

The electrophoretic mobility-shift assay is probably the most popular procedure used for the identification of sequence-specific DNA-binding proteins (Carey 1991). Originally, this assay was developed for kinetic analyses of highly purified prokaryotic regulatory proteins (Fried and Crother 1981, 1984a,b; Garner and Revzin 1981). By including nonspecific competitor DNA (such as salmon sperm DNA, calf thymus DNA, or *Escherichia coli* DNA), it was demonstrated that the assay could be used for the analysis of crude protein preparations (Strauss and Varshavsky 1984). Now it is common for these heterologous nonspecific competitor DNAs to be substituted by synthetic polymers (such as poly[dI-dC]·poly[dI-dC], poly[dA-dT]·poly[dA-dT], and others) (Carthew et al. 1985; Singh et al. 1986). On the basis of these improvements, the assay provides a simple, rapid, and sensitive method for detecting DNA-binding proteins in crude nuclear extracts and it is often used during the initial stages of the analysis of DNA-binding sites within a promoter of interest. Over the past few years, the protein-binding sites within many plant promoters have been characterized (Green et al. 1987; Stougaard et al. 1987; Deikman and Fischer 1988; Giuliano et al. 1988; Datta and Cashmore 1989; de Brujin et al. 1989; Holdsworth and Laties 1989; Lam and Chua 1989; Staiger et al. 1989; Schindler and Cashmore 1990; Buzby et al. 1990; Manzara et al. 1991; Petersen et al. 1991; Granell et al. 1992; Williams et al. 1992).

The electrophoretic mobility-shift assay is based on the observation that the mobility of a DNA fragment through a nondenaturing,

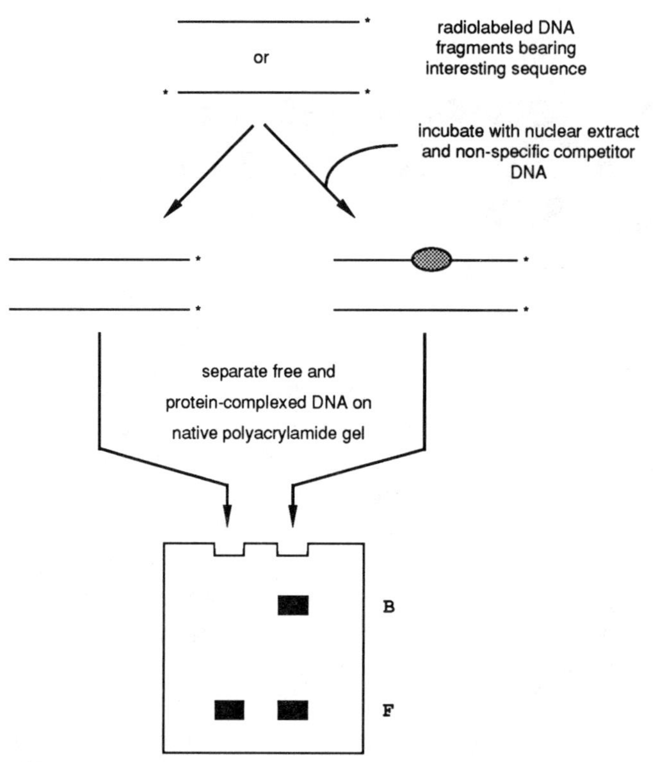

Figure 1
Schematic representation of the electrophoretic mobility-shift assay. The DNA fragment is radiolabeled on either one or both ends and incubated with a nuclear protein extract. Free (F) and protein-complexed (B) DNA fragments are separated on a native (nondenaturing), low-ionic-strength polyacrylamide gel. Protein-complexed DNA fragments appear as retarded bands on the autoradiograph.

low-ionic strength polyacrylamide gel is retarded upon association with a DNA-binding protein (Fig. 1). The protein-DNA interactions are stabilized by the low ionic strength of the buffer and the "caging effect" of the gel matrix (Fried and Crother 1981). However, the optimal binding and electrophoresis conditions must be determined empirically for each protein and DNA fragment. Factors influencing the results include pH and salt concentrations of the binding buffer, nature and amount of the nonspecific competitor DNA, amount and quality of the protein extract, incubation time, gel composition, and electrophoresis buffer.

The sensitivity of the electrophoretic mobility-shift assay increases significantly when small DNA fragments (i.e., <100 bp) are used. However, DNA fragments up to 400 bp can be analyzed and often multiple protein-DNA complexes mediated by different proteins can be obtained. The location of the individual DNA-binding sites of these proteins can be narrowed down by using subfragments of the

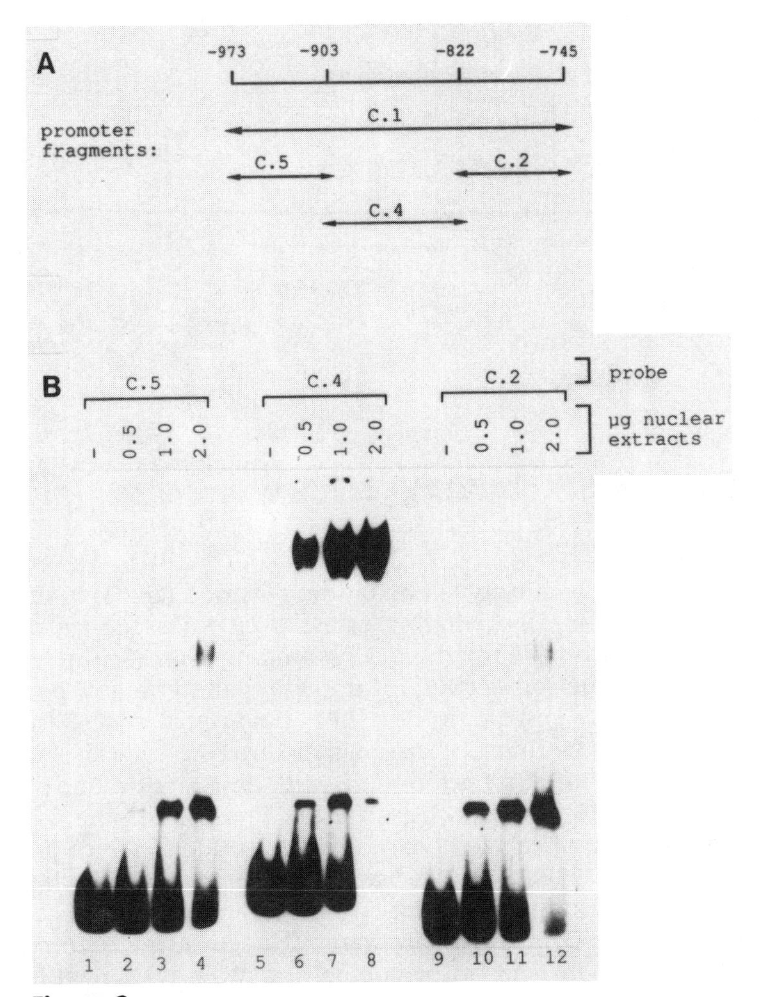

Figure 2
Identification of nuclear activities that interact with distinct DNA fragments derived from the *Nicotiana plumbaginifolia cab-E* promoter. (*A*) The *cab-E* promoter region extending from –745 to –973 is shown. Individual DNA fragments derived from this region are designated C.1, C.2, C.4, and C.5 (Schindler and Cashmore 1990). (*B*) Electrophoretic mobility-shift assay employing the radiolabeled promoter fragments C.2, C.4, and C.5. The fragments were incubated with increasing amounts of plant nuclear extract in the presence of 2 µg of poly(dI-dC)·poly(dI-dC). The amount of protein used in each reaction is indicated above the lanes. Lanes 1, 5, and 9 contain the free DNA probe only. The protein-DNA complexes are designated b1, b2 and b3.

larger promoter fragment (Fig. 2). Furthermore, during the initial stages of the analysis, it is important to perform titration experiments employing various amounts of protein so that proteins present in very low abundance can be identified (see Fig. 2).

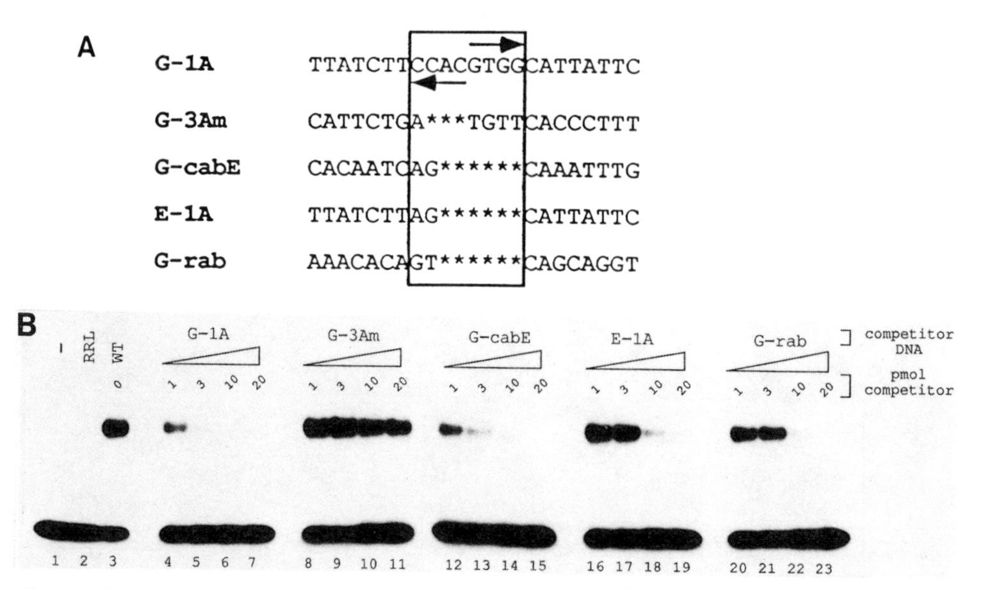

Figure 3
Determination of high- and low-affinity G-box binding factor 1 (GBF1) binding sites using the competitive DNA-binding assay. (*A*) DNA sequence of oligonucleotides employed in the competitive DNA-binding assay (Schindler et al. 1992). The G-1A oligonucleotide contains the 8-bp palindromic G-box (arrows), which represents a high-affinity GBF1 binding site. G-3Am represents a mutant binding site that is not recognized by GBF1. G-cabE, E-1A, and G-rab carry DNA motifs that are bound with various affinities by GBF1 (boxed nucleotides). Nucleotides identical to the palindrome are marked by asterisks. (*B*) The radiolabeled G-1A oligonucleotide (10 fmol per reaction) was incubated with GBF1 that had been synthesized in vitro using the rabbit reticulocyte lysate (RRL) system (lanes 3–23). Varying amounts of various unlabeled competitor DNA fragments were included in the binding reactions (lanes 4–23). The sequences of the competitor DNAs are shown in panel A. Lane 1 shows the free, radiolabeled G-1A oligonucleotide and lane 2 shows rabbit reticulocyte lysate to which no RNA had been added incubated with radiolabeled G-1A oligonucleotide.

The specificity of the protein-DNA interaction is determined by competitive binding studies in which various unlabeled DNA fragments are included in the binding reactions. Note that all DNA-binding proteins will interact with any DNA fragment if this fragment is present in vast excess. Hence, specificity is commonly demonstrated by the 100- to 1000-fold higher affinity of a protein for its specific DNA-binding site relative to other DNA sequences. Titration experiments similar to the one outlined in Figure 3 can be performed in order to determine the relative affinity of a protein for various DNA fragments.

REFERENCES

Buzby, J.S., T. Yamada, and E.M. Tobin. 1990. A light-regulated DNA-binding activity interacts with a conserved region of a *Lemna gibba rbcS* promoter. *Plant Cell* **2:** 805–814.

Carey, J. 1991. Gel retardation. *Methods Enzymol.* **208:** 103–117.

Carthew, R.W., L.A. Chodosh, and P.A. Sharp. 1985. An RNA polymerase II transcription factor binds to an upstream element in the adenovirus major late promoter. *Cell* **43:** 439–448.

Datta, N. and A.R. Cashmore. 1989. Binding of a pea nuclear protein to promoters of certain photoregulated genes is modulated by phosphorylation. *Plant Cell* **1:** 1069–1077.

de Brujin, F.J., G. Felix, B. Grunenberg, H.J. Hoffmann, B. Metz, P. Ratet, A. Simons-Schreier, L. Szabados, P. Welters, and J. Schell. 1989. Regulation of plant genes specifically induced in nitrogen-fixing nodules: Role of *cis*-acting elements and *trans*-acting factors in leghemoglobin gene expression. *Plant Mol. Biol.* **13:** 319–325.

Deikman, J. and R.L. Fischer. 1988. Interaction of a DNA-binding factor with the 5′-flanking region of an ethylene-responsive fruit ripening gene from tomato. *EMBO J.* **7:** 3315–3320.

Fried, M. and D.M. Crother. 1981. Equilibria and kinetics of *lac* repressor-operator interactions by polyacrylamide gel electrophoresis. *Nucleic Acids Res.* **9:** 6505–6525.

———. 1984a. Kinetics and mechanism in the reaction of gene regulatory proteins with DNA. *J. Mol. Biol.* **172:** 241–262.

———. 1984b. Equilibrium studies of the cyclic AMP receptor protein-DNA interaction. *J. Mol. Biol.* **172:** 263–282.

Garner, M.M. and A. Revzin. 1981. A gel electrophoresis method for quantifying the binding of proteins to specific DNA regions: Application to components of the *Escherichia coli* lactose operon regulatory system. *Nucleic Acids Res.* **9:** 3047–3060.

Giuliano, G., E. Pichersky, V.S. Malik, M.P. Timko, P.A. Scolnik, and A.R. Cashmore. 1988. An evolutionarily conserved protein binding sequence upstream of a plant light-regulated gene. *Proc. Natl. Acad. Sci.* **85:** 7089–7093.

Granell, A., J.G. Pereto, U. Schindler, and A.R. Cashmore. 1992. Nuclear factors binding to the extensin promoter exhibit differential activity in carrot protoplasts and cells. *Plant Mol. Biol.* **18:** 739–748.

Green, P.J., S.A. Kay, and N.-H. Chua. 1987. Sequence-specific interactions of a pea nuclear factor with light-responsive elements upstream of the *rbcS-3A* gene. *EMBO J.* **6:** 2543–2549.

Holdsworth, M.J. and G.G. Laties. 1989. Site-specific binding of a nuclear factor to the carrot extensin gene is influenced by both ethylene and wounding. *Planta* **179:** 17–23.

Lam, E. and N.-H. Chua. 1989. ASF-2: A factor that binds to the cauliflower mosaic virus 35S promoter and a conserved GATA motif in *Cab* promoters. *Plant Cell* **1:** 1147–1156.

Manzara, T., P. Carrasco, and W. Gruissem. 1991. Developmental and organ-specific changes in promoter DNA-protein interactions in the tomato *rbcS* gene family. *Plant Cell* **3:** 1305–1316.

Petersen, T.J., L.J. Arwood, S. Spiker, M.J. Guiltinan, and W.F. Thompson. 1991. High mobility group chromosomal proteins bind to AT-rich tracts flanking plant genes. *Plant Mol. Biol.* **16:** 95–104.

Sambrook, J., E.F. Fritsch, and T. Maniatis. 1989. *Molecular cloning: A laboratory manual*, 2nd edition. Cold Spring Harbor Laboratory Press, Cold Spring Harbor, New York.

Schindler, U. and A.R. Cashmore. 1990. Photoregulated gene expression may involve ubiquitous DNA binding proteins. *EMBO J.* **9:** 3415–3427.

Schindler, U., W.B. Terzaghi, H. Beckmann, T. Kadesch, and A.R. Cashmore. 1992. DNA binding site preferences and transcriptional activation properties of the *Arabidopsis* transcription factor GBF1. *EMBO J.* **11:** 1275–1289.

Singh, H., R. Sen, D. Baltimore, and P.A. Sharp. 1986. A nuclear factor that binds to a conserved sequence motif in transcriptional control elements of immunoglobulin genes. *Nature* **319:** 154–158.

Staiger, D., H. Kaulen, and J. Schell. 1989. A CACGTG motif of the *Antirrhinum majus* chalcone synthase promoter is recognized by an evolutionarily conserved nuclear protein. *Proc. Natl. Acad. Sci.* **86:** 6930–6934.

Stougaard, J., N.N. Sandal, A. Gron, A. Kuehle, and K.A. Marcker. 1987. 5′ analysis of the soyabean leghaemoglobin lbc_3 gene: Regulatory elements required for promoter activity and organ specificity. *EMBO J.* **6:** 3565–3569.

Strauss, F. and A. Varshavsky. 1984. A protein binds to a satellite DNA repeat at three specific sites that would be brought into mutual proximity by DNA folding in the nucleosome. *Cell* **37:** 889–901.

Williams, M.E., R. Foster, and N.-H. Chua. 1992. Sequences flanking the hexameric G-box core CACGTG affect the specificity of protein binding. *Plant Cell* **4:** 485–496.

• *TIMETABLE*

Electrophoretic Mobility-shift Assay to Characterize Protein/DNA-binding Sites

STEPS	*TIME REQUIRED*
Preparation of Radiolabeled DNA Probe	
Pour nondenaturing polyacrylamide gel and let gel polymerize	1 hour
Perform restriction digest	1–2 hours
Label DNA	30 minutes–1 hour
Separate labeled DNA fragment by PAGE	2–3 hours
Expose wet gel to X-ray film, develop autoradiograph, and excise labeled band	30 minutes
Elute labeled DNA fragment from gel	overnight
Recover labeled DNA fragment	1 hour
Protein/DNA-binding Reaction	
Pour nondenaturing polyacrylamide gel and let gel polymerize	1 hour
Set up binding reactions and incubate	1 hour
Load and run gel	3–4 hours
Dry gel	1 hour
Expose gel to X-ray film	5–24 hours

Preparation of Radiolabeled DNA Probe

In this procedure, a DNA fragment with a recessed 3′ end is labeled by the fill-in reaction, which is catalyzed by the Klenow fragment of *Escherichia coli* DNA polymerase.

MATERIALS AND EQUIPMENT

Ammonium persulfate (APS) (10%)
N,N,N′,N′,tetramethylethylenediamine (TEMED)
DNA fragment of interest subcloned into a plasmid (Sambrook et al. 1989)
Restriction enzymes (appropriate for the plasmid used)
Restriction digest buffer (as recommended by the supplier of the restriction enzymes used)
[α-^{32}P]dATP (~10 μCi/μl) (Amersham Life Science, Inc.)
dCTP (2 mM)
dGTP (2 mM)
dTTP (2 mM)
DNA polymerase (Klenow) (5 units/μl) (Promega Corporation)
X-ray film (e.g., Kodak-XAR; Eastman Kodak)
Ethanol (100% and 80%)

REAGENTS

Acrylamide (38%)/bisacrylamide (2%) stock solution
TBE (10x)
Loading buffer (10x)
TBE (0.5x) (prepared from 10x stock solution)
Elution buffer
TE
(For recipes, see Preparation of Reagents, pp. 277–278)

SAFETY NOTES

- Acrylamide and bisacrylamide are potent neurotoxins and are absorbed through the skin. Their effects are cumulative. Wear gloves and a mask when weighing acrylamide and bisacrylamide. Polyacrylamide is considered to be nontoxic, but it should be treated with care because it may contain small quantities of unpolymerized material.

- Wear gloves when handling radioactive substances. Consult the local safety office for further guidance in the appropriate use of radioactive materials.

PROCEDURE

Preparation of Polyacrylamide Gel

1. Pour a nondenaturing polyacrylamide gel using the procedure described on pp. 275–276 as a guide. Use 10x TBE to give a final TBE concentration in the gel of 0.5x. The polyacrylamide concentration you require will depend on the size of the DNA fragment you are using. Pour a 4–5% polyacrylamide gel for DNA fragments of approximately 200 bp and a 6–8% polyacrylamide gel for DNA fragments smaller than 100 bp.

Labeling of DNA

1. Digest 10 μg of plasmid DNA carrying the DNA fragment of interest with two restriction enzymes that cut at both ends of the DNA insert. Take care that at least one of these enzymes creates a 5′ overhang. Perform the digest in a total volume of 50 μl for at least 1 hour and under the buffer conditions recommended by the supplier of the enzymes.

2. Add to the digestion mixture:

[α-^{32}P]dATP (~10 μCi/μl) (provided the 5′ overhang carries T-residues)	10 μl
dCTP (2 mM)	1 μl
dGTP (2 mM)	1 μl
dTTP (2 mM)	1 μl
DNA polymerase (Klenow) (5 units/μl)	1 μl

Incubate the reaction for 30 minutes at room temperature.

Note: (i) The DNA fragment can be radiolabeled to a very high specific activity if more than one radiolabeled nucleotide is incorporated. For example, an *Eco*RI site (GAATTC) can be filled in with two A residues. Try to avoid using nucleotides that are incorporated at the most 3′ end; in other words, do not use [α-^{32}P]dTTP to fill in an *Eco*RI site. (ii) You can also incorporate two different radiolabeled nucleotides; for example, a *Hin*dIII site (AAGCTT) can be filled with [α-^{32}P]dATP and [α-^{32}P]dGTP. (iii) The DNA fragment can also be radiolabeled at the 5′ end using T4 polynucleotide kinase and [γ-^{32}P]ATP (Sambrook et al. 1989). However, this method is more laborious than the fill-in reaction described here since it requires removal of the 5′ phosphate residue prior to the labeling reaction. However, if synthetic oligonucleotides are used for DNA-binding studies, the phosphorylation reaction is usually the method of choice.

3. Stop the labeling reaction by adding 7 μl of 10x loading buffer.

Isolation of Radiolabeled Probe

1. Load the reaction mixture onto the nondenaturing gel prepared on p. 269.

2. Run the gel in 0.5x TBE at 10–15 V/cm for 2–3 hours.

3. Remove the glass plates from the electrophoresis tank and pry them apart so that the gel stays attached to one of the plates. Cover the gel with Saran® Wrap.

4. Expose the gel to X-ray film for 1 minute at room temperature. Take care that after developing the autoradiograph the film can be reoriented correctly with the gel.

   Note: We usually position the glass plate supporting the gel in one corner of a light-tight drawer and then place the X-ray film onto the gel.

5. Cover the autoradiograph with Saran® Wrap. Unwrap the gel and place the glass plate supporting the gel onto the autoradiograph. Using the autoradiograph as a template, take a razor blade and excise the gel slice containing the radiolabeled DNA fragment. Cut the slice into small pieces (~1 mm^2) and transfer them to a microfuge tube.

6. Add 1 ml of elution buffer to the gel pieces and incubate them overnight at 37°C with constant rotation or shaking.

7. Transfer the eluate into two clean microfuge tubes (500 μl in each) and add 1 ml of 100% ethanol to each tube. Incubate the tubes for 10 minutes at –80°C and pellet the DNA precipitates by spinning in a microfuge for 15 minutes at room temperature.

8. Remove the supernatants and wash the pellets with 80% ethanol. Spin again for 5 minutes at room temperature. Remove the super-

natants with a drawn-out pasteur pipette and dry the pellets in a SpeedVac® or in a desiccator.

9. Determine the amount of Cerenkov radiation in the samples using a scintillation counter and dissolve the pellets in TE so that the concentration of the radiolabeled probe is approximately 10^4 cpm/μl.

Protein/DNA-binding Reaction

MATERIALS AND EQUIPMENT

Ammonium persulfate (APS) (10%)
N,N,N′,N′,tetramethylethylenediamine (TEMED)
DNA probe (~10^4 cpm/μl) (see pp. 268–271)
Protein extract containing DNA-binding proteins of interest
Poly(dI-dC)·poly(dI-dC) (1 μg/μl)
X-ray film

REAGENTS

Acrylamide (38%)/bisacrylamide (2%) stock solution
TGE (10x)
Binding buffer (10x) (10x BB)
Loading buffer
TGE (1x) (prepared from 10x stock solution)
(For recipes, see Preparation of Reagents, pp. 277–278)

SAFETY NOTES

- Acrylamide and bisacrylamide are potent neurotoxins and are absorbed through the skin. Their effects are cumulative. Wear gloves and a mask when weighing acrylamide and bisacrylamide. Polyacrylamide is considered to be nontoxic, but it should be treated with care because it may contain small quantities of unpolymerized material.
- Wear gloves when handling radioactive substances. Consult the local safety office for further guidance in the appropriate use of radioactive materials.

PROCEDURE

Preparation of Polyacrylamide Gel

1. Pour a 5% nondenaturing polyacrylamide gel using the procedure described on pp. 275–276. Use 10x TGE to give a final TGE concentration in the gel of 1x.

Binding Reaction

1. Set up the binding reactions using the sample experiments in the following table as a guide. Start each reaction by adding the protein extract. Mix gently by pipetting up and down or by tapping the bottom of the tube.

Tube No.	DNA probe (μl)	Poly(dI-dC)·poly(dI-dC) (μl)	10 x BB (μl)	H_2O (μl)	Protein extract (μl)
1	1	2	3	24	0
2	1	2	3	23	1
3	1	2	3	21	3
4	1	2	3	19	5
5	1	2	3	14	10
6	1	0	3	22	4
7	1	0.5	3	21.5	4
8	1	1	3	21	4
9	1	3	3	19	4
10	1	5	3	17	4

Note: (i) The protein extract can be derived from plant tissue, *Escherichia coli* cells expressing a recombinant DNA-binding protein, or any other source. For the preparation of nuclear protein extracts from tomato cotyledons, see Section 12. (ii) The DNA-binding activity may vary dramatically among different protein preparations. Therefore, the binding activity in each preparation should be titrated as described in the table above (tubes 1–5) and in Figure 2.

2. Incubate the binding reactions for 30 minutes at room temperature.

   Note: The optimal incubation time and incubation temperature may vary depending on the nature of the protein.

3. Add 3 μl of loading buffer to each sample and load the samples onto the 5% nondenaturing polyacrylamide gel.

4. Run the gel in 1x TGE at 5–10 V/cm for 3–4 hours, depending on the length of the DNA fragment you are using. We routinely use 7 V/cm.

5. Remove the glass plates from the electrophoresis tank. Remove the spacers, and pry the plates apart so that the gel stays attached to one of the plates. Place a dry sheet of Whatman 3MM paper onto the gel, then carefully peel off the paper so that the gel stays at-

tached to it. Cover the gel with Saran® Wrap and dry it on a gel dryer for 1 hour at 80°C.

6. Expose the dried gel to X-ray film overnight at –80°C.

7. Analyze the data.

   Note: (i) If most or all of the radiolabeled DNA fragments remain at the top of the gel, try increasing the amount of nonspecific competitor DNA in the binding reaction and/or decreasing the polyacrylamide concentration and the cross-linkage (i.e., the ratio of acrylamide:bisacrylamide) of the gel matrix. (ii) If no protein-DNA complex can be visualized, try increasing the amount of protein and/or lowering the amount of nonspecific competitor DNA in the binding reaction. Alternatively, you could choose another type of nonspecific competitor.

Preparation of a Nondenaturing Polyacrylamide Gel

MATERIALS AND EQUIPMENT

Gel casting chamber (Life Technologies, Inc.)
Comb and spacers
Ammonium persulfate (APS) (10%)
N,N,N′,N′,tetramethylethylenediamine (TEMED)

REAGENTS

Acrylamide (38%)/bisacrylamide (2%) stock solution
TGE (10x)
TGE (1x) (prepared from 10x stock solution)
(For recipes, see Preparation of Reagents, pp. 277–278)

SAFETY NOTE

- Acrylamide and bisacrylamide are potent neurotoxins and are absorbed through the skin. Their effects are cumulative. Wear gloves and a mask when weighing acrylamide and bisacrylamide. Polyacrylamide is considered to be nontoxic, but it should be treated with care because it may contain small quantities of unpolymerized material.

PROCEDURE

1. Assemble the gel casting chamber, which consists of two glass plates (19.5 x 19.5 cm and 19.5 x 17 cm) with spacers (1.5 mm thick) inserted on three sides.

2. Prepare a 5% nondenaturing acrylamide gel solution by mixing the following components in the order listed:

Acrylamide (38%)/bisacrylamide (2%) stock solution	6.25 ml
TGE (10x)	5 ml
APS (10%)	500 μl

Adjust the volume to 50 ml with H_2O. Add 50 μl of TEMED. Mix well, pour the solution into the gel casting chamber, and insert the comb.

Note: (i) The final concentration of TGE in this gel is 1x and the gel should be run in 1x TGE. However, electrophoretic mobility-shift assays are often run in other buffer systems, such as 0.25–0.5x TBE or TE, and the buffer composition of the gel should be altered accordingly. (ii) The polyacrylamide concentration (4–8%) and the cross-linkage (the ratio of acrylamide:bisacrylamide) of the gel matrix can influence the migration of the protein-DNA complex. The sharpness of the retarded band depends on the properties of the DNA-binding protein and/or the length of the DNA fragment. We routinely use the system described above for small DNA fragments (<100 bp) and synthetic oligonucleotides. When we are working with longer DNA fragments or high-molecular-weight proteins, we prefer to use 4% gels (i.e., containing 28% acrylamide and 0.2% bisacrylamide) run in 0.25x TBE.

3. After the gel has polymerized (~30 minutes), remove the comb and the bottom spacer, attach the glass plates to the electrophoresis tank, and fill the upper and lower reservoir of the tank with 1x TGE. Flush out the wells and remove air bubbles that may have been trapped underneath the gel.

4. Store the gel at room temperature until you are ready to load your samples. Use the gel the day it is prepared.

• *PREPARATION OF REAGENTS*

Acrylamide (38%)/bisacrylamide (2%) stock solution

38% Acrylamide
2% Bisacrylamide

Filter through Whatman 3MM paper and store light tight at 4°C.

SAFETY NOTE

- Acrylamide and bisacrylamide are potent neurotoxins and are absorbed through the skin. Their effects are cumulative. Wear gloves and a mask when weighing acrylamide and bisacrylamide. Polyacrylamide is considered to be nontoxic, but it should be treated with care because it may contain small quantities of unpolymerized material.

Binding buffer (10x)

100 mM Tris-HCl (pH 7.5)
400 mM NaCl
40% Glycerol
10 mM EDTA
1 mM 2-Mercaptoethanol

Store at 4°C.

SAFETY NOTE

- 2-Mercaptoethanol may be fatal if swallowed and is harmful if inhaled or absorbed through the skin. High concentrations are extremely destructive to the mucous membranes, upper respiratory tract, skin, and eyes. Use only in a chemical fume hood. Gloves and safety glasses should be worn.

Elution buffer

0.5 M Ammonium acetate
1 mM EDTA

Store at room temperature.

Loading buffer (10x)

30% Glycerol
0.25% Bromophenol blue
0.25% Xylene cyanol

Store at room temperature.

TBE (10x)

(1 liter)

Tris base	108 g
Boric acid	55 g
0.5 M EDTA (pH 8.0)	40 ml

Adjust the volume to 1 liter with H_2O. Store at room temperature.

TE

10 mM Tris-HCl (pH 8.0)
1 mM EDTA (pH 8.0)

Store at room temperature.

TGE (10x)

0.25 M Tris base
1.9 M Glycine
0.01 M EDTA (pH 8.3)

Store at room temperature.

• *Section 14*

Characterization of Protein/DNA-binding Sites Using the Methylation Interference Assay

Following the identification of a protein that interacts with a given promoter fragment (e.g., using the electrophoretic mobility-shift assay described in Section 13), it is often desirable to determine the exact sequence of the DNA bound by the protein using a more refined method. Many procedures for characterizing protein/DNA-binding sites have been described in the literature, including methylation and ethylation interference assays (Maxam and Gilbert 1980; Siebenlist and Gilbert 1980; Hendrickson and Schleif 1985; Wissmann and Hillen 1991) and footprinting techniques employing DNase I (Galas and Schmitz 1978), 1,10-phenanthroline-copper ion (Kuwabara and Sigman 1987; Sigman et al. 1991), or hydroxyl radicals (Dixon et al. 1991). The methylation interference assay and the DNase I footprinting technique are the most commonly used methods.

The methylation interference assay essentially represents a modification of the chemical degradation method developed by Maxam and Gilbert for the analysis of DNA sequences (Maxam and Gilbert 1980). The major difference relative to a standard sequencing reaction is that the modified DNA is used in a protein-binding reaction prior to the cleavage of the DNA backbone (Fig. 1).

The identification of protein-binding sites using the methylation interference assay is based on the observation that protein-DNA interactions may be inhibited upon methylation of specific guanosine (G) or adenosine (A) residues (Fig. 1). A DNA fragment, radiolabeled at only one end, is used as a substrate for protein binding. The methylating agent dimethyl sulfate (DMS) promotes the methylation of the *N7* position of G residues in the major groove of DNA and, to some extent, the methylation of the *N3* position of A residues (Maxam and Gilbert 1980; Siebenlist and Gilbert 1980). Under optimal conditions, only one G or A residue should be methylated per DNA strand. Methyl groups impair protein-DNA interactions either by steric hindrance or by disrupting important protein-DNA contacts. After separating and isolating "free" and "protein-bound" DNA fragments,

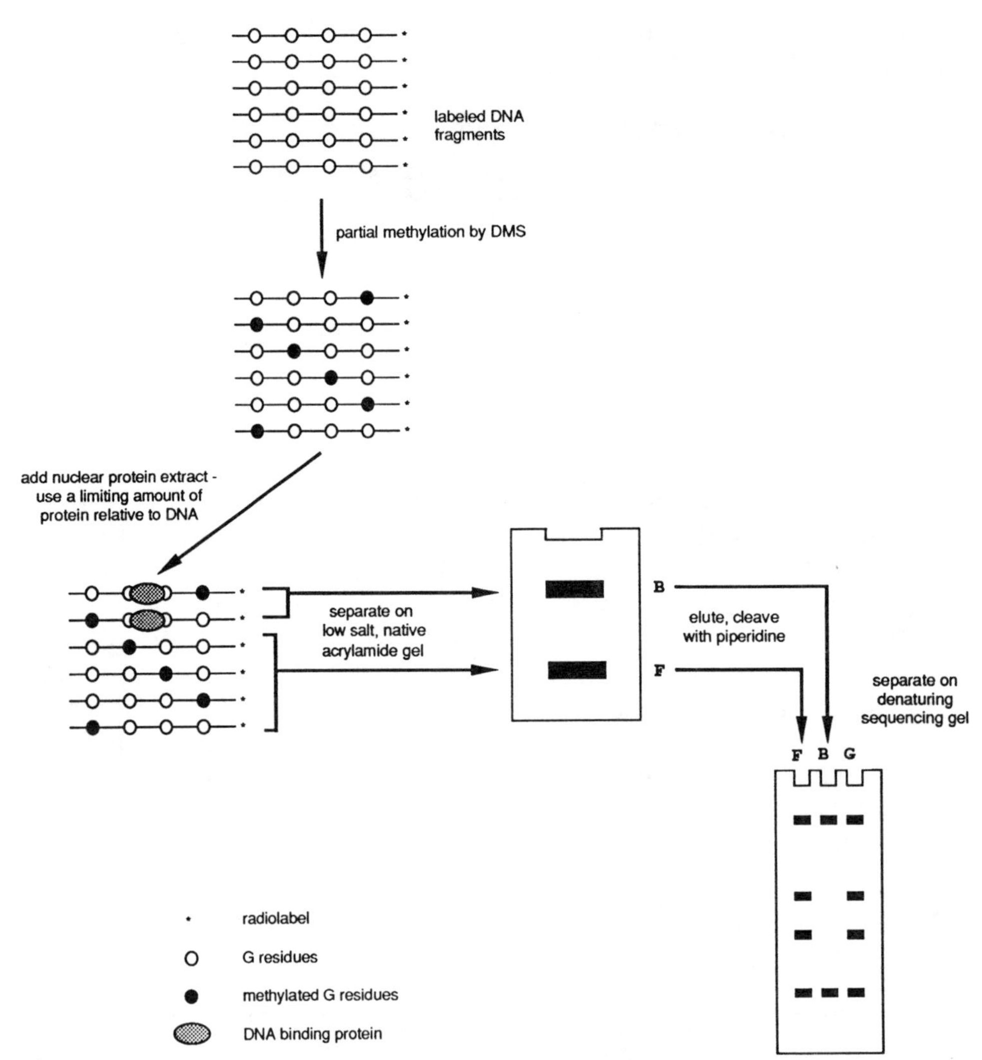

Figure 1

Schematic representation of the methylation interference assay. DNA fragments labeled at one end are partially methylated with dimethyl sulfate (DMS) so that each fragment carries approximately one methylguanosine residue. The DNA is incubated with protein extract. "Protein-bound" (B) and "free" (F) DNA fragments are separated using the electrophoretic mobility-shift assay and the radiolabeled DNA bands are eluted from the gel. The backbone of the DNA is cleaved with piperidine (cleavage occurs only at methylguanosine residues) and the cleavage products are analyzed on a denaturing polyacrylamide gel. Guanosine residues that interfere with protein binding when methylated are missing in the protein-bound fraction. G designates the marker lane on the denaturing gel, which contains a G-specific cleavage reaction (Maxam and Gilbert 1980).

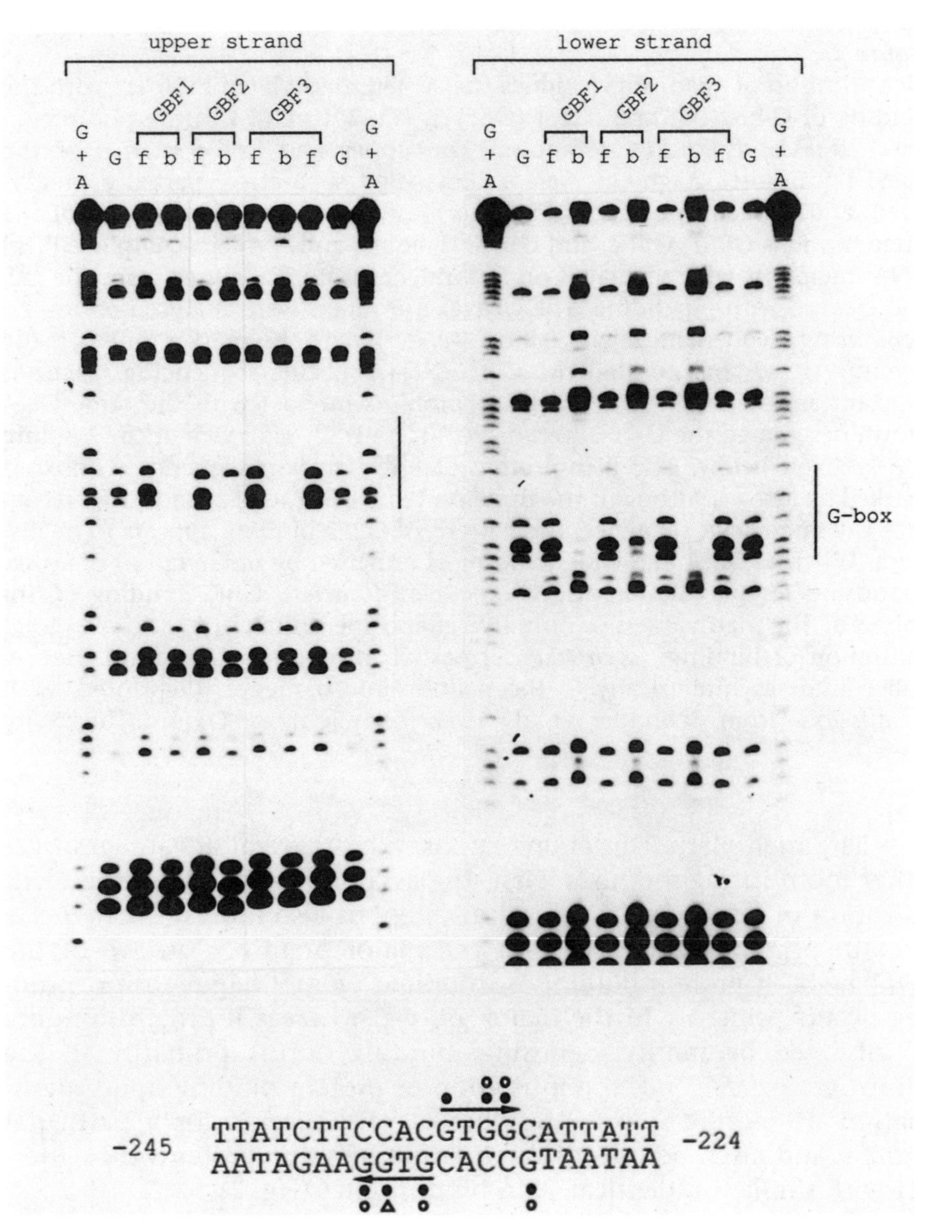

Figure 2 (*See following page for legend.*)

the fragments are cleaved with piperidine and the cleavage products are analyzed by denaturing polyacrylamide gel electrophoresis (PAGE). The protein-bound fraction is depleted of those G residues that interfere with protein binding when methylated (Figs. 1, 2).

Figure 2
Identification of guanosine residues that when methylated interfere with the binding of G-box binding factor (GBF) 1, GBF2, or GBF3 to the photoregulated *Arabidopsis RbcS1A* promoter. The upper and lower strands of the *RbcS1A* promoter fragment were radiolabeled separately, partially methylated, and a fraction of each DNA sample was incubated with each of the three proteins GBF1, GBF2, and GBF3. "Free" (f) and "protein-complexed" (b) DNA fragments were separated on a nondenaturing gel, eluted from the gel, and cleaved with piperidine. The cleavage products were analyzed on an 8% denaturing (sequencing) gel. G and G+A designate the marker lanes on the denaturing gel that contain the G- and G+A-specific sequencing reactions (Maxam and Gilbert 1980). All three proteins interact with the same DNA motif, designated the G-box (*vertical bar*). The DNA sequence of the binding site is given below. The palindromic core-recognition sequence (G-box) is marked by arrows. Although the three proteins GBF1, GBF2, and GBF3 recognize the same DNA sequence, the pattern of GBF2 binding appears to be distinct. Whereas GBF1 and GBF3 binding is impaired by methylation of all six guanosine residues within the G-box (*closed circles*), GBF2 binding is impaired by the methylation of only five guanosine residues ([*open circles*] total inhibition of binding; [*open triangle*] partial inhibition of binding). Hence, GBF2 binds asymmetrically to the palindromic sequence. (Reprinted, with permission, from Schindler et al. 1992, copyright by Oxford University Press.)

The methylation interference assay has several advantages over other footprinting methods. First, the assay not only allows the identification of a specific DNA-binding site, but also indicates whether a protein primarily interacts with the major or minor groove of the DNA helix. If protein binding is inhibited by methylguanosine, binding occurs primarily in the major groove, whereas if protein binding is inhibited by methyladenosine, binding occurs primarily in the minor groove. Second, the inhibition of protein binding upon methylation of specific G or A residues is very specific for a particular protein and often permits discrimination between proteins that interact with similar or identical DNA-binding sites (Fig. 2).

REFERENCES

Dixon, W.J., J.J. Hayes, J.R. Levin, M.F. Weidner, B.A. Dombroski, and T.D. Tullius. 1991. Hydroxyl radical footprinting. *Methods Enzymol.* **208:** 380–413.

Galas, D.J. and A. Schmitz. 1978. DNase footprinting: A simple method for the detection of protein-DNA binding specificity. *Nucleic Acids Res.* **5:** 3157–3170.

Hendrickson, W. and R. Schleif. 1985. A dimer of AraC protein contacts three adjacent major groove regions of the *araI* DNA site. *Proc. Natl. Acad. Sci.* **82:** 3129–3133.

Kuwabara, M. and D. Sigman. 1987. Footprinting DNA-protein complexes in situ following gel retardation assays using 1,10-phenanthroline-copper ion *E. coli* RNA polymerase-*lac* promoter complexes. *Biochemistry* **26:** 7234–7238.

Maxam, A. and W. Gilbert. 1980. Sequencing end-labeled DNA with base-specific chemical cleavages. *Methods Enzymol.* **65:** 499–560.

Sambrook, J., E.F. Fritsch, and T. Maniatis. 1989. *Molecular cloning: A laboratory manual,* 2nd edition. Cold Spring Harbor Laboratory Press, Cold Spring Harbor, New York.

Schindler, U., A.E. Menkens, H. Beckmann, J.R. Ecker, and A.R. Cashmore. 1992. Heterodimerization between light-regulated and ubiquitously expressed *Arabidopsis* GBF bZIP proteins. *EMBO J.* **11:** 1261–1273.

Siebenlist, U. and W. Gilbert. 1980. Contacts between *E. coli* RNA polymerase and an early promoter of phage T7. *Proc. Natl. Acad. Sci.* **77:** 122–126.

Sigman, D.S., M.D. Kuwabara, C.-H. Chen, and T.W. Brucie. 1991. Nuclease activity of 1,10-phenanthroline-copper in study of protein-DNA interactions. *Methods Enzymol.* **208:** 414–432.

Wissmann, A. and W. Hillen. 1991. DNA contacts probed by modification protection and interference studies. *Methods Enzymol.* **208:** 365–379.

• TIMETABLE
Characterization of Protein/DNA-binding Sites Using the Methylation Interference Assay

STEPS	TIME REQUIRED
Preparation of Radiolabeled DNA Probe	
See Section 13, pp. 268–271	~24 hours
Methylation Reaction	
Methylate labeled DNA	1 hour
Protein/DNA-binding Reaction	
Pour nondenaturing polyacrylamide gel and let gel polymerize	1 hour
Set up binding reactions and incubate	1 hour
Load and run gel	3–4 hours
Expose wet gel to X-ray film	1–14 hours
Elution of Radiolabeled DNA Bands	
Excise labeled bands from gel	30 minutes
Elute "free" and "protein-bound" DNA from gel	2–4 hours
Cleavage of DNA Fragments and Analysis by Denaturing PAGE	
Pour denaturing polyacrylamide (sequencing) gel and let gel polymerize	1–2 hours
Cleave DNA with piperidine	30 minutes

STEPS	***TIME REQUIRED***
Cleavage of DNA Fragments and Analysis by Denaturing PAGE (continued)	
Lyophilize DNA	2–5 hours
Prepare samples and load gel	1 hour
Run gel	2–3 hours
Dry gel	1 hour
Expose gel to X-ray film	5–24 hours

Preparation of Radiolabeled DNA Probe

A DNA fragment can be radiolabeled at either end using different protocols. A recessed 3′ end can be labeled by the fill-in reaction, which is catalyzed by the Klenow fragment of *Escherichia coli* DNA polymerase I. For a description of this procedure, including a list of materials, equipment, and reagents, see Section 13, pp. 268–271. The 5′ end can be labeled by phosphorylation with bacteriophage T4 polynucleotide kinase in the presence of [γ-^{32}P]ATP (see Sambrook et al. 1989). 3′ labeling is the method of choice if the DNA fragment is flanked by sites for restriction enzymes that create an overhang at the 5′ end and an overhang or a blunt end at the 3′ end. Since 3′ overhangs and blunt ends are not substrates for the Klenow enzyme, the fragment can be released by restriction digest and immediately radiolabeled; hence, a subsequent restriction digest is unnecessary. In contrast, if the DNA fragment is flanked only by sites for restriction enzymes that create 5′ overhangs or if the kinase reaction is used, the plasmid must first be linearized using one restriction enzyme, then labeled, and the DNA fragment is subsequently released using the second restriction enzyme (Sambrook et al. 1989).

Methylation Reaction

MATERIALS

Carrier DNA (1 μg/μl) (any plasmid DNA will suffice)
DNA probe (~10^4 cpm/μl; end-labeled at one end only) (see p. 286)
Dimethyl sulfate (DMS)
Formic acid (10%)
Sodium acetate (0.3 M; pH 7.0)
Ethanol (100%) (ice-cold)

REAGENTS

DMS buffer
DMS stop mix
(For recipes, see Preparation of Reagents, pp. 297–299)

SAFETY NOTES

- Wear gloves when handling radioactive substances. Consult the local safety office for further guidance in the appropriate use of radioactive materials.
- DMS is toxic and a carcinogen and should be handled in a chemical fume hood. Be aware that DMS can penetrate rubber and vinyl gloves. Tips and tubes that are contaminated with DMS should be rinsed with 5 M NaOH and discarded in separate waste containers. Solutions containing DMS should be discarded in a waste bottle containing 5 M NaOH.

PROCEDURE

1. Set up the following reactions in microfuge tubes at room temperature (you need six tubes per DNA fragment—three for the "start mix" and three for the "stop mix"):

	Tube 1 G-Reaction	Tube 2 G-Reaction for Binding	Tube 3 A+G-Reaction
Start mix			
DMS buffer	200 μl	200 μl	–
H_2O	–	–	18 μl
Carrier DNA	1 μl	1 μl	1 μl
Probe DNA	1 μl	10 μl	2 μl

	Tube 4 G-Stop	Tube 5 G-Stop	Tube 6 A+G-Stop
Stop mix			
DMS stop mix	50 μl	50 μl	–
Sodium acetate (0.3 M; pH 7.0)	–	–	200 μl
Ethanol (100%; ice-cold)	1 ml	1 ml	1 ml

Note: The A+G-specific reaction is optional, but it usually helps to localize the exact DNA-binding site (Fig. 2).

2. Initiate the reactions by adding 1 μl of DMS to tubes 1 and 2 and 3 μl of 10% formic acid to tube 3.

   Note: (i) The methylation reaction is the most crucial step during this procedure—work fast but do not rush. (ii) The amount of DMS required and the incubation time and temperature are critical and should be determined empirically for each DNA fragment. (iii) The carrier DNA should not compete for the DNA-binding protein.

3. Incubate tubes 1 and 2 for 4 minutes at 16°C and incubate tube 3 for 15 minutes at 37°C. Stop the reactions by adding the appropriate stop mixes (i.e., add the contents of tubes 4, 5, and 6 to tubes 1, 2, and 3, respectively).

4. Shake the tubes vigorously and place them in a dry ice/ethanol bath for 6 minutes.

5. Spin the tubes in a microfuge for 7 minutes at 4°C.

6. Aspirate the supernatants with a drawn-out pasteur pipette, and wash the pellets with 1 ml of 100% ethanol.

7. Spin the tubes for 3 minutes at room temperature.

8. Aspirate the supernatants.

9. Dry the pellets using a desiccator or SpeedVac®. Use the DNA pellet from tube 2 for the binding reaction (see pp. 289–290). Store tubes 1 and 3 at –20°C until they are required (see p. 294).

Protein/DNA-binding Reaction

MATERIALS AND EQUIPMENT

Ammonium persulfate (APS) (10%)
N,N,N′,N′,tetramethylethylenediamine (TEMED)
Methylated, labeled DNA (from the "G-reaction for binding" [tube 2], see pp. 287–288)
Poly(dI-dC)·poly(dI-dC) (1 μg/μl)
Protein extract containing DNA-binding proteins of interest
X-ray film (Kodak-XAR; Eastman Kodak)

REAGENTS

Acrylamide (38%)/bisacrylamide (2%) stock solution
TGE (10x)
Binding buffer (10x)
Loading buffer
TGE (1x) (prepared from 10x stock solution)
(For recipes, see Preparation of Reagents, pp. 297–299)

SAFETY NOTES

- Acrylamide and bisacrylamide are potent neurotoxins and are absorbed through the skin. Their effects are cumulative. Wear gloves and a mask when weighing acrylamide and bisacrylamide. Polyacrylamide is considered to be nontoxic, but it should be treated with care because it may contain small quantities of unpolymerized material.
- Wear gloves when handling radioactive substances. Consult the local safety office for further guidance in the appropriate use of radioactive materials.

PROCEDURE

Preparation of Polyacrylamide Gel

1. Pour a 5% nondenaturing polyacrylamide gel using the procedure described in Section 13, pp. 275–276. Use 10x TGE to give a final TGE concentration in the gel of 1x.

Binding Reaction

1. Dissolve the dry, methylated, labeled DNA pellet (tube 2, see pp. 287–288) in 23 μl of H_2O.

2. Add:

Poly(dI-dC)·poly(dI-dC) (1 μg/μl)	2 μl
Protein extract	10–20 μl
Binding buffer (10x)	5 μl

   Add H_2O to give a final volume of 50 μl.

   Note: (i) This DNA-binding reaction represents a scale-up of the standard binding reaction outlined in Section 13, pp. 272–274, except that methylated DNA is used. (ii) The protein concentration should be limiting and therefore the binding activity should be titrated before performing the assay. The amount of protein extract required depends on the relative proportion of protein-DNA complexes obtained. Usually a 5–10-fold scale-up of the protein extract is sufficient. (iii) The nonspecific competitor DNA should not be scaled-up to the same extent as the protein, since some carrier DNA (see p. 287) is already present in the reaction.

3. Incubate the binding reaction for 30 minutes at room temperature.

4. Add 3 μl of loading buffer to the sample and load it onto the 5% nondenaturing polyacrylamide gel.

5. Separate the "free" and "protein-bound" DNA fragments by running the gel in 1x TGE at 10 V/cm for 2–3 hours.

6. Remove the glass plates from the electrophoresis tank and pry them apart so that the gel stays attached to one of the plates. Cover the gel with Saran® Wrap.

7. Expose the gel to X-ray film for 1–2 hours at room temperature. Take care that after developing the autoradiograph the film can be reoriented correctly with the gel.

   Note: We usually position the glass plate supporting the gel in one corner of a light-tight drawer and then place the X-ray film onto the gel.

Elution of Radiolabeled DNA Bands

MATERIALS AND EQUIPMENT

Agarose
NA45 membrane (0.5 x 0.5 cm) (Schleicher & Schuell, Inc.)
Phenol/chloroform (1:1; equilibrated with 1 M Tris-HCl [pH 8.0])
Salmon sperm DNA (5 μg/μl; sonicated)
Ethanol (100% and 80%)

REAGENTS

TBE (1x) (prepared from 10x stock solution)
Elution buffer
(For recipes, see Preparation of Reagents, pp. 297–299)

SAFETY NOTES

- Wear gloves when handling radioactive substances. Consult the local safety office for further guidance in the appropriate use of radioactive materials.
- Phenol is highly corrosive and can cause severe burns. Wear gloves, protective clothing, and safety glasses when handling it. All manipulations should be carried out in a chemical fume hood. Any areas of skin that come in contact with phenol should be rinsed with a large volume of water or PEG 400 and washed with soap and water; do not use ethanol!
- Chloroform is irritating to the skin, eyes, mucous membranes, and respiratory tract. It should only be used in a chemical fume hood. Gloves and safety glasses should also be worn. Chloroform is a carcinogen and may damage the liver and kidneys.

PROCEDURE

1. Cover the autoradiograph with Saran® Wrap and place it onto a disposable white surface. Unwrap the gel, and place the glass plate supporting the gel onto the autoradiograph. Align the corners of the glass plate and the autoradiograph. Using the autoradiograph as a template, take a razor blade and excise the poly-

acrylamide gel slices containing the radiolabeled DNA fragments (both free and protein-bound fractions).

2. Prepare a 1.5% agarose gel solution by dissolving 1.5 g of agarose in 100 ml of 1x TBE and boiling in a microwave oven. Cool the solution to 60°C and pour into a gel casting tray. A minigel (Bio-Rad Laboratories) of 6.5 x 10 x 0.5 cm is usually sufficient for this purpose. The gel will set in approximately 30 minutes.

3. Using a scalpel, cut small holes (0.5 x 1 cm) in the agarose gel and fill them with 1x TBE.

4. Place a piece of NA45 membrane (0.5 x 0.5 cm) in each hole and ensure that the membrane extends toward the anode (see Fig. 3).

5. Using forceps, place one polyacrylamide gel slice in each hole; fold the gel slices over if they do not fit. Apply a voltage gradient so that the DNA migrates toward the NA45 membrane (i.e., ~10 V/cm for 30–40 minutes). Electrophoresis is carried out in 1x TBE buffer. During the elution, you can follow the migration of the DNA by turning off the power, removing a gel slice with forceps, and monitoring it with a Geiger counter.

6. Using forceps, transfer the membrane pieces into microfuge tubes and add 100–200 μl of elution buffer to each tube. Make sure the membrane pieces are completely submerged.

7. Incubate the tubes at 68°C for 30–60 minutes.

8. Transfer the eluates to fresh microfuge tubes and add an equal volume of phenol/chloroform (1:1) to each tube. Vortex vigorously.

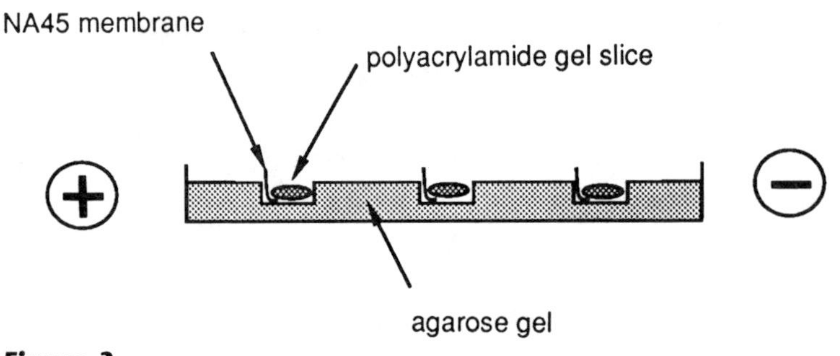

Figure 3
Schematic representation of the agarose gel used for the elution of the DNA fragment.

9. Separate the phases by spinning the tubes in a microfuge for 5 minutes at room temperature. Transfer the upper (aqueous) phases into fresh microfuge tubes.

10. Add 5 μg of sonicated salmon sperm DNA to each tube. Precipitate the DNA by adding two volumes of 100% ethanol to each tube. Incubate the tubes for 10 minutes at –80°C.

11. Pellet the precipitated DNA by spinning the tubes in a microfuge for 15 minutes at 4°C.

12. Using a drawn-out pasteur pipette, remove the supernatants. Monitor the labeled DNA in the pellets with a Geiger counter.

13. Wash each pellet in 500 μl of 80% ethanol. Vortex, and spin again for 5 minutes at 4°C.

14. Discard the supernatants and dry the pellets in a desiccator or SpeedVac®.

Cleavage of DNA Fragments and Analysis by Denaturing PAGE

MATERIALS

Dimethyldichlorosilane
Ethanol (100%)
Ammonium persulfate (APS) (10%)
N,N,N',N',tetramethylethylenediamine (TEMED)
Methylated, labeled DNA:
  Sample 1—free DNA eluted from the nondenaturing gel (pp. 291–293)
  Sample 2—protein-bound DNA eluted from the nondenaturing gel (pp. 291–293)
  Sample 3—DNA from the "G-reaction" (tube 1, pp. 287–288)
  Sample 4—DNA from the "A+G-reaction" (tube 3, pp. 287–288)
Piperidine (10% in H_2O) (freshly prepared)

REAGENTS

Denaturing gel stock solution
TBE (1x) (prepared from 10x stock solution)
Sequencing dye
(For recipes, see Preparation of Reagents, pp. 297–299)

SAFETY NOTES

- The denaturing gel stock solution contains acrylamide and bisacrylamide, which are potent neurotoxins that are absorbed through the skin. Their effects are cumulative. Wear gloves and a mask when weighing acrylamide and bisacrylamide. Polyacrylamide is considered to be nontoxic, but it should be treated with care because it may contain small quantities of unpolymerized material.
- Wear gloves when handling radioactive substances. Consult the local safety office for further guidance in the appropriate use of radioactive materials.
- Piperidine is corrosive to the eyes, skin, respiratory tract, and gas-

trointestinal tract. It reacts violently with acids and oxidizing agents. Keep piperidine away from heat or flames. Wear gloves and safety glasses when handling piperidine. Avoid breathing the vapors.

PROCEDURE

Preparation of Denaturing Polyacrylamide (Sequencing) Gel

1. Wash and siliconize two glass plates (33 x 42 cm and 33 x 39 cm) and then treat them with 100% ethanol. Assemble the gel casting chamber, which consists of the two glass plates with spacers (0.2 mm thick) inserted at three sides.

2. Add 200 µl of 10% APS and 200 µl of TEMED to 100 ml of denaturing gel stock solution. Mix well and pour immediately into the gel casting chamber. Insert the comb and let the gel polymerize for 1–2 hours.

3. Remove the comb from the gel and attach the glass plates to an electrophoresis tank. Fill the upper and lower reservoirs with 1x TBE. Using a pasteur pipette, flush the wells with 1x TBE. Pre-run the gel at 40–50 W for 1 hour. It is important that the gel is warmed before the samples are loaded.

Cleavage Reaction

1. Add 100 µl of 10% piperidine to each dry DNA pellet (i.e., samples 1–4).

2. Incubate the tubes for 30 minutes at 90°C.

3. Using a needle, puncture a hole in the cap of each microfuge tube. Lyophilize the cleaved DNA samples in a SpeedVac® for 2–5 hours, or until they are completely dry.

   Note: Lyophilization time will depend on the SpeedVac® and the number of samples.

4. Using a scintillation counter, determine the amount of Cerenkov radiation present in each DNA sample.

5. Dissolve each DNA sample in sequencing dye to give a final concentration of 5000 cpm/µl in samples 1–3 and 10^4 cpm/µl in sample 4.

6. Denature the DNA by heating the tubes for 10 minutes at 100°C and then chill them on ice.

7. Apply 5 μl of each sample onto the denaturing polyacrylamide gel that has been pre-run at 40–50 W for 1 hour.

   Note: It is important to load an equal number of counts from samples 1, 2, and 3 (i.e., 2.5 x 10^4 cpm per lane). You should load twice as many counts from sample 4 because more signals are usually obtained from this sample (see Fig. 2).

8. Run the gel at 40–50 W for 2–3 hours, depending on the length of the DNA fragment.

9. Disconnect the glass plates from the electrophoresis tank and remove the spacers. Pry the glass plates apart so that the gel stays attached to one of the plates. Carefully place a sheet of Whatman 3MM paper onto the gel. Slowly peel off the paper so that the gel stays attached to it. Cover the gel with Saran® Wrap and dry it for 1 hour at 80°C. Visualize the DNA bands by exposing the dried gel to X-ray film overnight at –80°C.

• *PREPARATION OF REAGENTS*

Acrylamide (38%)/bisacrylamide (2%) stock solution

38% Acrylamide
2% Bisacrylamide

Filter through Whatman 3MM paper and store light tight at 4°C.

SAFETY NOTE

- Acrylamide and bisacrylamide are potent neurotoxins and are absorbed through the skin. Their effects are cumulative. Wear gloves and a mask when weighing acrylamide and bisacrylamide. Polyacrylamide is considered to be nontoxic, but it should be treated with care because it may contain small quantities of unpolymerized material.

Binding buffer (10x)

100 mM Tris-HCl (pH 7.5)
400 mM NaCl
40% Glycerol
10 mM EDTA
1 mM 2-Mercaptoethanol

Store at 4°C.

SAFETY NOTE

- 2-Mercaptoethanol may be fatal if swallowed and is harmful if inhaled or absorbed through the skin. High concentrations are extremely destructive to the mucous membranes, upper respiratory tract, skin, and eyes. Use only in a chemical fume hood. Gloves and safety glasses should be worn.

Denaturing gel stock solution
(500 ml)

7 M Urea
8% Acrylamide (38%)/bisacrylamide (2%)
1x TBE

Adjust the volume to 500 ml with H_2O. Filter the solution through Whatman No. 1 paper. This stock solution can be stored for several weeks in a light-tight bottle at 4°C.

DMS buffer

50 mM Sodium cacodylate (pH 8.0)
1 mM EDTA

Store at 4°C.

DMS stop mix

1.5 M Sodium acetate (pH 7.0)
1 M 2-Mercaptoethanol

Store at 4°C.

Elution buffer

1.5 M NaCl
1 mM EDTA

Store at room temperature.

Loading buffer (10x)

30% Glycerol
0.25% Bromophenol blue
0.25% Xylene cyanol

Store at room temperature.

Sequencing dye

90% Formamide
1 mM EDTA (pH 8.0)
0.02% Bromophenol blue
0.02% Xylene cyanol FF

Store at –20°C.

TBE (10x)
(1 liter)

Tris base	108 g
Boric acid	55 g
0.5 M EDTA (pH 8.0)	40 ml

Adjust the volume to 1 liter with H_2O. Store at room temperature.

TGE (10x)

0.25 M Tris base
1.9 M Glycine
0.01 M EDTA (pH 8.3)

Store at room temperature.

• Section 15
Isolation of cDNAs Encoding Sequence-specific DNA-binding Proteins Using In Situ Screening

Sequence-specific DNA-binding proteins play a critical role in the regulation of transcription. The isolation of recombinant cDNAs encoding these proteins greatly facilitates the biochemical analysis of their structural and functional properties. Genes encoding such DNA-binding proteins have been isolated using classical genetics (Ludwig et al. 1989; Schmidt et al. 1990; Sommer et al. 1990; Yanofsky et al. 1990; Vollbrecht et al. 1991) and molecular biochemical approaches, including the screening of recombinant cDNA libraries with antibodies (Walter et al. 1985; Weinberger et al. 1985; Landschulz et al. 1988) or DNA probes (Walter et al. 1985; Kadonaga et al. 1987; Bodner et al. 1988). These molecular procedures usually require the biochemical isolation of the protein, which is dependent upon the availability of large amounts of starting material. Although the purification of these proteins has been greatly facilitated by the development of DNA-affinity matrices (Chodosh et al. 1986; Kadonaga and Tjian 1986; Rosenfeld and Kelly 1986), the procedure is still very laborious and/or expensive.

To date, with the exception of proteins belonging to the high-mobility class (Grasser and Feix 1991), no sequence-specific DNA-binding protein from any plant species has been purified to homogeneity for the purpose of isolating the corresponding gene. In contrast, the in situ screening procedure has been used successfully in many laboratories and has greatly facilitated the isolation of many sequence-specific DNA-binding proteins from various plant species (Katagiri et al. 1989; Tabata et al. 1989, 1991; Dehesh et al. 1990; Guiltinan et al. 1990; Lam et al. 1990; Scharf et al. 1990; Singh et al. 1990; Aeschbacher et al. 1991; Oeda et al. 1991; Weisshaar et al. 1991; Gilmartin et al. 1992; Perisic and Lam 1992; Schindler et al. 1992).

The in situ screening protocol, first developed in the laboratories of McKnight (Vinson et al. 1988) and Sharp (Singh et al. 1988), does not require the purification of the protein of interest, but depends on the availability and quality of an appropriate recombinant cDNA library constructed for expression in *Escherichia coli*, which is screened with a radiolabeled DNA probe. The procedure is schematically outlined in Figure 1. The cDNA is cloned into a bacteriophage λ expres-

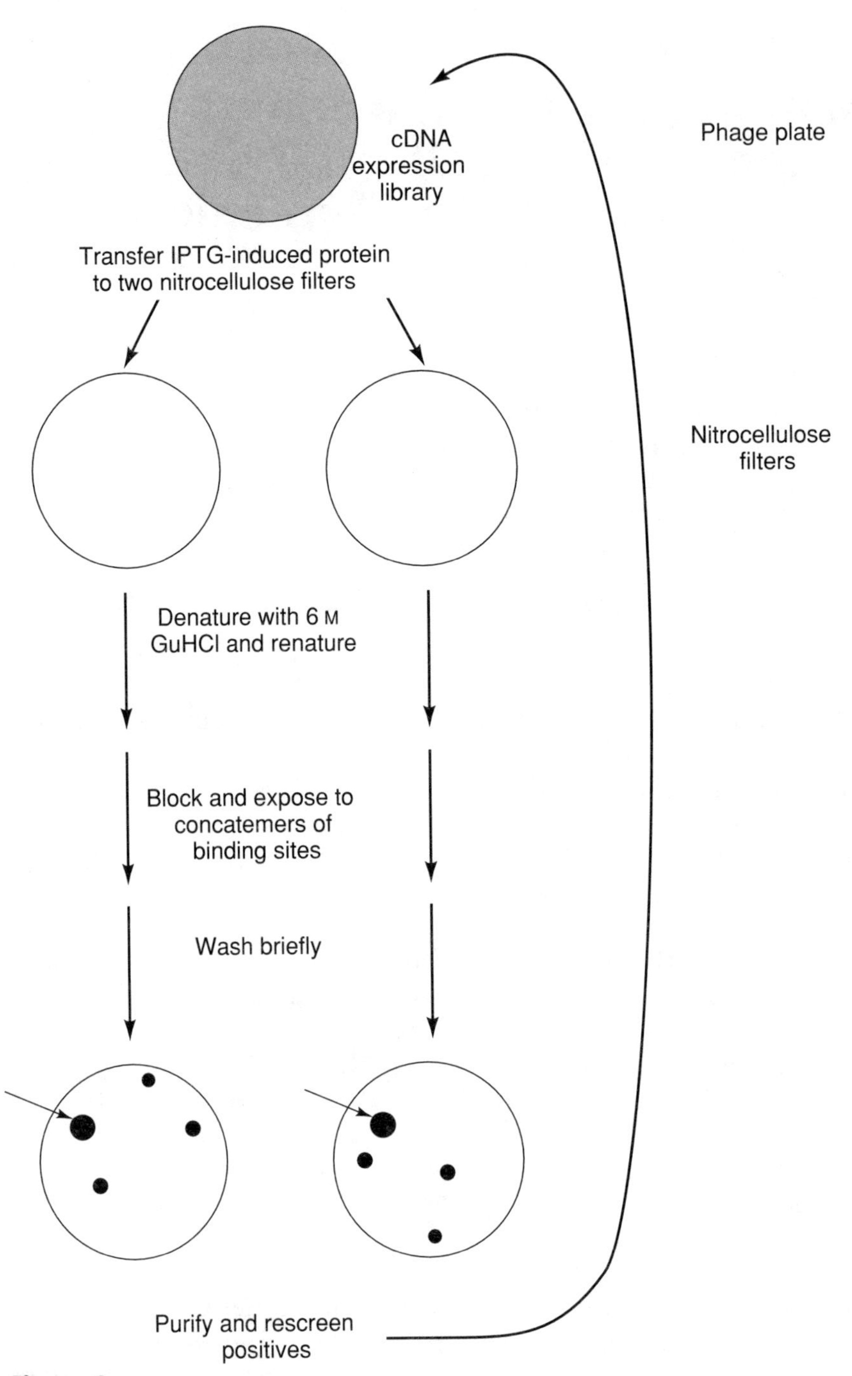

Figure 1
Schematic representation of the in situ screening procedure. The exact details of the procedure are described in the text.

sion vector (e.g., λgt11) downstream of the β-galactosidase promoter, which can be induced by isopropylthio-β-D-galactoside (IPTG). *E. coli* cells infected with the bacteriophage produce high levels of recombinant protein (as a β-galactosidase fusion protein) after being exposed to an IPTG-soaked nitrocellulose filter. Upon lysis of the bacterial cells, the recombinant protein adheres to the filter and can be subjected to various treatments. Denaturation with guanidine HCl and subsequent renaturation can greatly enhance the signal obtained from a phage expressing a specific protein (Vinson et al. 1988), although successful in situ screening experiments can be performed even if this treatment is omitted (Singh et al. 1988). The enhanced binding that occurs after a denaturation/renaturation cycle is most likely due to the fact that proteins present as insoluble precipitates regain their DNA-binding specificity and become more accessible to the radiolabeled DNA-binding site. Nonspecific association of the radiolabeled DNA probe is significantly reduced by blocking the filters using a 5% nonfat dry milk solution. Subsequently, the filters are exposed to the probe. Although probes with single DNA-binding sites can be used, it has been demonstrated that the signal can be greatly enhanced by using probes that carry multiple copies of a particular binding site (concatemers) (Singh et al. 1988; Vinson et al. 1988). One possible explanation for this effect is that a multisite probe can be bound by several proteins, which may increase the stability of the protein-DNA complex. The background signal, and therefore the chance of isolating clones encoding nonspecific DNA-binding proteins, can be reduced by including nonspecific competitor DNA such as poly(dI-dC)·poly(dI-dC) or sheared salmon sperm DNA (see Section 13) in the binding reaction (Miyamoto et al. 1988; Singh et al. 1988, 1989; Staudt et al. 1988; Vinson et al. 1988).

Although genes encoding very rare proteins can be isolated using this procedure, it should be kept in mind that this method requires that the protein has certain biochemical properties; for example, the protein must be functionally active when it is expressed in *E. coli*. Hence, heterodimeric proteins that require the presence of two or more different polypeptides for DNA binding or proteins that only interact with DNA after posttranslational modifications, such as phosphorylation (for reviews, see Berk 1989; Hunter and Karin 1992), cannot be identified in this screen. Furthermore, the protein must survive the harsh treatment during the denaturation/renaturation cycle and must renature appropriately in order to interact with its cognate binding site. An additional requirement is that the interaction between the protein and its DNA-binding sites must be sufficiently stable to be identified in this assay. Therefore, prior to screening for a particular protein, various parameters for binding, such as pH and salt concentrations, should be tested using crude nuclear extracts and the electrophoretic mobility-shift assay.

Despite all these potential difficulties, the in situ screening method has become one of the most powerful tools for isolating genes encoding sequence-specific DNA-binding proteins.

REFERENCES

Aeschbacher, R.A., M. Schrott, I. Protrykus, and M.W. Saul. 1991. Isolation and molecular characterization of *PosF21*, an *Arabidopsis thaliana* gene which shows characteristics of a b-Zip class transcription factor. *Plant J.* **1:** 303–316.

Berk, A.J. 1989. Regulation of eukaryotic transcription factors by post-translational modification. *Biochim. Biophys. Acta* **1009:** 103–109.

Bodner, M., J.L. Castrillo, L.E. Theill, T. Deerinck, M. Ellisman, and M. Karin. 1988. The pituitary-specific transcription factor GHF-1 is a homeobox containing protein. *Cell* **55:** 505–518.

Chodosh, L.A., R.W. Carthew, and P.A. Sharp. 1986. A single polypeptide possesses the binding and transcription activities of the adenovirus major late transcription factor. *Mol. Cell. Biol.* **6:** 4723–4733.

Dehesh, K., W.B. Bruce, and P.H. Quail. 1990. A *trans*-acting factor that binds to a GT-motif in a phytochrome gene promoter. *Science* **250:** 1397–1399.

Gilmartin, P.M., J. Memelink, K. Hiratsuka, S.A. Kay, and N.-H. Chua. 1992. Characterization of a gene encoding a DNA-binding protein with specificity for a light-responsive element. *Plant Cell* **4:** 839–849.

Grasser, K.D. and G. Feix. 1991. Isolation and characterization of maize cDNAs encoding a high mobility group protein displaying a HMG-box. *Nucleic Acids Res.* **19:** 2573–2577.

Guiltinan, M.J., W.R. Marcotte, and R.S. Quatrano. 1990. A plant leucine zipper protein that recognizes an abscisic acid response element. *Science* **250:** 267–271.

Hunter, T. and M. Karin. 1992. The regulation of transcription by phosphorylation. *Cell* **70:** 375–387.

Kadonaga, J.T. and R. Tjian. 1986. Affinity purification of sequence-specific DNA binding proteins. *Proc. Natl. Acad. Sci.* **83:** 5889–5893.

Kadonaga, J.T., K.R. Carner, F.R. Masiarz, and R. Tjian. 1987. Isolation of cDNA encoding transcription factor Sp1 and functional analysis of the DNA binding domain. *Cell* **55:** 1079–1090.

Katagiri, F., E. Lam, and N.-H. Chua. 1989. Two tobacco DNA-binding proteins with homology to the nuclear factor CREB. *Nature* **340:** 727–730.

Lam, E., Y. Kano-Murakami, P. Gilmartin, B. Niner, and N.-H. Chua. 1990. A metal-dependent DNA-binding protein interacts with a constitutive element of a light-responsive promoter. *Plant Cell* **2:** 857–866.

Landschulz, W.H., P.F. Johnson, E.Y. Adashi, B.J. Graves, and S.L. McKnight. 1988. Isolation of a recombinant copy of the gene encoding C/EBP. *Genes Dev.* **2:** 786–800.

Ludwig, S.R., L.F. Habera, S.L. Dellaporta, and S. Wessler. 1989. *Lc*, a member of the maize *R* gene family responsible for tissue-specific anthocyanin production, encodes a protein similar to transcriptional activators and contains the *myc*-homology region. *Proc. Natl. Acad. Sci.* **86:** 7092–7096.

Miyamoto, M., T. Fujita, Y. Kimura, M. Maruyama, H. Harada, Y. Sudo, T. Miyata, and T. Taniguchi. 1988. Regulated expression of a gene encoding a nuclear factor, IRF-1, that specifically binds to IFN-β gene regulatory elements. *Cell* **54:** 544–551.

Oeda, K., J. Salinas, and N.-H. Chua. 1991. A tobacco bZip transcription activator (TAF-1) binds to a G-box-like motif conserved in plant genes. *EMBO J.* **10:** 1793–1802.

Perisic, O. and E. Lam. 1992. A tobacco DNA binding protein that interacts with a light-responsive box II element. *Plant Cell* **4:** 831–838.

Rosenfeld, P.J. and T.J. Kelly. 1986. Purification of nuclear factor I by DNA recognition site affinity chromatography. *J. Biol. Chem.* **261:** 1398–1408.

Sambrook, J., E.F. Fritsch, and T. Maniatis. 1989. *Molecular cloning: A laboratory manual*, 2nd edition. Cold Spring Harbor Laboratory Press, Cold Spring Harbor, New York.

Scharf, K.D., S. Rose, W. Zott, R. Schöff, and L. Nover. 1990. Three tomato genes code for heat stress transcription factors with a region of remarkable homology to the DNA binding domain of the yeast HSF. *EMBO J.* **9:** 4495–4501.

Schindler, U., A.E. Menkens, H. Beckmann, J.R. Ecker, and A.R. Cashmore. 1992. Heterodimerization between light-regulated and ubiquitously expressed *Arabidopsis* GBF bZIP proteins. *EMBO J.* **11:** 1261–1273.

Schmidt, R.J., F.A. Burr, M.J. Aukerman, and B. Burr. 1990. Maize regulatory gene *opaque-2* encodes a protein with a "leucine-zipper" motif that binds to zein DNA. *Proc. Natl. Acad. Sci.* **87:** 46–50.

Singh, H., J.H. LeBowitz, A.S. Baldwin, and P.A. Sharp. 1988. Molecular cloning of an enhancer binding protein: Isolation by screening of an expression library with a recognition site DNA. *Cell* **52:** 415–423.

Singh, H., R.G. Clerc, and J.H. LeBowitz. 1989. Molecular cloning of sequence-specific DNA binding proteins using recognition site probes. *BioTechniques* **7:** 252–261.

Singh, K., E.S. Dennis, J.G. Ellis, D.J. Llewellyn, J.G. Tokuhisa, J.A. Wahleithner, and W.J. Peacock. 1990. OCSBF-1, a maize ocs enhancer binding factor: Isolation and expression during development. *Plant Cell* **2:** 891–903.

Sommer, H., J.P. Beltrán, P. Huijser, H. Pape, W.E. Lönnig, H. Saedler, and S.Z. Schwarz. 1990. *Deficiens*, a homeotic gene involved in the control of flower morphogenesis in *Antirrhinum majus*: The protein shows homology to transcription factors. *EMBO J.* **9:** 605–613.

Staudt, L.M., R.G. Clerc, H. Singh, J.H. LeBowitz, P.A. Sharp, and D. Baltimore. 1988. Cloning of a lymphoid specific cDNA encoding a protein binding the regulatory octamer DNA motif. *Science* **241:** 577–580.

Tabata, T., H. Takase, S. Takayama, K. Mikami, A. Nakatsuka, T. Kawata, T. Nakayama, and M. Iwabuchi. 1989. A protein that binds to a *cis*-acting element of wheat histone genes has a leucine zipper motif. *Science* **245:** 965–967.

Tabata, T., T. Nakayama, K. Mikami, and M. Iwabuchi. 1991. HBP-1a and HBP-1b: Leucine zipper-type transcription factors of wheat. *EMBO J.* **10:** 1459–1467.

Vinson, C.R., K.L. LaMarca, P.F. Johnson, W.H. Landschulz, and S.L. McKnight. 1988. In situ detection of sequence-specific DNA binding activity specified by a recombinant bacteriophage. *Genes Dev.* **2:** 801–806.

Vollbrecht, E., B. Veit, N. Sinha, and S. Hake. 1991. The developmental gene *Knotted-1* is a member of a maize homeobox gene family. *Nature* **350:** 241–243.

Walter, P., S. Green, G. Green, A. Krust, J.-M. Bornert, J.-M. Jeltsch, A. Staub, E. Jensen, G. Scrace, M. Waterfield, and P. Chambon. 1985. Cloning of the human estrogen receptor cDNA. *Proc. Natl. Acad. Sci.* **82:** 7889–7893.

Weinberger, C., S.M. Hollenberg, E.S. Ong, J.M. Harmon, S.T. Brower, J. Cidlowski, E.B. Thompson, M.G. Rosenfeld, and R.M. Evans. 1985. Identification of human glucocorticoid receptor complementary DNA clones by epitope selection. *Science* **228:** 740–742.

Weisshaar, B., G.A. Armstrong, A. Block, O. da Costa e Silva, and K. Hahlbrock. 1991. Light-inducible and constitutively expressed DNA-binding proteins recognizing a plant promoter element with functional relevance in light responsiveness. *EMBO J.* **10:** 1777–1786.

Yanofsky, M.F., H. Ma, J.L. Bowman, G.N. Drews, K.A. Feldmann, and E.M. Meyerowitz. 1990. The protein encoded by the *Arabidopsis* homeotic gene *agamous* resembles transcription factors. *Nature* **346:** 35–39.

• *TIMETABLE*
Isolation of cDNAs Encoding Sequence-specific DNA-binding Proteins Using In Situ Screening

STEPS	***TIME REQUIRED***
Preparation of Radiolabeled DNA Probe	
See Section 13, pp. 268–271	~24 hours
Preparation of Competent *E. coli* Cells	
Grow up bacteria in phage broth	~4 hours
Infection of *E. coli* and Induction of Protein Synthesis	
Prepare phage plates, dry them, and prewarm at 37°C	2–3 hours
Infect bacteria and plate onto phage plates	1 hour
Incubate plates at 42°C (prepare 1st set of IPTG-soaked nitrocellulose filters during this incubation)	4–5 hours
Place 1st set of filters onto plates	10 minutes
Incubate plates at 37°C (prepare 2nd set of IPTG-soaked nitrocellulose filters during this incubation)	4 hours
Remove 1st set of filters from plates and replace with 2nd set of filters	30 minutes
Incubate plates at 37°C (begin processing 1st set of filters during this incubation)	2–4 hours

STEPS	*TIME REQUIRED*
Denaturation, Renaturation, and Blocking	
Incubate filters in 6 M guanidine HCl	10 minutes
Wash filters	45 minutes
Incubate filters in BLOTTO buffer	1 hour
Binding Reaction	
Incubate filters with radiolabeled DNA probe (begin processing 2nd set of filters during this incubation)	1 hour
Wash filters	15 minutes
Expose both sets of filters to X-ray film	overnight
Identification of Positive Plaques	
Isolate positive plaques	(depends on number of positive plaques)
Let phage diffuse from agarose plugs	> 1 hour
Second and Third Round of Screening	
Perform second screen	2 days
Perform third screen	2 days

Preparation of Radiolabeled DNA Probe

DNA probes for in situ screening experiments can be labeled by the fill-in reaction as described in Section 13, pp. 268–271, except that a plasmid carrying multiple copies of the DNA-binding site of interest should be used (Vinson et al. 1988; Singh et al. 1989). The presence of multiple DNA-binding sites within one DNA fragment (concatenation) enhances the stability of the protein-DNA complex. Radiolabeled probes can also be prepared by nick translation or by labeling synthetic oligonucleotides at the 5′ end using bacteriophage T4 polynucleotide kinase (Sambrook et al. 1989). If the latter procedure is used, the oligonucleotides should be annealed and ligated in order to obtain multisite DNA probes (Singh et al. 1989).

Preparation of Competent *E. coli* Cells

All the steps described below should be performed under aseptic conditions using sterile equipment and reagents.

MATERIALS

E. coli cells—use the strain that is recommended as a host by the manufacturer of the λ expression vector you have chosen (e.g., CLONTECH Laboratories, Inc. or Stratagene)

MEDIA

Phage broth
(For recipe, see Preparation of Media, p. 322)

REAGENTS

TM buffer
(For recipe, see Preparation of Reagents, pp. 323–324)

PROCEDURE

1. Inoculate 2 ml of phage broth with *E. coli* cells and shake overnight at 37°C (overnight culture).

2. Inoculate 100 ml of phage broth with 1 ml of the overnight culture of *E. coli* cells and shake at 37°C until an OD_{600} of 0.4 is reached.

3. Spin the cells in a Beckman tabletop centrifuge at 3000 rpm for 5 minutes at 4°C and resuspend the bacteria in 10 ml of TM buffer. The cells can be stored for up to 5 days at 4°C without significant loss of competence.

Infection of *E. coli* and Induction of Protein Synthesis

MATERIALS AND EQUIPMENT

cDNA library in a bacteriophage λ expression vector (phage stock)
Competent *E. coli* cells (see p. 310)
Nitrocellulose filters (132 mm) (Schleicher and Schuell, Inc.)
Isopropylthio-β-D-galactoside (IPTG) (10 mM)

MEDIA

Phage broth supplemented with 15 g of agar/liter
Top agarose
(For recipes, see Preparation of Media, p. 322)

REAGENTS

SM buffer
(For recipe, see Preparation of Reagents, pp. 323–324)

PROCEDURE

1. Prepare the phage plates for the first screen by adding 70 ml of phage broth supplemented with 15 g of agar/liter to each of several 150-mm petri dishes. The exact number of phage plates required will depend on the design of your experiment. Dry the plates before use by placing them uncovered in a fume hood for 2–3 hours.

2. Dilute the phage stock to 5 x 10^6 pfu/ml (pfu = plaque-forming units) with SM buffer.

   Note: (i) Sequence-specific DNA-binding proteins are usually encoded by very low-abundant mRNAs. The chance of isolating a clone that expresses a sequence-specific DNA-binding protein increases significantly if you use an unamplified cDNA library for this procedure. (ii) The titer of the phage stock is determined by plating various dilutions of the original stock and counting the number of plaques formed.

3. Melt the top agarose in a microwave and keep it at 50°C until you are ready to use it.

4. Add 600 μl of competent *E. coli* cells to a 15-ml screw-cap tube, infect with 4–6 μl of diluted phage stock, and let the phage absorb for 20 minutes at room temperature. Prepare the desired number of samples for your experiment.

5. Add 10 ml of melted top agarose to each sample of phage-infected *E. coli* cells. Invert the tubes 2–3 times and pour the contents onto 150-mm phage plates that have been prewarmed for 30 minutes at 37°C. Tilt each plate to all sides so that the entire surface is covered with top agarose. Avoid air bubbles.

   Note: (i) Be sure to use freshly prepared phage plates that have been dried in an incubator at 37°C. Prewarming the plates significantly increases the chance that plaques will be visible after 4 hours. (ii) The size of the plaques depends on the nature of the λ expression vector. Some vectors, like λgt11, give very small plaques and 3×10^4 pfu can be applied per plate, whereas other vectors, like λZAP, give larger plaques and only 2×10^4 pfu can be applied per plate.

6. Let the top agarose solidify at room temperature (this takes ~5 minutes). Incubate the plates upside down for 4 hours at 42°C.

7. During this incubation, mark and number the 132-mm nitrocellulose filters using a ballpoint pen.

   Note: Each filter should have at least three marks so that it can be oriented relative to the appropriate plate when you are identifying positive plaques (see step 1, p. 318).

8. Soak the nitrocellulose filters in 10 mM IPTG for several minutes. Let the filters air dry, but do not let them dry out completely.

   Note: IPTG-soaked filters are required to induce the expression of recombinant protein.

9. After 4 hours at 42°C, the plaques on the phage plates should just be visible. If plaques have not appeared, incubate the plates a little longer. The plaques should certainly be visible after 5 hours. When you can discern the plaques, place a nitrocellulose filter on top of each plate with the marks facing toward the top agarose. Using a waterproof marker, label the bottom of each plate by tracing the marks on the nitrocellulose filter. Incubate for 4 hours at 37°C.

10. Repeat steps 7 and 8 to prepare a second set of IPTG-soaked nitrocellulose filters.

11. Using forceps, carefully remove the first set of filters from the plates and place them onto Whatman 3MM paper with the plaque-exposed sides up. To prevent peeling off the top agarose layer, cool the plates for 5–10 minutes at 4°C before you attempt to remove the filters.

12. Place the second set of filters onto the plates. Again, label the bottom of each plate by tracing the marks on the filter. Incubate the plates for 2–4 hours at 37°C.

13. During the incubation with the second set of filters, start processing the first set (see pp. 314–315). After you have removed the second set of filters for processing, store the plates at 4°C until you are ready to identify positive plaques (see p. 318).

Denaturation, Renaturation, and Blocking

MATERIALS

Plaque-exposed nitrocellulose filters (see pp. 312–313)

REAGENTS

Denaturation buffer
Dignam buffer
Binding buffer
BLOTTO (bovine lacto transfer technique optimizer) buffer
(For recipes, see Preparation of Reagents, pp. 323–324)

SAFETY NOTE

- The denaturation buffer contains guanidine HCl, which is irritating to the mucous membranes, the upper respiratory tract, the skin, and the eyes. Avoid breathing the dust. Wear gloves and safety glasses when handling this compound.

PROCEDURE

The following denaturation and renaturation steps can be carried out at room temperature provided the buffers are kept at 4°C.

1. Place the plaque-exposed nitrocellulose filters into denaturation buffer (4 filters per 100 ml) and shake gently for 10 minutes.

2. Remove 50 ml of the denaturation buffer, add 50 ml of Dignam buffer, and shake gently for 5 minutes.

3. Repeat step 2 four times.

4. Transfer the filters to fresh Dignam buffer and shake gently for 5 minutes.

5. Rinse the filters in binding buffer.

6. Place the filters in BLOTTO buffer (4 filters per 100 ml) and shake gently for 1 hour at 4°C. This step will prevent nonspecific binding of the DNA probe to the filters.

7. Rinse the filters briefly in binding buffer.

   Note: At this stage, depending on the stability of the protein, the filters often can be kept at 4°C for several hours.

Binding Reaction

MATERIALS AND EQUIPMENT

Processed, plaque-exposed nitrocellulose filters (see pp. 314–315)
Radiolabeled DNA probe (~10^4 cpm/μl; concatenated and end-labeled at one end only) (see p. 309)
X-ray film

REAGENTS

Binding buffer
(For recipe, see Preparation of Reagents, pp. 323–324)

SAFETY NOTE

- Wear gloves when handling radioactive substances. Consult the local safety office for further guidance in the appropriate use of radioactive materials.

PROCEDURE

1. Place the filters into binding buffer supplemented with the radiolabeled DNA probe (5–10 x 10^5 cpm/ml; 4 filters per 100 ml). Shake gently for at least 1 hour at 4°C.

   Note: This is a good time to start processing the second set of plaque-exposed nitrocellulose filters.

2. Wash the filters by shaking gently in binding buffer (4 filters per 100 ml) for 5 minutes. Repeat this step once or twice. Monitor the amount of radioactivity present on the filters using a Geiger counter. These washing steps can be performed at room temperature provided the binding buffer is kept at 4°C.

   Note: High nonspecific background is usually caused by inefficient blocking of the filters (see step 6, p. 315). Washing for longer than 30 minutes will not reduce the nonspecific background. To minimize nonspecific background, include a nonspecific competitor DNA in the binding reaction (see p. 303 and Singh et al. 1989).

3. Place the filters onto Whatman 3MM paper and remove any excess liquid. Wrap the filters with Saran® Wrap.

4. Expose both sets of filters to X-ray film overnight at –80°C.

5. Develop the autoradiographs.

Identification of Positive Plaques

MATERIALS

Autoradiographs of nitrocellulose filters (see p. 317)
Phage plates (see p. 313)

REAGENTS

SM buffer
(For recipe, see Preparation of Reagents, pp. 323–324)

PROCEDURE

1. Align the autoradiographs of each set of nitrocellulose filters with the marks on the bottom of the corresponding phage plate. Phage-expressing DNA-binding proteins will give signals on the autoradiographs of both the first and second filters (Fig. 2).

2. Remove the positive plaques with the wide end of a pasteur pipette and place each agarose plug into 500 μl of SM buffer. Let the phage diffuse for several hours at 4°C.

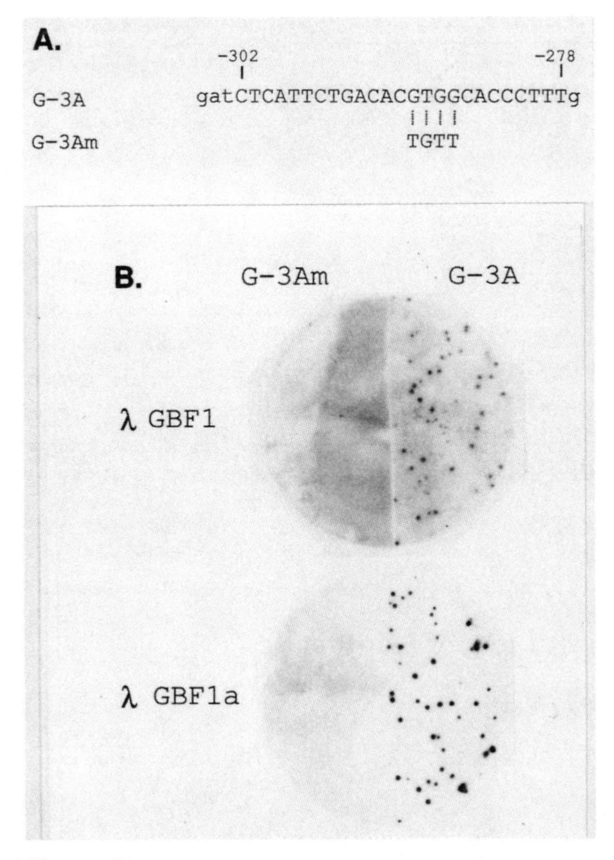

Figure 2
DNA-binding specificity of two DNA-binding proteins expressed from *Arabidopsis* cDNA clones. The two phage, λGBF1 and λGBF1a, express DNA-binding proteins that specifically interact with the G-3A oligonucleotide (right half of both filters) but not with the G-3Am oligonucleotide, a mutant DNA element that carries four base-pair substitutions (left half of both filters) (Schindler et al. 1992).

Second and Third Round of Screening

MATERIALS AND EQUIPMENT

Phage stocks prepared from positive plaques (see p. 318)
Competent *E. coli* cells (see p. 310)
Nitrocellulose filters (82 mm) (Schleicher and Schuell, Inc.)
Isopropylthio-β-D-galactoside (IPTG) (10 mM)

MEDIA

Phage broth supplemented with 15 g of agar/liter
Top agarose
(For recipes, see Preparation of Media, p. 322)

REAGENTS

SM buffer
(For recipe, see Preparation of Reagents, pp. 323–324)

PROCEDURE

1. Prepare the phage plates for the second and third screens by adding 30 ml of phage broth supplemented with 15 g of agar/liter to each of several 90-mm petri dishes. The exact number of phage plates required will depend on the design of your experiment. Dry the plates before use by placing them uncovered in an incubator for 2–3 hours at 37°C.

2. Determine the titer of the phage stocks as follows:

   a. Prepare 10^{-2}, 10^{-3}, and 10^{-4} dilutions of each phage stock in SM buffer.

   b. Mix 1 μl of each dilution with 250 μl of competent *E. coli* cells in a 5-ml screwcap tube and incubate for 20 minutes at room temperature.

c. Melt the top agarose in a microwave and let it cool to 50°C.

d. Add 4 ml of top agarose to each bacteria/phage mix and pour the entire contents of each tube onto a 90-mm phage plate.

e. Let the top agarose solidify for 5 minutes at room temperature and incubate the plates overnight at 37°C. The titer is determined by counting the number of plaques that form.

3. Perform the second screen using the procedure described for the first screen (see pp. 311–313), except infect 250-μl aliquots of competent *E. coli* cells with 100–300 pfu and plate onto 90-mm phage plates using 4 ml of top agarose. Use 82-mm nitrocellulose filters for the 90-mm phage plates. The number of positives per plate should increase in the second screen.

4. Remove the positive plaques with the narrow end of a pasteur pipette—try to pick single plaques. Place each agarose plug into 100 μl of SM buffer and let the phage diffuse for several hours at 4°C. Determine the titer of the phage stocks (see step 2).

5. Perform the third screen as described in step 3, except use two different DNA probes to investigate the binding specificity of the expressed protein. To achieve this, cut the plaque-exposed filters in half, incubate one half with the wild-type probe and the other half with a probe containing a mutant DNA-binding site. Sequence-specific DNA-binding proteins will interact with the wild-type but not with the mutant probe (Fig. 2).

• *PREPARATION OF MEDIA*

Phage broth

(1 liter)

Yeast extract	5 g
Bacto tryptone	10 g
NaCl	5 g
1 M Tris-HCl (pH 7.5)	10 ml
1 M $MgSO_4$	10 ml
Maltose	2.5 g

Adjust the volume to 1 liter with H_2O. Autoclave.

Top agarose

Phage broth supplemented with 6 g of agarose/liter.

• PREPARATION OF REAGENTS

Binding buffer

10 mM Tris-HCl (pH 7.5)
40 mM NaCl
1 mM EDTA
1 mM Dithiothreitol (DTT)

DTT should be added just before use.

BLOTTO buffer

5% Nonfat dry milk (Carnation)
50 mM Tris-HCl (pH 7.5)
40 mM NaCl
1 mM EDTA

Denaturation buffer

6 M Guanidine HCl in Dignam buffer

SAFETY NOTE

- Guanidine HCl is irritating to the mucous membranes, the upper respiratory tract, the skin, and the eyes. Avoid breathing the dust. Wear gloves and safety glasses when handling this compound.

Dignam buffer

20 mM HEPES (pH 8.0)
0.1 mM KCl
0.2 mM EDTA
0.5 mM Dithiothreitol (DTT)
0.5 mM Phenylmethylsulfonyl fluoride (PMSF)

DTT and PMSF should be added just before use.

SAFETY NOTE

- PMSF is extremely destructive to the mucous membranes of the respiratory tract, the eyes, and the skin. It may be fatal if inhaled, swallowed, or absorbed through the skin. It is a highly toxic cholinesterase inhibitor. It should be used in a chemical fume hood. Gloves and safety glasses should be worn.

SM buffer

(1 liter)

NaCl	5.8 g
$MgSO_4$	2 g
1 M Tris-HCl (pH 7.5)	50 ml
2% Gelatin	5 ml

Adjust the volume to 1 liter with H_2O. Autoclave.

TM buffer

10 mM Tris-HCl (pH 7.5)
10 mM $MgSO_4$

• *Section 16*
Phosphorylation Studies of DNA-binding Proteins

Regulation of gene expression is accomplished, in part, by the action of transcription factors, some of which are highly specific DNA-binding proteins that recognize defined sequence elements within promoters. Upon binding to a promoter, these transcription factors are presumed to interact with the transcriptional initiation complex and modify its activity in a positive or negative manner, thereby effecting stimulation or suppression of transcription (for review, see Ptashne 1988; Dynan 1989; Latchman 1990). The activity of transcription factors, that is, their ability to affect the rate of transcriptional initiation, is subject to regulation by biochemical mechanisms similar to those identified for regulatory enzymes. Such mechanisms include ligand binding (Evans 1988; Beato 1989), posttranslational modifications (Yamamoto et al. 1988; Manak et al. 1990; Wegner et al. 1992), and de novo synthesis (Curran and Franza 1988). As a consequence, regulation of transcription can become integrated into an appropriate signal transduction pathway. The regulatory reaction can result in translocation of newly synthesized transcription factor to the nucleus, modulation of the DNA-binding activity of the transcription factor, or modulation of its ability to interact with the initiation complex. DNA-binding proteins with lower sequence specificity than transcription factors, such as high-mobility-group proteins or histones, may also be targets for regulatory interactions (see, e.g., Nissen et al. 1991) since they can affect the rate of transcription via the higher-order structure of the DNA template.

In this section, we focus on studying the effects of phosphorylation on the DNA-binding properties of specific DNA-binding proteins. Protein phosphorylation constitutes a major mechanism of regulatory posttranslational modification, and its role in the regulation of transcription was reviewed recently (Hunter and Karin 1992).

Once a DNA-binding protein is isolated and identified in a specific binding assay, a preliminary evaluation of the possible regulatory role of phosphorylation can be performed by subjecting the DNA-binding protein preparation to phosphatase treatment (Datta and Cashmore 1989; Harrison et al. 1991). Since this approach is indirect and harbors many potential pitfalls, several distinct control experiments should be performed in parallel (e.g., studies of the effects of different phosphatases, pH values, and phosphatase inhibitors). In particular, spurious effects may be observed as a result of the change in binding conditions, a reversal of the effect of endogenous inhibitors, and proteolytic degradation by endogenous or exogenous proteases.

After this preliminary experiment, more rigorous in vitro and in vivo studies have to be completed to demonstrate that the activity of a DNA-binding protein is actually regulated by phosphorylation and that this effect is biologically relevant. These studies should test the criteria proposed by Krebs and Beavo (1979), which state that: (i) the protein must be phosphorylated in vitro to a substantial molar amount and at a significant rate by a particular kinase and be dephosphorylated by a phosphoprotein phosphatase; (ii) the functional properties of the factor must undergo meaningful changes in correlation with the degree of phosphorylation; (iii) phosphorylation and dephosphorylation must occur in vivo and be accompanied by changes in functional properties; and (iv) the cellular levels of effectors of the protein kinase or phosphatase or the degree of their activation must correlate with the extent of in vivo phosphorylation of the factor.

The following protocols outline basic experiments that should be performed in the initial stages of the characterization of the effect of phosphorylation on DNA-binding proteins. These experiments will provide information regarding the first and second of the above criteria. They were originally developed for the study of phosphorylation of the *Arabidopsis* transcription factor GBF1 (G-box binding factor) by casein kinase II (Klimczak et al. 1992), but the conditions used are quite generic and, if necessary, could easily be adapted to any other factor/protein kinase pair by appropriate modification of the buffer components.

REFERENCES

Affolter, M., A. Percival-Smith, M. Müller, W. Leupin, and W.J. Gehring. 1990. DNA binding properties of the purified *Antennapedia* homeodomain. *Proc. Natl. Acad. Sci.* **87:** 4093–4097.

Beato, M. 1989. Gene regulation by steroid hormones. *Cell* **56:** 335–344.

Datta, N. and A.R. Cashmore. 1989. Binding of a pea nuclear protein to promoters of certain photoregulated genes is modulated by phosphorylation. *Plant Cell* **1:** 1069–1077.

Cao, Z., R.M. Umek, and S.L. McKnight. 1991. Regulated expression of three C/EBP isoforms during adipose conversion of 3T3-L1 cells. *Genes Dev.* **5:** 1538–1552.

Curran, T. and B.R. Franza. 1988. Fos and Jun: The AP-1 connection. *Cell* **55:** 395–397.

Dynan, W.S. 1989. Modularity in promoters and enhancers. *Cell* **58:** 1–4.

Evans, R.M. 1988. The steroid and thyroid hormone receptor superfamily. *Science* **240:** 889–895.

Harrison, M.J., M.A. Lawton, C.J. Lamb, and R.A. Dixon. 1991. Characterization of a nuclear protein that binds to three elements within the silencer region of a bean chalcone synthase. *Proc. Natl. Acad. Sci.* **88:** 2515–2519.

Hunter, T. and M. Karin. 1992. The regulation of transcription by phosphorylation. *Cell* **70:** 375–387.

Klimczak, L.J., U. Schindler, and A.R. Cashmore. 1992. DNA binding activity of the *Arabidopsis* G-box binding factor GBF1 is stimulated by phosphorylation by casein kinase II from broccoli. *Plant Cell* **4:** 87–98.

Krebs, E.G. and J.A. Beavo. 1979. Phosphorylation-dephosphorylation of enzymes. *Annu. Rev. Biochem.* **48:** 923–947.

Latchman, D.S. 1990. Eukaryotic transcription factors. *Biochem. J.* **270:** 281–289.

Laemmli, U.K. 1970. Cleavage of structural proteins during the assembly of the head of bacteriophage T4. *Nature* **227:** 680–685.

Manak, J.R., N. deBisschop, R.M. Kris, and R. Prywes. 1990. Casein kinase II enhances the DNA binding activity of serum response factor. *Genes Dev.* **4:** 955–967.

Marais, R.M., J.J. Hsuan, C. McGuigan, J. Wynne, and R. Treisman. 1992. Casein kinase II phosphorylation increases the rate of serum response factor-binding site exchange. *EMBO J.* **11:** 97–105.

Nissen, M.S., T.A. Langan, and R. Reeves. 1991. Phosphorylation by cdc2 kinase modulates DNA binding activity of high mobility group I nonhistone chromatin protein. *J. Biol. Chem.* **266:** 19945–19952.

Ptashne, M. 1988. How eukaryotic transcriptional activators work. *Nature* **335:** 683–689.

Wegner, M., Z. Cao, and M.G. Rosenfeld. 1992. Calcium-regulated phosphorylation within the leucine zipper of C/EBPβ. *Science* **256:** 370–373.

Yamamoto, K.K., G.A. Gonzalez, W.H. Biggs III, and M.R. Montminy. 1988. Phosphorylation-induced binding and transcriptional efficacy of a nuclear factor CREB. *Nature* **334:** 494–498.

• *TIMETABLE*
Phosphorylation Studies of DNA-binding Proteins

STEPS	*TIME REQUIRED*
Determination of Stoichiometry of Phosphorylation	
Set up reaction mixture	15 minutes
Perform phosphorylation assays	30 minutes
Spot half of each sample onto filters, treat filters with TCA, and wash[1]	40 minutes
Pour SDS-polyacrylamide gel and let gel polymerize[1]	30 minutes
Treat remaining half of each sample with TCA and prepare samples for SDS-PAGE	1 hour
Run gel	40 minutes
Expose wet gel to X-ray film	3 hours
Excise labeled bands	20 minutes
Solubilize labeled protein from gel	6–12 hours
Count samples	20 minutes
Measurement of Kinetics of Phosphoprotein Formation	
Set up reaction mixture	15 minutes
Perform phosphorylation assays	30 minutes
Spot half of each sample onto filters, treat filters with TCA, and wash[2]	40 minutes

STEPS	*TIME REQUIRED*
Measurement of Kinetics of Phosphoprotein Formation (continued)	
Pour SDS-polyacrylamide gel and let gel polymerize[2]	30 minutes
Treat remaining half of each sample with TCA and prepare samples for SDS-PAGE	1 hour
Run gel	40 minutes
Expose wet gel to X-ray film	3 hours
Excise labeled bands	20 minutes
Solubilize labeled protein from gel	6–12 hours
Count samples	20 minutes
Mobility-shift Assay with Radiolabeled Phosphoprotein and Unlabeled DNA Probe	
Set up reaction mixture	15 minutes
Perform phosphorylation reaction	30 minutes
Perform binding assay[3]	40 minutes
Pour nondenaturing gel and let gel polymerize[3]	30 minutes
Run gel	3 hours
Expose gel to X-ray film	overnight
Mobility-shift Assay with Unlabeled Phosphoprotein and Radiolabeled DNA Probe	
Set up reaction mixture	15 minutes
Perform phosphorylation reaction	30 minutes

STEPS	TIME REQUIRED
Mobility-shift Assay with Unlabeled Phosphoprotein and Radiolabeled DNA Probe (continued)	
Perform binding assay[4]	90 minutes
Pour nondenaturing gel and let gel polymerize[4]	30 minutes
Run gel	3 hours
Expose gel to X-ray film	overnight

[1, 2, 3, 4]Each of these sets of steps can be performed in parallel.

Determination of Stoichiometry of Phosphorylation and Measurement of Kinetics of Phosphoprotein Formation

A rigorous biochemical study of phosphorylation requires at least 20–50 μg of the pure protein of interest. Since plant transcription factors are very low in abundance, it is almost impossible to obtain sufficient material from plant tissues. Therefore, such studies are limited by the availability of a cloned gene that can be overexpressed. Bacterial expression systems are preferred because they generate unmodified (in particular, unphosphorylated) proteins and this is not necessarily the case with eukaryotic systems. To demonstrate phosphorylation, the protein expressed in bacteria is incubated with an appropriate protein kinase (relatively crude preparations may sometimes be sufficient) in the presence of $MgCl_2$ and [γ-^{32}P]ATP. The proteins in the assay mixture are precipitated with trichloroacetic acid and analyzed by SDS-polyacrylamide gel electrophoresis (Laemmli 1970) to identify a phosphoprotein of appropriate molecular mass.

Here we describe procedures for determining the stoichiometry of phosphorylation close to saturation and for measuring the kinetics of formation of a phosphorylated protein. These procedures were originally devised to study the phosphorylation of GBF1 by casein kinase II. It is not possible to give universal protocols for the isolation and purification of various protein kinases and DNA-binding proteins as strategies have to be developed individually for each particular protein. Protocols for the expression of GBF1 in *Escherichia coli*, purification of GBF1, and partial purification of nuclear casein kinase II from broccoli can be found in Klimczak et al. 1992.

MATERIALS AND EQUIPMENT

Tris-HCl (50 mM; pH 8.0)/$MgCl_2$ (10 mM)
Adenosine-5′-triphosphate (ATP) (10 mM) (prepared from 100 mM stock solution)
Purified DNA-binding protein of interest (1 mg/ml)
[γ-^{32}P]ATP (10 μCi/μl; 3000 mCi/mmol)
Protein kinase (appropriate for phosphorylating DNA-binding protein

of interest) (30 milliunits; 1 milliunit incorporates 1 pmole of phosphate into protein per minute)
Trichloroacetic acid (TCA) (10%)/sodium pyrophosphate (1%)
Minigel electrophoresis apparatus (Bio-Rad Laboratories or Hoefer Scientific Instruments)
Tris-HCl (1 M; pH 6.8 and pH 8.8)
Sodium dodecyl sulfate (SDS) (10%)
Ammonium persulfate (APS) (50%)
N,N,N',N',tetramethylethylenediamine (TEMED)
Glycerol
Ethanol (100%)
Acetone (90% and 100%)
Hydrogen peroxide (30%)

REAGENTS

TCA (100%)
Acrylamide/bisacrylamide (30%)
Denaturing sample buffer (5x)
Running buffer (for denaturing gel) (10x)
(For recipes, see Preparation of Reagents, pp. 347–348)

SAFETY NOTES

- Wear gloves when handling radioactive substances. Consult the local safety office for further guidance in the appropriate use of radioactive materials.
- TCA is highly caustic. Always wear gloves when handling this compound.
- Acrylamide and bisacrylamide are potent neurotoxins and are absorbed through the skin. Their effects are cumulative. Wear gloves and a mask when weighing acrylamide and bisacrylamide. Polyacrylamide is considered to be nontoxic, but it should be treated with care because it may contain small quantities of unpolymerized material.
- Wear a mask when weighing SDS.
- Hydrogen peroxide is a strong oxidant. Always wear gloves and safety glasses when handling this compound. Avoid contact with combustible materials—the drying of concentrated hydrogen peroxide on clothing or other combustible materials may cause a fire.

PROCEDURE

Determination of Stoichiometry of Phosphorylation

1. Set up the following reaction mixture on ice:

Tris-HCl (50 mM; pH 8.0)/$MgCl_2$ (10 mM)	250 µl
ATP (10 mM)	1 µl
Purified DNA-binding protein (1 mg/ml)	4 µl
[γ-^{32}P]ATP (13 µCi)	1.3 µl

2. Divide the reaction mixture into two 125-µl portions, designated **portion A** and **portion B**. The latter portion will be used in the kinetics experiment (see p. 334).

3. Transfer six 20-µl aliquots of **portion A** into clean tubes and start the phosphorylation assays by adding 0, 0.5, 1, 2, 3, and 4 milli-units of protein kinase to the tubes. Incubate for 30 minutes at room temperature.

4. Spot 10 µl of each phosphorylation assay mixture onto a 2 x 2-cm square of Whatman 3MM filter paper. Wash the filters in 10% TCA/1% sodium pyrophosphate for 10 minutes (use 5 ml of solution per filter). Repeat the wash twice with fresh solution.

   Note: This is a convenient time to prepare the SDS-polyacrylamide gel (see p. 335).

   Rinse the filters in 100% ethanol. Determine the counts incorporated into phosphoprotein in each sample by counting the Cerenkov radiation emitted by each filter.

5. Precipitate the remaining 10 µl of each phosphorylation assay mixture by adding 1.1 µl of 100% TCA to each tube. Incubate on ice for 20 minutes. Collect the protein precipitates by spinning in a microfuge at 10,000 rpm for 10 minutes at room temperature. Rinse the pellets with 90% acetone and then with 100% acetone. Air dry the pellets for 5 minutes. Resuspend each pellet in 8 µl of H_2O, add 2 µl of 5x denaturing sample buffer, and boil for 5 minutes.

6. Load the samples onto the SDS-polyacrylamide gel (see p. 335) and run the gel in 1x running buffer (prepared from 10x stock solution) at 160 V for 40 minutes. Wrap the wet gel with Saran® Wrap and expose it to X-ray film for approximately 3 hours.

7. Using the autoradiograph as a template, excise the labeled bands that correspond to the DNA-binding protein. Solubilize the DNA-

binding protein by incubating the gel pieces in five volumes of 30% hydrogen peroxide for 6–12 hours at 37°C. Determine the counts incorporated into the DNA-binding protein by counting the Cerenkov radiation emitted by each sample.

Note: SDS-PAGE is not necessary if the DNA-binding protein is the only phosphorylated substrate in the reaction mixture. However, when there is significant phosphorylation of contaminating proteins in the DNA-binding protein or protein kinase preparations, the phosphorylated DNA-binding protein must be isolated to obtain a precise determination of phosphate incorporation.

8. Determine the total number of counts in the assay using two different methods:

   a. Add 9 µl of assay buffer (50 mM Tris-HCl [pH 8.0]/10 mM $MgCl_2$) to 1 µl of **portion A** and spot the mixture onto a 2 x 2-cm square of Whatman 3MM filter paper. Do not treat the filter with TCA. Count the Cerenkov radiation emitted by the filter.

   b. Add 1 µl of **portion A** to an appropriate volume of 30% hydrogen peroxide and count the Cerenkov radiation.

9. Calculate the molar incorporation of phosphate into protein. In this experiment, each assay contains 800 pmoles of ATP and, assuming the DNA-binding protein of interest is GBF1 (M.W. 37,000), 8 pmoles (0.3 µg) of protein. One per cent of radioactivity incorporated into protein corresponds to 1 mole of phosphate per mole of protein.

   Note: For examples of biochemical calculations, see pp. 344–346.

Measurement of Kinetics of Phosphoprotein Formation

1. Add 18 milliunits of protein kinase to **portion B** (see p. 333) and incubate at room temperature. Remove 20-µl aliquots at intervals of 0, 5, 10, 20, 25, and 30 minutes. *Immediately* terminate the phosphorylation reaction at each time point by spotting 10 µl of each aliquot onto a 2 x 2-cm square of Whatman 3MM filter paper and treating with TCA (see p. 333, step 4) and precipitating the remaining 10 µl with 1.1 µl of 100% TCA (see p. 333, step 5).

2. Process the samples from **portion B** exactly as those from **portion A** (see pp. 333–334, steps 4–8).

3. Determine the rate of phosphoprotein formation.

Preparation of SDS-Polyacrylamide Gel for Electrophoresis under Denaturing Conditions

It is convenient to pour the gel while you are washing the filters after the phosphorylation reaction (see p. 333, step 4).

1. Prepare a 10% acrylamide resolving gel solution by mixing the following components in the order listed:

Acrylamide/bisacrylamide (30%)	3.38 ml
Tris-HCl (1 M; pH 8.8)	3.80 ml
SDS (10%)	0.10 ml

   Adjust the volume to 10 ml with H_2O. Add 12 μl of 50% APS and 10 μl of TEMED. Mix well. (This is sufficient for two minigels approximately 8.3 x 7.4 x 0.075 cm in size.) Pour the solution into the gel casting chamber so that it fills approximately 80% of the space between gel plates. Overlay the gel with H_2O and wait 10 minutes for the gel to polymerize. Remove the H_2O overlay.

2. Prepare a 4% acrylamide stacking gel solution by mixing the following components in the order listed:

Acrylamide/bisacrylamide (30%)	0.27 ml
Tris-HCl (1 M; pH 6.8)	0.25 ml
Glycerol	0.20 ml
SDS (10%)	20 μl

   Adjust the volume to 2 ml with H_2O. Add 2.4 μl of 50% APS and 2 μl of TEMED. Pour the solution onto the resolving gel to approximately 5 mm below the edge of the shorter gel plate and insert the comb. The gel can be stored for several hours at room temperature or for several days in a sealed plastic bag at 4°C until ready to use.

Binding of Phosphorylated Proteins to DNA: Mobility-shift Assay with Radiolabeled Phosphoprotein and Unlabeled DNA Probe

Once it has been established that a DNA-binding protein is phosphorylated to a substantial molar amount and at a significant rate, the effect of phosphorylation on the properties of the protein should be investigated. In particular, it is necessary to determine whether the phosphorylated protein retains its ability to bind to its cognate DNA element in a specific manner. A simple way to do this is to perform a "reverse-labeling" gel mobility-shift assay—that is a DNA-binding assay in which the protein is radiolabeled by phosphorylation, but the DNA fragment is unlabeled. Such an experiment also provides confirmation that the phosphorylated species is indeed the DNA-binding protein itself, and not, for example, a protein contaminant that has a high affinity for the protein kinase used.

MATERIALS AND EQUIPMENT

DNA-binding protein phosphorylated to saturation (for the phosphorylation reaction, follow the procedure outlined on p. 333)
Poly(dI-dC)·poly(dI-dC) (referred to hereafter as poly[dI-dC])
Unlabeled DNA probe (e.g., a G-box-containing oligonucleotide)
Salmon sperm DNA (sonicated)
Vertical gel unit (Life Technologies, Inc.)
Ammonium persulfate (APS) (50%)
N,N,N′,N′,tetramethylethylenediamine (TEMED)

REAGENTS

Mobility-shift buffer prepared without poly(dI-dC)
Native sample buffer
Acrylamide/bisacrylamide (30%)
Running buffer (for nondenaturing gel) (10x and 1x)
(For recipes, see Preparation of Reagents, pp. 347–348)

SAFETY NOTES

- Wear gloves when handling radioactive substances. Consult the local safety office for further guidance in the appropriate use of radioactive materials.
- Acrylamide and bisacrylamide are potent neurotoxins and are absorbed through the skin. Their effects are cumulative. Wear gloves and a mask when weighing acrylamide and bisacrylamide. Polyacrylamide is considered to be nontoxic, but it should be treated with care because it may contain small quantities of unpolymerized material.

PROCEDURE

Mobility-shift Assay

1. Add 2 μl of the DNA-binding protein phosphorylated to saturation to 110 μl of mobility-shift buffer prepared without poly(dI-dC).

2. Transfer 20 μl to a clean tube—this will serve as the control sample (sample 1). Add poly(dI-dC) to the remaining 92 μl to give a final concentration of 0.07 μg/ml. Remove four 20-μl aliquots (samples 2–5).

3. Add 0, 2, and 10 ng of unlabeled DNA probe to samples 2, 3, and 4, respectively. Add 0.1 μg of sonicated salmon sperm DNA to sample 5. Incubate samples 1–5 for 30 minutes at room temperature.

   Note: This is a convenient time to prepare the nondenaturing polyacrylamide gel (see p. 338).

4. Add 5 μl of native sample buffer to each sample. Load the samples onto the nondenaturing polyacrylamide gel and run in 1x running buffer at no more than 100 V for 3 hours. Wrap the wet gel with Saran® Wrap and expose it to X-ray film overnight. For increased resolution, dry the gel onto Whatman 3MM paper and expose the dried gel to X-ray film using a Kodak intensifying screen.

   Note: The results from sample 1 will show the migration of the radioactively labeled protein alone, in the absence of any DNA. Samples 2–5 will show the protein-DNA complexes. In samples 2–4, a wide diffuse band of unspecific complexes with poly(dI-dC) will be replaced by a sharp band corresponding to the complex with the specific unlabeled DNA probe as more oligonucleotide is added. Sample 5 will show a different diffuse band, the position of which depends on the average size of the sonicated sperm DNA fragments.

Preparation of Nondenaturing Polyacrylamide Gel

It is convenient to pour the gel while you are incubating the binding assay.

1. Prepare a 4% nondenaturing acrylamide gel solution by mixing the following components in the order listed:

Acrylamide/bisacrylamide (30%)	6.6 ml
Running buffer (10x)	5 ml

   Adjust the volume to 50 ml with H_2O. Add 60 µl of 50% APS and 50 µl of TEMED.

2. Mix well and pour the solution into the gel casting chamber (14.5 x 17 x 0.2 cm) of the vertical gel unit. No stacking gel is required for this procedure as the separating gel has a self-stacking effect. Polymerization will take place in approximately 10 minutes. The gel can be stored for several weeks submerged in 1x running buffer at 4°C until ready to use.

Binding of Phosphorylated Proteins to DNA: Mobility-shift Assay with Unlabeled Phosphoprotein and Radiolabeled DNA Probe

The phosphorylated form of the DNA-binding protein may have an enhanced, reduced, or unaltered DNA-binding activity compared to the nonphosphorylated form. Therefore, the effect of phosphorylation on DNA binding should be analyzed quantitatively. Phosphorylation may stimulate or inhibit DNA-binding activity as a result of (i) modulation of the kinetics of the interaction between protein and DNA (i.e., an alteration in the ability of the protein to associate with or dissociate from the DNA fragment [the exchange rate]) or (ii) modulation of the equilibrium affinity (i.e., the ratio of the number of bound to unbound molecules at any particular concentration after an incubation time sufficient to reach equilibrium). These two possibilities can be distinguished in a simple experiment by following the binding reaction over an extended incubation period (see, e.g., Marais et al. 1992). The final data should be expressed as kinetic or equilibrium constants (see, e.g., Affolter et al. 1990; Cao et al. 1991; Nissen et al. 1991). The study of the conditions of the phosphorylation reaction can supply additional evidence that the reaction is mediated by the protein kinase in question, provided the kinase displays some characteristic biochemical property. Casein kinase II exhibits such a property in its ability to utilize GTP as a phosphate donor. The following experiment illustrates how the conditions of the phosphorylation reaction can affect the DNA-binding properties of the target protein.

MATERIALS AND EQUIPMENT

Purified DNA-binding protein of interest
Protein kinase (appropriate for phosphorylating DNA-binding protein of interest) (9 milliunits)
Adenosine-5′-triphosphate (ATP) (4 mM) (prepared from 100 mM stock solution)
Guanosine-5′-triphosphate (GTP) (4 mM) (prepared from 100 mM stock solution)
Cytosine-5′-triphosphate (CTP) (4 mM) (prepared from 100 mM stock solution)

Uridine-5′-triphosphate (UTP) (4 mM) (prepared from 100 mM stock solution)
EDTA (100 mM)
Radiolabeled DNA probe (see Section 13, pp. 268–271)
Vertical gel unit (Life Technologies, Inc.)
Ammonium persulfate (APS) (50%)
N,N,N′,N′,tetramethylethylenediamine (TEMED)

REAGENTS

Phosphorylation buffer
Mobility-shift buffer
Native sample buffer
Acrylamide/bisacrylamide (30%)
Running buffer (for nondenaturing gel) (10x and 1x)
(For recipes, see Preparation of Reagents, pp. 347–348)

SAFETY NOTES

- Wear gloves when handling radioactive substances. Consult the local safety office for further guidance in the appropriate use of radioactive materials.
- Acrylamide and bisacrylamide are potent neurotoxins and are absorbed through the skin. Their effects are cumulative. Wear gloves and a mask when weighing acrylamide and bisacrylamide. Polyacrylamide is considered to be nontoxic, but it should be treated with care because it may contain small quantities of unpolymerized material.

PROCEDURE

Phosphorylation of DNA-binding Protein

In this procedure, phosphorylated DNA-binding protein is generated in those samples that contain protein kinase and the appropriate phosphate donor (generally ATP, but for certain protein kinases also GTP). This experiment is designed for the study of systems in which phosphorylation stimulates DNA binding. Therefore, a low amount of DNA-binding protein is used, just enough to ensure that a faint retarded protein-DNA band is visible in the control assays. Upon phosphorylation of the protein, DNA-binding activity will greatly increase and it is important to include enough free DNA probe in the assays to ensure that the concentration of the probe is not rate limiting. The opposite is necessary for the study of systems in which phosphorylation inhibits DNA binding.

1. Add 800 ng of the DNA-binding protein to 100 μl of phosphorylation buffer. Transfer 10 μl to a tube containing 1 μl of 4 mM ATP—this is the minus-kinase control sample (sample 1). Add 9 milliunits of protein kinase to the remaining 90 μl, and transfer 10 μl to a clean tube—this is the minus-ATP control sample (sample 2).

2. Remove four 20-μl aliquots from the remaining 80 μl (samples 3–6). Add 2 μl of 4 mM ATP, 2 μl of 4 mM GTP, 2 μl of 4 mM CTP, and 2 μl of 4 mM UTP to samples 3, 4, 5, and 6, respectively.

3. Incubate samples 1–6 for 20 minutes at room temperature.

4. Stop the reactions by adding 1.5 μl of 100 mM EDTA to samples 1 and 2 and 3 μl of 100 mM EDTA to samples 3–6. The final concentration of EDTA in each tube will be approximately 15 mM. Store the samples on ice.

Effect of Phosphorylation on Binding Kinetics

Here we examine how phosphorylation affects the efficiency of DNA binding. Unlabeled phosphoprotein is incubated with radiolabeled DNA probe and the DNA-protein complexes are analyzed by a mobility-shift assay on a nondenaturing polyacrylamide gel. This experiment should be performed in parallel with the experiment to study the effect of different phosphate donors (see p. 342).

1. Add 2-μl aliquots of samples 3 and 5 (i.e., DNA-binding protein phosphorylated in the presence of ATP and CTP, respectively) to 20-μl aliquots of mobility-shift buffer containing premixed radiolabeled DNA probe and perform a "staggered-start" time course reaction by incubating for various time intervals at room temperature. The incubation time is calculated as the time that elapses between the addition of phosphoprotein to mobility-shift buffer and the start of electrophoresis. Since all samples are analyzed in parallel on one gel, this experiment cannot be done by starting all incubations simultaneously and removing samples at appropriate times—the binding reaction cannot be stopped easily without disrupting the bound complexes. Start the first pair of reactions 80 minutes before the start of the gel, the second pair 40 minutes before, the third pair 20 minutes before, and the fourth

pair 10 minutes before, so as to obtain samples at 80-, 40-, 20-, and 10-minute time points.

Note: It is convenient to prepare the nondenaturing polyacrylamide gel (see p. 343) during the binding reaction.

2. Add 5 μl of native sample buffer to each sample. Load the samples with the longest incubation times onto the gel first, so that the incubation time of the 10-minute sample is not increased too much. To obtain a 0-minute time point, add the phosphoprotein samples to the mobility-shift buffer on ice, mix, load onto the gel, and start the current immediately. Run the gel in 1x running buffer at no more than 100 V for 3 hours.

3. Dry the gel and expose it to X-ray film overnight.

   Note: This experiment will generate two series of data—the binding kinetics of the unphosphorylated protein (i.e., the time points obtained with the sample in which CTP was the phosphate donor) and the binding kinetics of the phosphorylated protein (i.e., the time points obtained with the sample in which ATP was the phosphate donor). Look at the binding kinetics of the unphosphorylated protein: (i) if the abundance of the retarded protein-DNA complex remains low even at very long incubation times, then phosphorylation of the DNA-binding protein affects the equilibrium affinity of the DNA-binding protein; (ii) if the initial low level of protein-DNA complex formation increases at longer incubation times and eventually reaches that seen with the phosphorylated protein, then phosphorylation of the DNA-binding protein affects the rate at which the protein binds to the DNA fragment.

Effect of Different Phosphate Donors

Here we examine how phosphorylation in the presence of different phosphate donors affects the efficiency of DNA binding. This experiment should be performed in parallel with the experiment to study the effect of phosphorylation on binding kinetics.

1. Add 2-μl aliquots of samples 1–6 (see page 341) to 20-μl aliquots of mobility-shift buffer containing premixed radiolabeled DNA probe and incubate for 30 minutes at room temperature.

   Note: (i) It is convenient to prepare the nondenaturing polyacrylamide gel (see p. 343) during the binding reaction. (ii) Only one time point is required for this experiment. Depending on the phosphate donor present, different amounts of the retarded protein-DNA complex will be generated. The control samples (1 and 2) will show that no stimulation of complex formation takes place when one component of the phosphorylation reaction (protein kinase or ATP) is omitted.

2. Add 5 μl of native sample buffer to each sample. Load the samples onto the nondenaturing polyacrylamide gel and run the gel in 1x running buffer at no more than 100 V for 3 hours.

3. Dry the gel and expose it to X-ray film overnight.

Preparation of Nondenaturing Polyacrylamide Gel

It is convenient to perform this procedure while you are incubating the binding assays.

1. Prepare a 4% nondenaturing acrylamide gel solution by mixing the following components in the order listed:

Acrylamide/bisacrylamide (30%)	6.6 ml
Running buffer (10x)	5 ml

   Adjust the volume to 50 ml with H_2O. Add 60 µl of 50% APS and 50 µl of TEMED.

2. Mix well and pour the solution into the gel casting chamber (14.5 x 17 x 0.2 cm) of the vertical gel unit. No stacking gel is required for this procedure as the separating gel has a self-stacking effect. Polymerization will take place in approximately 10 minutes. The gel can be stored for several weeks submerged in 1x running buffer at 4°C until ready to use.

Examples of Biochemical Calculations

QUESTIONS

(See facing page for answers.)

1. A phosphorylation assay is performed in 20 μl of 25 μM ATP and the total radioactivity in the assay is 500,000 cpm. 0.5 μg of a 65-kD phosphorylation product are labeled to 20,000 cpm. What is the stoichiometry of phosphorylation in mol phosphate/mol protein?

2. A 20-μl gel-shift assay contains 1 ng of a 200-bp DNA probe. A 44-kD DNA-binding protein is added to the assay. The dissociation constant of the DNA-protein complex is 10^{-10} M. How much of this protein (in pg) is bound to DNA, and how much is free in an assay that shows 50% of the probe retarded? Assume a 1:1 molar ratio of DNA:protein in the complex.

3. A 32-kD DNA-binding protein is purified on an affinity column. 3.5 ml of a solution containing 0.15 mg/ml of protein is loaded onto the column and 1 μl of this solution retards a 300-cpm band in a gel-shift assay that contains 10,000 cpm of the probe. The purified fraction contains a total of 0.5 μg of protein in 250 μl and 1 μl of this fraction retards 2500 cpm of the probe. Assume that the assay is set up to make the intensity of the retarded band proportional to the amount of DNA-binding protein added, and that the molar ratio of DNA:protein in the complex is 1:1.

   a. Calculate the yield (%) and purification (x-fold) obtained in this procedure.

   b. Estimate the total mass of the pure DNA-binding protein in the purified fraction (2 fmoles of DNA were used in a 20-μl assay). How pure is the protein (%)?

ANSWERS

1. The assay contains:

   20 μl x 25 μM = 500 pmol of ATP

   and

   0.5 μg/65,000 g = 7.7 pmol of protein.

   The counts incorporated are:

   20,000 cpm/500,000 cpm = 4% of total phosphate,

   which is

   0.04 x 500 pmol = 20 pmol phosphate.

   The molar ratio is then:

   20 pmol phosphate/7.7 pmol protein = 2.6 mol phosphate/mol protein

2. 1 ng of 200-bp DNA is:

   1 ng/(200 x 660 g) = 7.6 fmol.

   The DNA concentration is:

   7.6 fmol/20 μl = 380 pM.

   If 50% of the probe is retarded, then:

   [L] = [L·P] and K_d = [P]

   where [L] and [P] denote the molar concentrations of the DNA ligand and protein, respectively, and [L·P] denotes the molar concentration of the protein-DNA complex. 380 pM DNA is divided equally into 190 pM in free and bound fractions.

   Free protein is 10^{-10}M, that is:

   100 pM x 20 μl = 2 fmol

   which is

   2 fmol x 44,000 g = 88 pg.

   Bound protein is:

   190 pM x 20 μl x 44,000 g = 167 pg.

   This example illustrates that if the DNA concentration is greater than the K_d, the majority of the protein will be bound at below 50% retardation, and such an assay can be used to estimate the amount of the protein.

3. a. Total activity in fraction loaded is:

   300 cpm x (3.5 ml/1 μl) = 1,050,000 cpm;

   total activity in fraction eluted is:

   2500 cpm x (250 μl/1 μl) = 625,000 cpm;

   hence the yield is:

   625,000 cpm/1,050,000 cpm = 0.60 or 60%.

   Specific activity in fraction loaded is:

   (300 cpm/μl)/(0.15 μg/μl) = 2000 cpm/μg;

   specific activity in fraction eluted is:

   (2500 cpm/μl)/(0.5/250) μg/μl = 1,250,000 cpm/μg;

   hence the purification is

   (1,250,000 cpm/μg)/(2000 cpm/μg) = 625-fold.

   b. The proportion of the probe retarded is:

   2500 cpm/10,000 cpm = 0.25 or 25%;

   which is:

   0.25 x 2 fmol = 0.5 fmol.

   The molar ratio of DNA:protein is 1:1, therefore the amount of protein retarded is:

   0.5 fmol x 32,000 g = 16 pg in 1 μl;

   that is:

   16 pg x (250 μl/1 μl) = 4 ng in the whole purified fraction.

   The purity is then:

   4 ng/0.5 μg = 0.008 or 0.8%.

• PREPARATION OF REAGENTS

Acrylamide/bisacrylamide (30%)

Acrylamide	29 g
Bisacrylamide	1 g

Dissolve in H_2O and adjust the volume to 100 ml with H_2O. Store at 4°C.

SAFETY NOTE

- Acrylamide and bisacrylamide are potent neurotoxins and are absorbed through the skin. Their effects are cumulative. Wear gloves and a mask when weighing acrylamide and bisacrylamide. Polyacrylamide is considered to be nontoxic, but it should be treated with care because it may contain small quantities of unpolymerized material.

Denaturing sample buffer (5x)

50 mM Tris-HCl (pH 8.0)
25% 2-Mercaptoethanol
50% Glycerol
2.5 mg/ml Bromophenol blue
5% Sodium dodecyl sulfate (SDS)

SAFETY NOTES

- 2-Mercaptoethanol may be fatal if swallowed and is harmful if inhaled or absorbed through the skin. High concentrations are extremely destructive to the mucous membranes, upper respiratory tract, skin, and eyes. Use only in a chemical fume hood. Gloves and safety glasses should be worn.
- Wear a mask when weighing SDS.

Mobility-shift buffer

10 mM Tris-HCl (pH 8.0)
40 mM KCl
1 mM EDTA
0.3 mg/ml Bovine serum albumin
0.07 mg/ml Poly(dI-dC)

Store at –20°C.

Native sample buffer (5x)

50 mM Tris-HCl (pH 8.0)
50% Glycerol
2.5 mg/ml Bromophenol blue

Phosphorylation buffer

50 mM Tris-HCl (pH 8.0)
10 mM $MgCl_2$
10 mM NaF
50 μg/ml Bovine serum albumin

Store at –20°C.

Running buffer (for denaturing gel) (10x)

0.25 M Tris
1 M Glycine

To use, dilute the stock solution 10-fold with H_2O and add SDS to a final concentration of 0.1%. It is not necessary to adjust the pH; this buffer will have a pH of approximately 8.3. Store at room temperature.

Running buffer (for nondenaturing gel) (10x)

0.25 M Tris
1 M Glycine

It is not necessary to adjust the pH; this buffer will have a pH of approximately 8.3. Store at room temperature. To prepare a 1x solution, dilute the stock solution 10-fold with H_2O.

Trichloroacetic acid (TCA) (100% w/v)

Add 227 ml of H_2O to a 500-g bottle of TCA. This stock solution can then be used to prepare 10% TCA/1% sodium pyrophosphate (see p. 332).

SAFETY NOTE

- TCA is highly caustic. Always wear gloves when handling this compound.

• *Section 17*

In Vivo Genomic Footprinting by Ligation-mediated PCR: Comparison of the Tomato *RbcS3B* Promoter in Dark- and Light-grown Seedlings

In vivo genomic footprinting can reveal DNA-protein interactions that are sometimes difficult or impossible to demonstrate by in vitro DNase I protection assays. This is because it is often the modified form of a DNA-binding protein that binds DNA, and the modification can be lost during the preparation of a nuclear extract or other in vitro manipulations. In vivo genomic footprinting is also useful for confirming the results of DNase I protection assays and for validating or excluding DNA-protein interactions at specific sites. In addition, this technique has the potential to reveal changes in DNA structure that occur as a result of DNA-protein or DNA-protein-protein interactions and that are difficult to detect in vitro. This section describes a protocol for the use of in vivo genomic footprinting by ligation-mediated polymerase chain reaction (PCR) to detect differences in the DNA-protein interactions of tomato *RbcS* promoters from dark- and light-grown seedlings and to confirm the results of DNase I protection assays obtained in vitro (Manzara et al. 1991, 1993; Carrasco et al. 1993) (see also Section 12).

The basis of the in vivo genomic footprinting technique is that dimethyl sulfate (DMS) can penetrate tissues quickly and methylate guanosine (G) residues in the DNA in vivo at sites that are not protected by proteins. The methylated G residues in the purified genomic DNA can subsequently be cleaved using Maxam-Gilbert chemistry (Church and Gilbert 1984). The principle of in vivo genomic footprinting by ligation-mediated PCR, which was developed by Mueller and Wold (1989), is similar to that of the original procedure (Church and Gilbert 1984), but it has advantages in that it eliminates much of the genomic DNA handling and fractionation and increases the signal-to-noise ratio, thus shortening the exposure time necessary to detect a good signal. Most of the advantages of this procedure have been discussed in detail by Mueller and Wold (1989; 1991a).

Figure 1 shows a schematic representation of the in vivo genomic footprinting by ligation-mediated PCR technique. Genomic DNA from DMS-treated tissue is cleaved with piperidine (A), and sub-

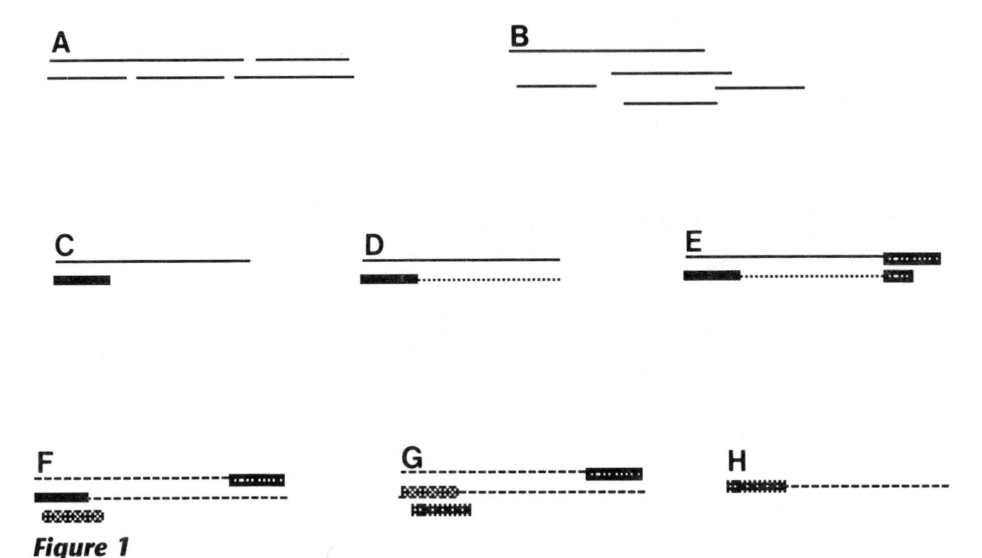

Figure 1
Schematic representation of the in vivo genomic footprinting by ligation-mediated PCR technique. For details, see text.

sequent denaturation of the cleaved DNA generates a population of single-stranded DNA fragments ending in a G residue (B). The target sequence (i.e., the promoter region for which in vivo DNA-protein interactions are to be determined) is hybridized to a sequence-specific primer (C) and extended with Sequenase®, creating a population of blunt-ended fragments 3′ distal to the primer (D). To amplify the sequence of interest, an asymmetric double-stranded linker-primer is ligated to the blunt-ended fragments generated in the Sequenase® reaction (E). PCR is carried out using a second sequence-specific primer (which hybridizes close to or overlaps with the primer used in the Sequenase® reaction) and the long strand of the linker-primer (F). Following PCR, a third sequence-specific primer (which can be the same as that used for the PCR reaction) is labeled with [γ-^{32}P]ATP and used for one new cycle of denaturation, hybridization, and extension with *Taq* polymerase (G). The labeled DNA fragments (H) are separated on a sequencing gel. The design of the primers is critical for the success of the experiment. If possible, primers should be gene specific, otherwise undesired products could be obtained after the Sequenase® or PCR reactions. For a detailed discussion of the design of primers and their specificity, see Mueller and Wold (1991b).

The following experimental procedure was specifically designed to study a segment of the tomato *RbcS3B* promoter that contains several DNA sequence motifs with putative functions in the light-induced regulation of transcription of this gene. Transcription of the tomato *RbcS3B* gene in germinating seedlings is strictly regulated by light (Sugita and Gruissem 1987; Wanner and Gruissem 1991). In vitro DNase I footprinting studies of the tomato *RbcS3B* promoter have indicated that extensive sequences within the promoter are protected from digestion by DNA-protein interactions (Manzara et al. 1991; Carrasco et al. 1993). However, the footprinting studies did not reveal any differences in DNA binding between nuclear protein extracts from seedlings grown in the dark and those from seedlings illuminated for six hours, even though there is considerable light-induced transcription of the *RbcS3B* gene after six hours of illumination (Wanner and Gruissem 1991). This result could indicate that: (i) the DNA-binding proteins are already present in dark-grown seedlings; (ii) protein modifications are required for differential binding, and constitutive binding occurs if these modifications are lost in vitro; or (iii) DNA-protein complexes are already formed in dark-grown seedlings, and modification of existing complexes and/or protein-protein interactions may occur in a secondary, light-dependent reaction. The data generated by in vivo genomic footprinting for this segment of the tomato *RbcS3B* promoter confirm the results obtained by DNase I footprinting (Manzara et al. 1991; Carrasco et al. 1993).

In vivo genomic footprinting can clarify some of the questions raised above by providing information about the temporal and spatial interactions of proteins with DNA. The technique is limited, however, and does not provide information on the length of a DNA sequence protected by a protein. Also, protection against G methylation is not always complete. This could result from prolonged DMS treatment, in which case the experimental conditions may have to be adjusted. Alternatively, if a whole organ is treated with DMS, then methylation of genomic DNA may vary between tissues in which genes are differentially expressed. Under these circumstances, incomplete methylation protection may indicate differences in DNA-protein interactions that reflect the transcriptional status of the gene of interest in different tissues. These limitations should be considered when using the techniques described here.

REFERENCES

Carrasco, P., T. Manzara, and W. Gruissem. 1993. Developmental and organ-specific changes in DNA-protein interactions in the tomato *rbcS3B* and *rbcS3C* promoter regions. *Plant Mol. Biol.* **21:** 1–15.

Church, G.M. and W. Gilbert. 1984. Genomic sequencing. *Proc. Natl. Acad. Sci.* **81:** 1991–1995.

Manzara, T., P. Carrasco, and W. Gruissem. 1991. Developmental and organ-specific changes in promoter DNA-protein interactions in the tomato *rbcS* gene family. *Plant Cell.* **3:** 1305–1316.

———. 1993. Developmental and organ-specific changes in DNA-protein interactions in the tomato *rbcS1*, *rbcS2* and *rbcS3A* promoter regions. *Plant Mol. Biol.* **21:** 69–88.

Mueller, P.R. and B. Wold. 1989. In vivo footprinting of a muscle specific enhancer by ligation mediated PCR. *Science* **246:** 780–786.

———. 1991a. Ligation-mediated PCR for genomic sequencing and footprinting. *Current protocols in molecular biology* (ed. F.M. Ausubel et al.), pp. 15.5.1–15.5.15. Wiley, New York.

———. 1991b. Ligation-mediated PCR: Applications to genomic footprinting. *Methods* **2:** 20–31.

Sambrook, J., E.F. Fritsch, and T. Maniatis. 1989. *Molecular cloning: A laboratory manual,* 2nd edition. Cold Spring Harbor Laboratory Press, Cold Spring Harbor, New York.

Sugita, M. and W. Gruissem. 1987. Developmental, organ-specific, and light-dependent expression of the tomato ribulose-1,5-bisphosphate carboxylase small subunit gene family. *Proc. Natl. Acad. Sci.* **84:** 7104–7108.

Wahl, G.M., S.L. Berger, and A.R. Kimmel. 1987. Molecular hybridization of immobilized nucleic acids: Theoretical concepts in practical considerations. *Methods Enzymol.* **152:** 399–407.

Wanner, L.A. and W. Gruissem. 1991. Expression dynamics of the tomato *rbcS* gene family during development. *Plant Cell.* **3:** 1289–1303.

• *TIMETABLE*

In Vivo Genomic Footprinting by Ligation-mediated PCR

STEPS	***TIME REQUIRED***
DMS Treatment of Cotyledons from Tomato Seedlings	
Grow tomato seedlings in dark	7 days
Expose some seedlings to light	6 hours
Harvest cotyledons from dark- and light-grown seedlings	30 minutes
Treat cotyledons with DMS	10 minutes
Minipreparation of Genomic DNA	
Extract DNA from DMS-treated cotyledons	3–4 hours
Resuspend DNA in TE	several hours or overnight
Treat DNA with RNase	1 hour
Re-extract DNA	2 hours
Resuspend DNA in TE	several hours or overnight
Run agarose gel to determine quality of DNA	3 hours
In Vitro Methylation of Genomic DNA	
Prepare genomic DNA methylated in vitro to use as control	1–2 hours

STEPS	*TIME REQUIRED*
Chemical Cleavage of Methylated DNA with Piperidine	
Incubate methylated DNA samples with piperidine	30 minutes
Dry samples and wash three times with H_2O	5–6 hours
First-strand Synthesis	
Denature cleaved, methylated DNA	2 minutes
Hybridize primer to methylated DNA fragments	30 minutes
Perform first-strand synthesis reaction to yield blunt-ended DNA fragments	20 minutes
Linker-primer Ligation	
Ligate linker-primer to blunt-ended DNA fragments	overnight
PCR Amplification	
Perform 20 PCR cycles	3–4 hours
Labeling of DNA and Final PCR Cycle	
Prepare labeled primer	2 hours
Perform final PCR cycle	1 hour
Sequencing Gel Analysis	
Pour gel	1 hour
Load and run gel	3–4 hours
Dry gel and expose gel to X-ray film	overnight[1]

[1]Overnight exposure may be too long. Therefore, try exposing gel for only a few hours first.

DMS Treatment of Cotyledons from Tomato Seedlings

MATERIALS

Tomato seeds (e.g., VFNT Cherry or other cultivars)
Clorox bleach (5% in H_2O)
Dimethyl sulfate (DMS) (0.1%)
Liquid nitrogen

SAFETY NOTES

- DMS is toxic and a carcinogen and should be handled in a chemical fume hood. Be aware that DMS can penetrate rubber and vinyl gloves. Tips and tubes that are contaminated with DMS should be rinsed with 5 M NaOH and discarded in separate waste containers. Solutions containing DMS should be discarded in a waste bottle containing 5 M NaOH.
- The temperature of liquid nitrogen is –185°C. Handle with great care; always wear gloves and a face protector.

PROCEDURE

1. Sterilize the tomato seeds by washing them in 5% Clorox bleach for 20 minutes. Rinse the seeds thoroughly in sterile H_2O. Spread the sterilized seeds on three layers of wet, sterile Whatman 3MM paper in two large dishpans and cover with aluminum foil. Allow the seeds to germinate by placing them in the dark for 7 days at 25°C. After 7 days, expose the etiolated seedlings in one dishpan to light (500–1000 $\mu E \cdot m^{-2} \cdot sec^{-1}$) for 6 hours.

2. Harvest the cotyledons from the dark- and light-grown seedlings (2–5 g from each dishpan). Use a razor blade to shear off the cotyledons quickly; you will also harvest part of the hypocotyl.

   Note: If possible, harvest the cotyledons from the dark-grown seedlings under a green safe light. This will avoid rapid protein modifications that may occur in white light. Green safe light does not activate the photochrome system.

3. Place each batch of cotyledons directly into 0.1% DMS in a vacuum flask. Slowly apply a vacuum and allow the DMS to infiltrate the tissue for 1–2 minutes. Gently release the vacuum. Quickly wash each batch of cotyledons once with ice-cold H_2O, and immediately transfer them to liquid nitrogen. Store them at –80°C.

   Note: Again, it is preferable to treat the cotyledons from the dark-grown seedlings with DMS under a green safe light.

Minipreparation of Genomic DNA

MATERIALS

DMS-treated cotyledons from dark- and light-grown tomato seedlings (see pp. 355–356)
Liquid nitrogen
Proteinase K (10 mg/ml in grinding buffer)
Phenol/chloroform/isoamyl alcohol (12:12:1) (preequilibrate the phenol with grinding buffer)
Sodium acetate (3 M; pH 5.5)
Ethanol (100%)
RNase A (DNase-free; heat-treated) (5 mg/ml in TE [pH 7.0])

REAGENTS

Grinding buffer
TE (pH 7.0)
(For recipes, see Preparation of Reagents, pp. 376–378)

SAFETY NOTES

- The temperature of liquid nitrogen is –185°C. Handle with great care; always wear gloves and a face protector.
- Phenol is highly corrosive and can cause severe burns. Wear gloves, protective clothing, and safety glasses when handling it. All manipulations should be carried out in a chemical fume hood. Any areas of skin that come in contact with phenol should be rinsed with a large volume of water or PEG 400 and washed with soap and water; do not use ethanol!
- Chloroform is irritating to the skin, eyes, mucous membranes, and respiratory tract. It should only be used in a chemical fume hood. Gloves and safety glasses should also be worn. Chloroform is a carcinogen and may damage the liver and kidneys.

PROCEDURE

Prepare genomic DNA from DMS-treated cotyledons harvested from both dark- and light-grown tomato seedlings.

1. Using a mortar and pestle, grind 1–2 g of cotyledons briefly in liquid nitrogen. Add 5 ml of grinding buffer and continue grinding until the tissue is homogenized into a liquid suspension.

2. Pour the suspension into a 50-ml Falcon tube. Rinse the mortar with 5 ml of grinding buffer and combine the rinse with the suspension. Add 10 mg/ml proteinase K to give a final concentration of 0.1 mg/ml. Rock gently for 1 hour at room temperature.

3. Add 5 ml of phenol/chloroform/isoamyl alcohol (12:12:1) and mix gently by inversion several times. Separate the phases by spinning in an IEC centrifuge at maximum speed (~4000*g*) for 5 minutes at room temperature.

4. Transfer the aqueous (upper) phase to a clean tube with a wide-bore pipette, and repeat the phenol/chloroform/isoamyl alcohol extraction.

5. Transfer the aqueous phase to a 30-ml Corex tube, and add 1 ml of 3 M sodium acetate (pH 5.5) and 10 ml of 100% ethanol. Mix gently until fibrous material precipitates. Spin at 10,000 rpm for 10 minutes at 4°C, and drain the tube.

6. Resuspend the pellet in 0.5 ml of TE (pH 7.0)—this may take several hours.

7. Add 5 μl of 5 mg/ml RNase A and incubate overnight at 4°C or for 30 minutes at 37°C.

   Note: The RNase digestion is critical because the genomic DNA preparation is contaminated with RNA, which can contribute to high background levels in the footprinting experiment.

8. Transfer the mixture to a microfuge tube and extract the DNA once with phenol/chloroform/isoamyl alcohol (12:12:1) (see step 3) and precipitate it with 50 μl of 3 M sodium acetate (pH 5.5) and 800 μl of 100% ethanol (see step 5). Collect the DNA by spinning in a microfuge at 10,000 rpm for 10 minutes at 4°C.

9. Resuspend the DNA pellet in 50 μl of TE (pH 7.0) by incubating overnight at 4°C.

10. Determine the DNA concentration by reading the OD_{260}.

11. Check the quality of the DMS-treated genomic DNA by agarose gel electrophoresis (Sambrook et al. 1989). You should expect to see a broad band of DNA of approximately 40 kb.

    Note: The DMS-treated genomic DNA will appear slightly degraded, but this is normal.

12. To prepare samples for cleavage with piperidine (see step 1, p. 362), precipitate 3 μg of the DNA with 100% ethanol. Collect the DNA by spinning at 10,000 rpm for 10 minutes at 4°C. Dry the DNA pellet in a SpeedVac® for 5–10 minutes.

In Vitro Methylation of Genomic DNA

When performing an in vivo genomic footprinting experiment, it is a good idea to use genomic DNA methylated in vitro as a control. Here we describe the standard Maxam-Gilbert procedure for treatment of DNA with DMS in vitro.

MATERIALS

Genomic DNA purified from seedlings that have not been treated with DMS (use the procedure described on pp. 357–359)
Dimethyl sulfate (DMS) (ice-cold)
Ethanol (100% and 70%; ice-cold)
Sodium acetate (0.3 M; pH 5.2)

REAGENTS

DMS buffer
DMS stop solution
(For recipes, see Preparation of Reagents, pp. 376–378)

SAFETY NOTE

- DMS is toxic and a carcinogen and should be handled in a chemical fume hood. Be aware that DMS can penetrate rubber and vinyl gloves. Tips and tubes that are contaminated with DMS should be rinsed with 5 M NaOH and discarded in separate waste containers. Solutions containing DMS should be discarded in a waste bottle containing 5 M NaOH.

PROCEDURE

1. Mix 5 μl of genomic DNA (3 μg) with 200 μl of DMS buffer and chill to 0°C. Add 1 μl of ice-cold undiluted DMS.

   Note: Alternatively, cloned DNA can be used as a control for base modifications. Approximately 0.01 fmol of cloned DNA will give bands on the sequencing gel that are similar in intensity to those obtained with 3 μg of genomic DNA.

2. Incubate for 10 minutes at room temperature.

3. Add 50 μl of DMS stop solution and 750 μl of ice-cold 100% ethanol.

4. Incubate for 5 minutes at –70°C. Spin in a microfuge for 5 minutes at 4°C.

5. Resuspend the pellet in 250 μl of 0.3 M sodium acetate (pH 5.2). Add 750 μl of ice-cold 100% ethanol, mix well, and incubate for 5 minutes at –70°C. Spin in a microfuge for 5 minutes at 4°C.

6. Wash the pellet with 500 μl of ice-cold 70% ethanol. Spin in a microfuge for 5 minutes at 4°C. Remove the ethanol and dry the methylated DNA in a SpeedVac® for 5–10 minutes.

Chemical Cleavage of Methylated DNA with Piperidine

MATERIALS

Methylated DNA:
- Sample 1—genomic DNA from DMS-treated cotyledons harvested from dark-grown tomato seedlings (pp. 357–359)
- Sample 2—genomic DNA from DMS-treated cotyledons harvested from light-grown tomato seedlings (pp. 357–359)
- Sample 3—genomic DNA treated with DMS in vitro (pp. 360–361)

Piperidine (0.1 M)

SAFETY NOTE

- Piperidine is corrosive to the eyes, skin, respiratory tract, and gastrointestinal tract. It reacts violently with acids and oxidizing agents. Keep piperidine away from heat or flames. Wear gloves and safety glasses when handling piperidine. Avoid breathing the vapors.

PROCEDURE

1. Resuspend 3 μg of each sample of methylated DNA in 100 μl of 0.1 M piperidine. Close the tubes tightly and incubate them for 30 minutes at 90°C.

2. Remove or puncture the caps and dry the samples in a SpeedVac®. This can take up to 2 hours and the samples should be checked every 15 minutes.

3. Resuspend each pellet in 20 μl of H_2O and vortex for 30 seconds. Dry the samples again in the SpeedVac® for 10–15 minutes.

4. Repeat step 3 twice.

5. Resuspend each DNA sample in a small volume of H_2O to give a concentration of approximately 1 μg/μl.

First-strand Synthesis

MATERIALS

Methylated DNA cleaved with piperidine (see p. 362)
Tomato *RbcS3B* primer I dissolved in H_2O or TE (pH 7.5). This primer can be synthesized using the published sequence or obtained from the laboratory of Dr. W. Gruissem (see Appendix 2).
Sequenase® 2.0 (from Sequenase® version 2.0 DNA sequencing kit [Amersham Life Science, Inc. US 70770]) (diluted 1:4 in TE [pH 7.5])
Tris-HCl (310 mM; pH 7.7)

REAGENTS

Sequenase® buffer (magnesium-free) (5x)
First-strand synthesis buffer
TE (pH 7.5)
(For recipes, see Preparation of Reagents, pp. 376–378)

PROCEDURE

1. Mix 3 µg of each sample of methylated, cleaved DNA with 0.3 pmol of tomato *RbcS3B* primer I and 3 µl of 5x magnesium-free Sequenase® buffer to give a final volume of 15 µl.

2. Denature the DNA by incubating for 2 minutes at 95°C.

3. Hybridize the primer to the DNA by incubating for 30 minutes at 40°C.

   Note: The hybridization temperature depends on the T_m of the primer. Hybridization should be performed approximately 5°C below the T_m of the primer. T_m is calculated according to the formula:

   $$T_m = 81.5 + 16.6(\log M) + 0.41(\%G\text{–}C) - 500/n$$

   where M is the molar concentration of monovalent cation (~0.05 in this experiment) and n is the length of the primer (Wahl et al. 1987). The hybridization temperature used in this experiment is specific for tomato *RbcS3B* primer I.

4. Transfer the hybridization mixtures to ice.

5. Add 7.5 µl of first-strand synthesis buffer and 1.5 µl of a 1:4 dilution of Sequenase® 2.0 in TE (pH 7.5) to each sample.

6. Incubate for 5 minutes at 40°C.

   Note: (i) Again, the incubation temperature depends on the T_m of the primer, and can be 40–45°C (for a more detailed discussion, see Mueller and Wold 1991b). (ii) This is a good time to prepare the ligase reaction mixture (see p. 365, step 1).

7. To terminate the first-strand synthesis reactions, incubate the tubes for 5 minutes at 60°C, remove the tubes from the heat, immediately add 6 µl of 310 mM Tris-HCl (pH 7.7), and incubate at 67°C for 10 minutes.

8. Transfer the tubes to ice. The blunt-ended DNA fragments are now ready for the linker-primer ligation reaction.

Linker-primer Ligation

MATERIALS

Linker-primer (20 μM in 250 mM Tris-HCl [pH 7.7]) (synthesize to specification)
T4 DNA ligase (3–9 "Weiss" units)
Blunt-ended DNA fragments (see p. 364)
Sodium acetate (3 M; pH 7.0)
Yeast tRNA (10 mg/ml)
Ethanol (100% and 70%; ice-cold)

REAGENTS

Ligase mix
Ligase dilution buffer (ice-cold)
(For recipes, see Preparation of Reagents, pp. 376–378)

PROCEDURE

1. Mix 5 μl of 20 μM linker-primer with 3–9 "Weiss" units of T4 DNA ligase and adjust the volume to 25 μl with ligase mix. Prepare 25 μl of this ligase reaction mixture per DNA sample. Keep the ligase reaction mixture on ice until you are ready to use it.

   Note: You should prepare the ligase reaction mixture while you are incubating the first-strand synthesis reaction (see p. 364)

2. Add 20 μl of ice-cold ligase dilution buffer to each DNA sample. Mix.

3. Add 25 μl of ligase reaction mixture to each sample.

4. Incubate the ligation reactions overnight at 16°C.

   Note: Blunt-end ligations are notoriously difficult, and therefore, shortcuts at this step are not recommended.

5. Terminate the ligation reactions by heating the samples for 10 minutes at 70°C.

6. Add 10 μl of 3 M sodium acetate (pH 7.0) and 1 μl of 10 mg/ml yeast tRNA to each sample.

7. To precipitate the DNA, add 220 μl of ice-cold 100% ethanol to each sample and incubate for 2 hours at –20°C (or for 20 minutes in dry ice). Spin at 10,000 rpm for 10 minutes at 4°C. Wash the DNA pellet with ice-cold 70% ethanol, spin briefly, and dry in a SpeedVac®.

8. Resuspend each DNA sample in 50 μl of H_2O.

PCR Amplification

MATERIALS AND EQUIPMENT

DNA samples resuspended in H_2O (see p. 366, step 8)
Bovine serum albumin (BSA) (DNase-free; 10 mg/ml)
dNTPs (mixture prepared from 100 mM stock solutions; Pharmacia Biotech, Inc.)
Tomato *RbcS3B* primer II dissolved in TE (pH 7.5). This primer can be synthesized using the published sequence or obtained from the laboratory of Dr. W. Gruissem (see Appendix 2).
Long strand of linker-primer dissolved in TE (pH 7.5)
Taq polymerase (Perkin Elmer Corp. N 8010060)
Mineral oil (Sigma)
PCR thermal cycler (Perkin Elmer Corp.)

REAGENTS

Taq buffer (10x)
TE (pH 7.5)
(For recipes, see Preparation of Reagents, pp. 376–378)

PROCEDURE

This reaction is critical and will determine the success or failure of the experiment. It is important to calculate the T_m of the primers that are used in this reaction (see p. 363) so as to achieve good priming and amplification with the appropriate PCR program. Also, it is important that the temperature calibration is correct.

1. Set up the following reaction mixture in a total volume of 100 µl:

DNA sample resuspended in H_2O	45 µl
Taq buffer (10x)	10 µl
BSA (10 mg/ml)	1 µl
dNTPs	20 nmol
Tomato *RbcS3B* primer II	10 pmol
Long strand of linker-primer	10 pmol
Taq polymerase	2.5 units

2. Cover each reaction mixture with mineral oil.

3. Heat for 4 minutes at 94°C.

4. Run 20 PCR cycles with the following program:

Process	**Time**	**Temperature**
Denaturation	1 minute	94°C
Annealing	1 minute	56°C
Extension	1 minute	74°C

Note: The annealing temperature depends on the T_m of the primers: use the lower of the two T_m values.

5. Spin each sample in a microfuge for 15 seconds at room temperature, and keep the reaction mixtures on ice.

Labeling of DNA and Final PCR Cycle

MATERIALS AND EQUIPMENT

Tomato *RbcS3B* primer II. This primer can be synthesized using the published sequence or obtained from the laboratory of Dr. W. Gruissem (see Appendix 2).
[γ-^{32}P]ATP (6000 Ci/mmol)
T4 polynucleotide kinase
EDTA (0.5 M)
NENSORB® nucleic acid purification cartridges (DuPont NEN® NPL-028)
PCR reaction mixtures (see p. 368, step 5)
Taq polymerase (Perkin Elmer Corp. N 8010060)
dNTPs (supplied in Sequenase® 2.0 kit)
PCR thermal cycler (Perkin Elmer Corp.)
Sodium acetate (3 M; pH 7.0)
Phenol/chloroform/isoamyl alcohol (12:12:1)
Ethanol (100% and 70%)
Sequenase® version 2.0 DNA sequencing kit (Amersham Life Science, Inc. US 70770)

REAGENTS

Kinase buffer (10x)
TE (pH 8.0)
Taq buffer (10x)
(For recipes, see Preparation of Reagents, pp. 376–378)

SAFETY NOTES

- Wear gloves when handling radioactive substances. Consult the local safety office for further guidance in the appropriate use of radioactive materials.
- Phenol is highly corrosive and can cause severe burns. Wear gloves, protective clothing, and safety glasses when handling it. All manipulations should be carried out in a chemical fume hood. Any areas of skin that come in contact with phenol should be rinsed with a large volume of water or PEG 400 and washed with soap and water; do not use ethanol!

- Chloroform is irritating to the skin, eyes, mucous membranes, and respiratory tract. It should only be used in a chemical fume hood. Gloves and safety glasses should also be worn. Chloroform is a carcinogen and may damage the liver and kidneys.

PROCEDURE

Preparation of Labeled Primer

1. Set up the following reaction mixture in a total volume of 30 μl:

Kinase buffer (10x)	3 μl
Tomato *RbcS3B* primer II (5 pmol/μl)	3–5 pmol
[γ-^{32}P]ATP (6000 Ci/mmol)	5 μCi
T4 polynucleotide kinase	30 units

   Adjust the volume to 30 μl with H_2O.

   Note: (i) You will need one 30-μl reaction mixture for each PCR reaction. (ii) In case primer II is not absolutely specific for the DNA sequence of interest, you can use a third primer (III) that overlaps primer II for the final PCR cycle. This will decrease the background of the PCR labeling step. High background should not be a problem in the experiment described here because tomato *RbcS3B* primer II is specific.

2. Incubate for 45 minutes at 37°C.

3. Stop the reaction by adding 1 μl of 0.5 M EDTA and 170 μl of TE (pH 8.0).

4. Separate the labeled primer from unincorporated [γ-^{32}P]ATP by chromatography on a NENSORB® nucleic acid purification cartridge according to the manufacturer's instructions.

   Note: A Sephadex™ G-10 spin column can be used instead of the NENSORB® nucleic acid purification cartridge.

Final PCR Cycle

1. Set up the following reaction mixture in a total volume of 100 μl:

PCR reaction mixture	50 μl
Labeled primer II	1–5 pmol
Taq polymerase	2.5 units
dNTPs	20 nmol
Taq buffer (10x)	5 μl

   Adjust the volume to 100 μl with H_2O.

2. Run the final PCR cycle with the following program:

Process	Time	Temperature
Denaturation	2 minutes	94°C
Annealing	2 minutes	56°C
Extension	2 minutes	74°C

Note: Again, the annealing temperature depends on the T_m of the primer. If you use a primer III that is different from primer II, you must calculate the T_m of the new primer.

3. Place the reaction mixture on ice.

4. Add 12 µl of 3 M sodium acetate (pH 7.0) and 120 µl of phenol/chloroform/isoamyl alcohol (12:12:1).

5. Mix thoroughly, and spin at room temperature to separate the phases.

6. Transfer the aqueous (upper) phase to a clean tube. Precipitate the DNA by adding two volumes of 100% ethanol. Pellet the DNA by spinning at 10,000 rpm for 10 minutes at 4°C, and wash the pellet with 70% ethanol.

7. Dry the DNA in a SpeedVac®, and resuspend the DNA pellet in 3–5 µl of sequencing dye solution (supplied with the Sequenase® 2.0 kit).

Sequencing Gel Analysis

MATERIALS AND EQUIPMENT

Sequencing gel apparatus and power supply
Urea
Ammonium persulfate (APS) (10%)
N,N,N',N',tetramethylethylenediamine (TEMED)
DNA samples (amplified by PCR and labeled) (see p. 371, step 7)

REAGENTS

Acrylamide (38%)/bisacrylamide (2%) stock solution
TBE (10x)
TBE (1x) (prepared from 10x stock solution)
(For recipes, see Preparation of Reagents, pp. 376–378)

SAFETY NOTE

- Acrylamide and bisacrylamide are potent neurotoxins and are absorbed through the skin. Their effects are cumulative. Wear gloves and a mask when weighing acrylamide and bisacrylamide. Polyacrylamide is considered to be nontoxic, but it should be treated with care because it may contain small quantities of unpolymerized material.

PROCEDURE

1. Assemble the sequencing gel casting chamber.

2. For a narrow plate, prepare a 5% acrylamide gel solution by mixing the following components in the order listed:

Acrylamide (38%)/bisacrylamide (2%) stock solution	6.25 ml
Urea	24 g
TBE (10x)	5 ml
APS (10%)	50 μl

Adjust the volume to 50 ml with H_2O and mix well. Add 50 μl of TEMED. Pour the solution into the gel casting chamber, and insert the comb. The gel should take approximately 20 minutes to polymerize.

3. After the gel has polymerized, remove the comb and attach the glass plates to the electrophoresis tank. Fill the upper and lower reservoirs of the tank with 1x TBE. Flush out the wells with 1x TBE.

4. Denature the DNA samples by heating them at 95°C for 5 minutes. Place the samples on ice.

5. Load the samples onto the gel. Run the gel at approximately 10 V/cm until the bromophenol blue dye in the sequencing dye solution (see step 7, p. 371) has reached the bottom of the gel.

6. Dry the gel and expose it to X-ray film.

   Note: You should be able to detect a good signal after overnight exposure to X-ray film in the presence of an intensifying screen. Alternatively, you can analyze the gel using a phosphorimager.

Expected Results

The results expected for the experiment described in this protocol are shown in Figure 2.

- The control lanes on the sequencing gel contain genomic DNA or cloned DNA methylated in vitro and they show all the G bands of the genomic sequence from the corresponding DNA strand.

- The lanes containing genomic DNA from DMS-treated tissues show that some of the methylated G bands present in the genomic sequence disappear or are present at a reduced intensity. This indicates protection from methylation as a result of protein-DNA interactions.

- The intensity of some methylated G bands may increase relative to the controls. This is because the proximity of a protein may cause neighboring bases to become more susceptible to methylation.

- Although the in vivo footprinting procedure uses the Maxam and Gilbert protocol for the detection of G residues, adenosine (A) residues are also detected by the autoradiography, indicating that A residues are also methylated to some degree. The A residues are usually protected from methylation, and therefore, A methylation may indicate that DNA-protein interactions cause torsional changes in the DNA that make A residues more accessible to DMS.

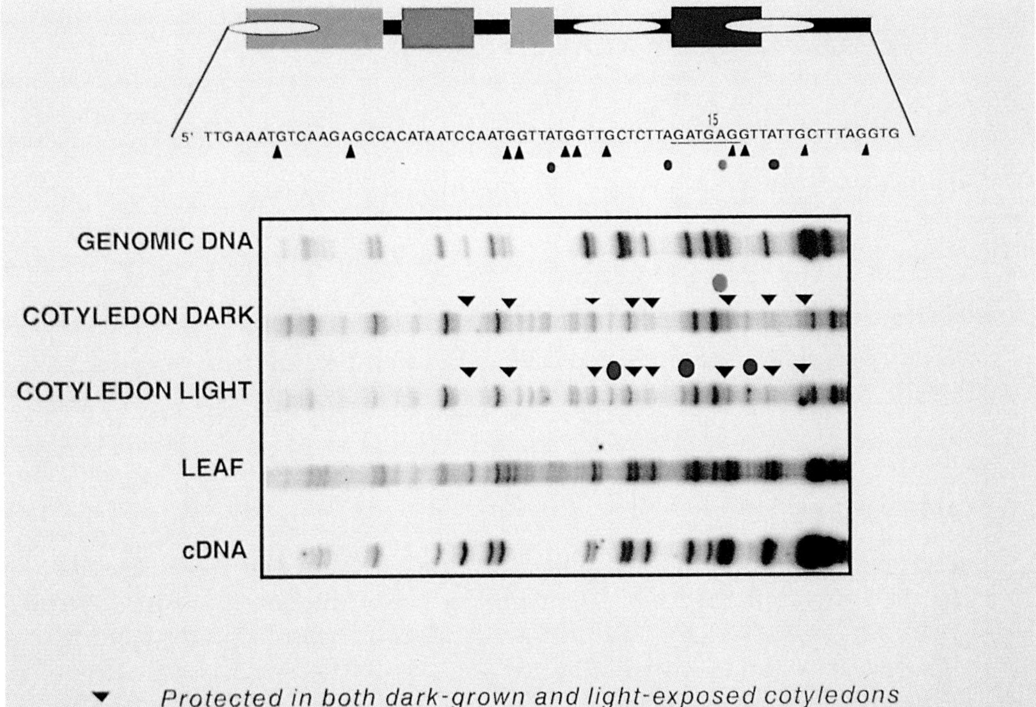

Figure 2

DNA-protein interactions detected by in vivo footprinting. In vivo analysis of the *RbcS3B* promoter during light-induced transcriptional activation in cotyledons. The control lanes contain genomic DNA and cloned DNA methylated in vitro and they show all the guanosine (G) bands of the genomic sequence. The black triangles indicate the G bands that are protected from methylation in both dark-grown and light-exposed cotyledons. The red circles indicate the methylated G bands that are enhanced in light-exposed cotyledons, and the green circles indicate the methylated G bands that are enhanced in dark-grown cotyledons.

• *PREPARATION OF REAGENTS*

Acrylamide (38%)/bisacrylamide (2%) stock solution
(100 ml)

Acrylamide (Bio-Rad Laboratories)	38 g
Bisacrylamide	2 g

Dissolve the components in distilled H_2O and adjust the volume to 100 ml. Store at 4°C in the dark.

SAFETY NOTE

- Acrylamide and bisacrylamide are potent neurotoxins and are absorbed through the skin. Their effects are cumulative. Wear gloves and a mask when weighing acrylamide and bisacrylamide. Polyacrylamide is considered to be nontoxic, but it should be treated with care because it may contain small quantities of unpolymerized material.

DMS buffer

50 mM Sodium cacodylate (pH 8.0)
1 mM EDTA

Store at –20°C.

DMS stop solution

1.5 M Sodium acetate (pH 7.0)
1 M 2-Mercaptoethanol
250 μg/ml Yeast tRNA

Store at –20°C.

SAFETY NOTE

- 2-Mercaptoethanol may be fatal if swallowed and is harmful if inhaled or absorbed through the skin. High concentrations are extremely destructive to the mucous membranes, upper respiratory tract, skin, and eyes. Use only in a chemical fume hood. Gloves and safety glasses should be worn.

First-strand synthesis buffer

20 mM $MgCl_2$
20 mM Dithiothreitol
0.02 mM dNTPs

Store at –20°C.

Grinding buffer

0.1 M Tris-HCl (pH 8.5)
50 mM EDTA
0.1 M NaCl
2% Sodium dodecyl sulfate (SDS)

Store at room temperature.

SAFETY NOTE

- Wear a mask when weighing SDS.

Kinase buffer (10x)

500 mM Tris-HCl (pH 7.5)
100 mM $MgCl_2$
50 mM Dithiothreitol
50 μg/ml Bovine serum albumin

Store at –20°C.

Ligase dilution buffer

17.5 mM $MgCl_2$
42.3 mM Dithiothreitol
125 μg/ml Bovine serum albumin

Store at –20°C.

Ligase mix

10 mM $MgCl_2$
20 mM Dithiothreitol
3 mM ATP
50 μg/ml Bovine serum albumin

Store at –20°C.

Sequenase® buffer (magnesium-free) (5x)

200 mM Tris-HCl (pH 7.7)
250 mM NaCl

Taq buffer (10x)

500 mM KCl
50 mM Tris-HCl (pH 8.9)
50 mM $MgCl_2$

TBE (10x)
(1 liter)

Tris base	108 g
Boric acid	55 g
0.5 M EDTA (pH 8.0)	40 ml

Adjust the volume to 1 liter with H_2O. Store at room temperature.

TE (pH 7.0)

10 mM Tris-HCl (pH 7.0)
1 mM EDTA

Store at room temperature.

TE (pH 7.5)

10 mM Tris-HCl (pH 7.5)
1 mM EDTA

Store at room temperature.

TE (pH 8.0)

10 mM Tris-HCl (pH 8.0)
1 mM EDTA

Store at room temperature.

• *Section 18*

Pulsed-field Gel Electrophoresis and Yeast Artificial Chromosome Analysis of the *Arabidopsis* Genome

Yeast artificial chromosomes (YACs) were conceived as tools for the analysis, mapping, and manipulation of mammalian genomes (Burke et al. 1987) (for review, see Schlessinger 1990). The main breakthrough associated with the use of YACs is the large increase in the size of DNA that can be cloned as a single segment. Before the development of YACs, the largest capacity vectors were cosmids, which can accommodate a maximum of 50 kb of DNA. The first YAC library (Burke et al. 1987) exceeded this by a factor of five, and more recently constructed libraries contain inserts averaging more than one megabase (Chumakov et al. 1992). The reason for this large capacity is that, unlike traditional vectors, the YAC cloning system places no theoretical constraints on the amount of DNA that can be carried by one clone.

As shown in Figure 1 (A), the YAC vector, in this case pYAC4, is propagated as a plasmid in *Escherichia coli* (*E. coli*). Digestion with *Eco*RI and *Bam*HI yields two vector arms that contain all the elements necessary for maintaining exogenous DNA as a chromosome in the yeast, *Saccharomyces cerevisiae*. These elements are telomeres, the selectable markers URA3 and TRP1 that confer uracil- and tryptophan-independent growth, a centromere, and an autonomous replicating sequence (origin of replication). The vector arms are ligated to large segments of genomic DNA (Fig. 1, B) and the ligated fragments are introduced into yeast cells where they replicate as authentic chromosomes. The large cloning capacity of YACs offers many advantages over conventional cloning systems. Perhaps the most important of these is that the size range of YACs bridges the gap between linkage studies and conventional cloning methods, greatly facilitating the identification of genes by positional cloning and permitting the assembly of contiguous long-range physical maps of overlapping clones. The many benefits of YACs apply equally to plant studies (Guzman and Ecker 1988; Ecker 1990; Ward and Jen 1990; Grill and Somerville 1991; Hwang et al. 1991) and the manipulation of large segments of plant DNA cloned as YACs has become an essential part of plant molecular genetics, particularly for the many investigators attempting to clone genes by chromosome walking. The effective use of YACs as research tools requires familiarity with a battery of methods, some unique to YACs and some modified from other areas of molecular biology. This section contains eight protocols that will familiarize an

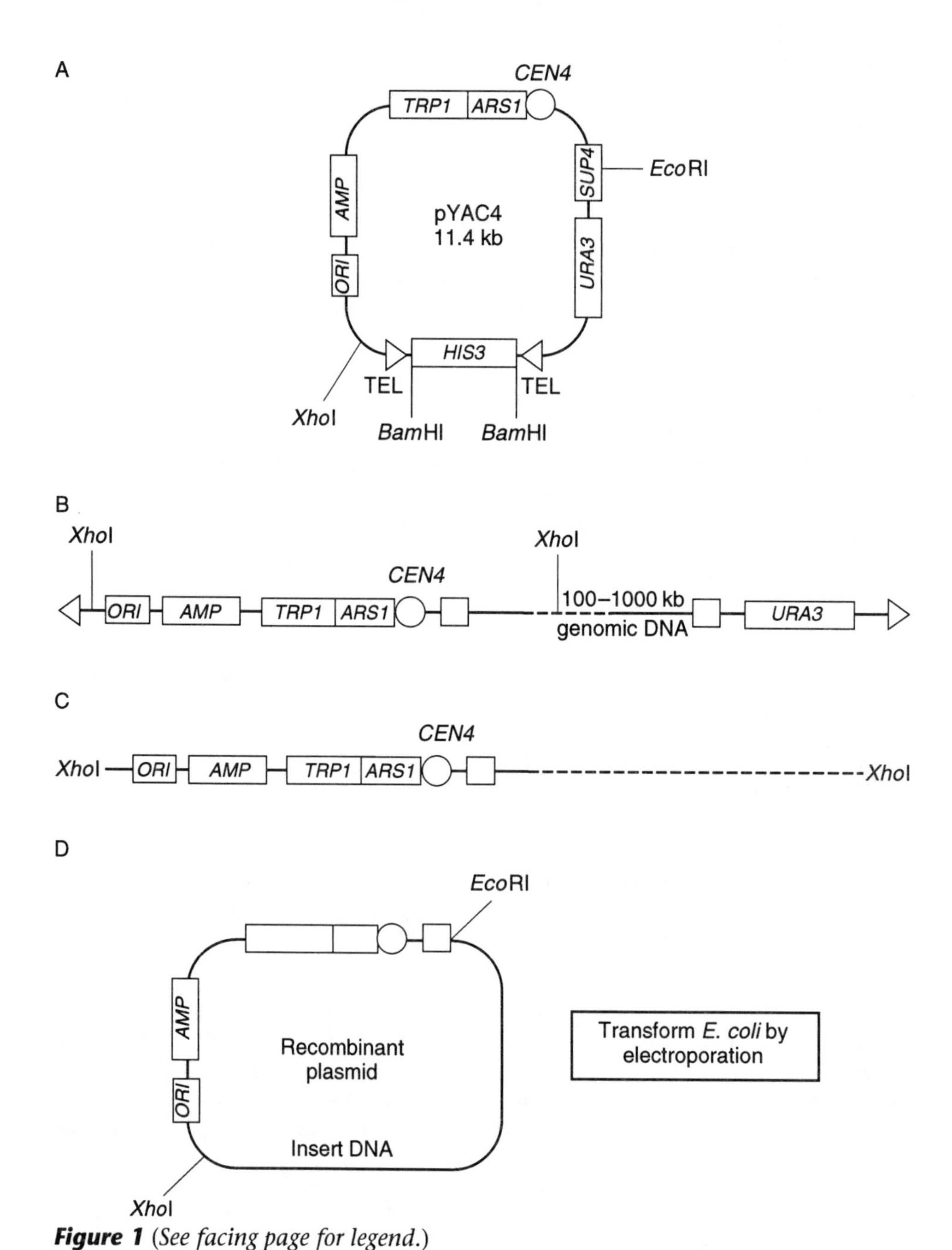

Figure 1 *(See facing page for legend.)*

investigator with these methods and provide the background necessary to incorporate work with YACs successfully into their research.

The Yeast DNA Minipreparation described on pp. 390–393 provides small amounts of DNA that are pure enough to be digested with restriction enzymes, ligate efficiently with T_4 DNA ligase, and serve as templates for the polymerase chain reaction (PCR). This DNA is required for several subsequent experiments. The protocol for the Recovery of YAC Insert Terminal Sequences by Plasmid Rescue in *E. coli* (pp. 394–398) is used to obtain plant DNA sequences that are immediately adjacent to the left arm of the YAC. These sequences are frequently required as probes to identify further YACs that overlap the original clone. The left arm of the YAC (Fig. 1, B–D) retains elements from pYAC4 for selection and maintenance in *E. coli* and thus provides a convenient way to obtain the terminal sequences of the insert cloned as a plasmid in *E. coli.* The right arm of the YAC lacks these elements, and Recovery of YAC Insert Terminal Sequences by Inverse PCR (pp. 399–404) describes an elegant solution to this problem, which can usually be relied on to provide a probe consisting of sequences from the right end of the YAC insert.

Screening a YAC Library for a Sequence-tagged Site by PCR (pp. 405–408) describes how a YAC library is screened by PCR to identify YACs that contain a particular sequence. A sequence-tagged site (STS) is a single-copy sequence in a genome that may be specifically amplified by PCR using a unique pair of primers (Olson et al. 1989) and is completely defined by the primer sequences and the size of the

Figure 1

Isolation of insert sequences adjacent to the left arm of the YAC by plasmid rescue in *E. coli.* (*A*) Structure of the plasmid pYAC4. Abbreviations: ORI, bacterial origin of replication; AMP, bacterial ampicillin resistance gene; TRP1, URA3 and HIS3, yeast selectable markers; ARS1, yeast autonomous replicating sequence (origin of replication); CEN4, centromere; SUP4, suppressor containing the *Eco*RI cloning site; TEL, sequence that forms a functional telomere after introduction into yeast. (*B*) Schematic representation of genomic DNA cloned as a yeast artificial chromosome. Digestion of the plasmid pYAC4 with *Bam*HI and *Eco*RI gives two vector arms between which large segments of genomic DNA are cloned. The left arm contains a bacterial ampicillin resistance gene and origin of replication, the TRP1 gene for selection in a *trp*⁻ host, the yeast autonomous replicating sequence (ARS), and the centromere CEN4, which causes the YAC to segregate correctly at mitosis. The right arm contains the URA3 gene for selection in a *ura*⁻ host. Both arms contain telomeres at their ends. (*C*) Structure of a left arm junction fragment after digestion with *Xho*I. (*D*) Structure of a recombinant plasmid after religation of the compatible *Xho*I termini. The plasmid retains the bacterial origin of replication and the ampicillin gene for selection in *E. coli.*

amplification product. This information can be used to screen YAC clones for the presence or absence of an STS simply by using the clones as templates for PCR and attempting amplification with the specific primer pair (Green and Olson 1989). Presence or absence of an amplification product indicates a positive or negative result in the screen. Although PCR screening is superior in many ways, conventional screening of YAC libraries by colony hybridization is still routinely carried out, and Gridding a YAC Library for Colony Hybridization (pp. 409–414) describes how the YAC clones are transferred from storage in microtiter plates onto a regular grid on nylon filters prior to such a hybridization experiment.

An important part of characterizing YACs is to determine the insert sizes. DNA molecules longer than 50–100 kb are easily broken by mechanical shear, so to analyze YACs larger than this it is necessary to immobilize the DNA in an agarose matrix (Schwartz and Cantor 1984). Preparation of High-molecular-weight Yeast DNA in Agarose Plugs (pp. 415–417) describes how yeast cells are embedded in low-melting-point agarose and then treated to obtain a relatively pure sample of high-molecular-weight chromosomal DNA that may be used to estimate the size of the YAC by pulsed-field gel electrophoresis. The electrophoresis itself is described in Analysis of YACs by Contour-clamped Homogeneous Electric Field Electrophoresis (pp. 418–420).

As explained previously, it is common to screen a YAC library for clones containing a particular sequence by hybridization. This is, in principle, straightforward, and is similar to any hybridization experiment in which filters bearing immobilized DNA are screened with a radioactive probe. Unlike DNA cloned in bacteria, however, YACs are maintained at only one copy per cell which makes yeast colony hybridization much less sensitive than bacterial colony hybridization. Screening a YAC Library by Hybridization (pp. 421–424) describes both the probe preparation and hybridization protocols we have found to be most successful for yeast colony hybridization.

ACKNOWLEDGMENTS

This work was supported in part by The National Center for Human Genome Research (grant #HG00322) to J.R.E. E.M. is supported by a Fulbright Foundation Fellowship from the Spanish MEC.

REFERENCES

Arveiler, B. and D.J. Porteous. 1991. Amplification of end fragments of YAC recombinants by inverse-polymerase chain reaction. *Technique* **3:** 24–28.

Burke, D.T., G.F. Carle, and M.V. Olson. 1987. Cloning of large segments of exogenous DNA into yeast by means of artificial chromosome vectors. *Science* **236:** 806–812.

Carle, G.F. and M.V. Olson. 1984. Separation of chromosomal DNA molecules from yeast by orthogonal-field-alternation gel electrophoresis. *Nucleic Acids Res.* **12:** 5647–5664.

Chu, G., D. Vollrath, and R.W. Davis. 1986. Separation of large DNA molecules by contour-clamped homogeneous electric fields. *Science* **234:** 1582–1585.

Chumakov, I., P. Rigault, S. Guillou, P. Ougen, A. Billaut, G. Guasconi, P. Gervy, I. LeGall, P. Soularue, L. Grinas, L. Bougueleret, C. Bellane-Chantelot, B. Lacroix, E. Barillot, P. Gesnouin, S. Pook, G. Vaysseix, G. Frelat, A. Schmitz, J. Sambucy, A. Bosch, X. Estivill, A. Weissenbach, A. Vignal, H. Reithman, D. Cox, D. Patterson, K. Gardiner, M. Hattori, Y. Sakaki, H. Ichikawa,M. Ohki, D. Le Paslier, R. Heilig, S. Antonarakis, and D. Cohen. 1992. Continuum of overlapping clones spanning the entire human chromosome 21q. *Nature* **359:** 380–387.

Deitrich, W., H. Katz, S.E. Lincoln, H.-S. Shin, J. Friedman, N.C. Dracopoli, and E.S. Lander. 1992. A genetic map of the mouse suitable for typing intraspecific crosses. *Genetics* **131:** 423–447.

Ecker, J.R. 1990. PFGE and YAC analysis of the *Arabidopsis* genome. *Methods* **1:** 186–194.

Feinberg, A.P. and B. Vogelstein. 1983. A technique for radiolabeling DNA restriction endonuclease fragments to high specific activity. *Anal. Biochem.* **132:** 6–13.

Green, E.D. and M.V. Olson. 1989. Systematic screening of yeast artificial chromosome libraries by use of the polymerase chain reaction. *Proc. Natl. Acad. Sci.* **87:** 1213–1217.

———. 1990. Chromosomal region of the cystic fibrosis gene in yeast artificial chromosomes: A model for human genome mapping. *Science* **250:** 94–98.

Grill, E. and C. Somerville. 1991. Construction and characterization of a yeast artificial chromosome library of *Arabidopsis* that is suitable for chromosome walking. *Mol. Gen. Genet.* **226:** 484–490.

Guzman, P. and J.R. Ecker. 1988. Development of large DNA methods for plants: Molecular cloning of large segments of *Arabidopsis* and carrot DNA into yeast. *Nucleic Acids Res.* **17:** 11091–11105.

Hwang, I., T. Kohchi, T., B.M. Hauge, H.M. Goodman, R. Schmidt, G. Cnops, C. Dean, S. Gibson, K. Iba, B. Lemieux, V. Arondel, L, Danhoff, and C. Somerville. 1991. Identification and map position of YAC clones comprising one-third of the *Arabidopsis* genome. *Plant J.* **1:** 367–374.

Jeffreys, A.J., V. Wilson, and S.L. Thein. 1985. Hypervariable "minisatellite" regions in human DNA. *Nature* **314:** 67–73.

Larin, Z., A.P. Monaco, and H. Lerach. 1991. Yeast artificial chromosome libraries from mouse and human DNA. *Proc. Natl. Acad. Sci.* **88:** 4123–4127.

Ochman, H., A.S. Gerber, and D.L. Hartl. 1988. Genetic applications of an inverse polymerase chain reaction. *Genetics* **120:** 621–623.

Olson, M.V., L. Hood, C. Cantor, and D. Botstein. 1989. A common language for physical mapping of the human genome. *Science* **245:** 1434–1435.

Philippsen, P., A. Stotz, and C. Scherf. 1991. DNA of *Saccharomyces cerevisiae*. *Methods Enzymol.* **194:** 169–182.

Ponce, M.R. and J.L. Micol. 1992. PCR amplification of long DNA fragments. *Nucleic Acids Res.* **20:** 623.

Sambrook, J., E.F. Frisch, and T. Maniatis. 1989. *Molecular cloning. A laboratory manual*, 2nd edition. Cold Spring Harbor Laboratory Press, Cold Spring Harbor, New York.

Schlessinger, D. 1990. Yeast artificial chromosomes: Tools for mapping and analysis of complex genomes. *Trends Genet.* **6:** 248–259.

Schwartz, D.C. and C.R. Cantor. 1984. Separation of yeast chromosome sized DNAs by pulsed field gradient gel electrophoresis. *Cell* **37:** 67–75.

Ward, E.R. and G.C. Jen. 1990. Isolation of single copy sequence clones from a yeast artificial chromosome library of randomly-sheared *Arabidopsis thaliana* DNA. *Plant Mol. Biol.* **14:** 561–568.

Weber, J.L. and P.E. May. 1989. Abundant class of human DNA polymorphisms which can be typed using the polymerase chain reaction. *Am. J. Human Genet.* **44:** 388–396.

• *TIMETABLE*
Pulsed-field Gel Electrophoresis and Yeast Artificial Chromosome Analysis of the *Arabidopsis* Genome

STEPS	***TIME REQUIRED***
Yeast DNA Minipreparation	
Grow yeast cells	2–3 days
Collect and wash cells	15 minutes
Digest cells with Zymolyase® 20T	30 minutes
Lyse spheroplasts with detergent	30 minutes
Precipitate proteins with potassium acetate	1 hour–overnight
Extract DNA from supernatant	45 minutes
Digest DNA with RNase	30 minutes
Precipitate DNA, wash, dry, and resuspend in TE	40 minutes
Recovery of YAC Insert Terminal Sequences by Plasmid Rescue in *E. coli*	
Digest DNA with *Xho*I	2–3 hours
Heat inactivate enzyme	15 minutes
Extract DNA	30 minutes

STEPS	*TIME REQUIRED*
Recovery of YAC Insert Terminal Sequences by Plasmid Rescue in *E. coli* (continued)	
Set up ligation reaction	15–30 minutes
Incubate ligation reaction	overnight
Extract DNA	30 minutes
Electroporate *E. coli* in presence of DNA	30 minutes
Grow transformed cells	1 hour
Plate cells on selective medium	15 minutes
Select for ampicillin-resistant colonies	overnight
Recovery of YAC Insert Terminal Sequences by Inverse PCR	
Digest DNA with restriction enzyme	2–3 hours
Extract DNA with phenol/chloroform or heat inactivate enzyme	15 minutes
Set up ligation reaction	15–30 minutes
Incubate ligation reaction	overnight
Set up PCR reaction	15–30 minutes
Run PCR reaction	3 hours–overnight
Run, stain, and view agarose minigel	2–3 hours
Screening a YAC Library for a Sequence-tagged Site by PCR	
Set up first set of PCR reactions (microtiter plate pools)	15–30 minutes

STEPS	***TIME REQUIRED***
Screening a YAC Library for a Sequence-tagged Site by PCR (continued)	
Run PCR reactions	3 hours–overnight
Run, stain, and view agarose gel	2–3 hours
Set up second set of PCR reactions (row and column pools)	15–30 minutes
Run PCR reactions	3 hours–overnight
Run, stain, and view agarose gel	2–3 hours
Gridding a YAC Library for Colony Hybridization	
Thaw frozen library and set up laminar flow hood	1 hour
Replicate library onto nylon filters	2 hours (for a 24-plate library)
Grow cells on nylon filters on YPD nutrient agar	overnight
Prepare Zymolyase® 20T plates for cell-wall digestion	30 minutes–1 hour
Transfer filters to Zymolyase®20T plates	30 minutes–1 hour
Digest cell walls	overnight
Lyse cells	1 hour
Air dry filters	30 minutes
Bake filters to immobilize DNA	2 hours

STEPS	*TIME REQUIRED*
Preparation of High-molecular-weight Yeast DNA in Agarose Plugs	
Grow yeast cells	2–3 days
Collect, wash, and resuspend cells	1 hour
Embed cells in agarose	30 minutes–1 hour
Digest with Zymolyase® 20T	overnight
Lyse cells	overnight
Analysis of YACs by Contour-clamped Homogeneous Electric Field Electrophoresis	
Melt agarose and cast gel	30 minutes
Load DNA samples	30 minutes
Equilibrate buffer to 10–12°C	15 minutes
Run gel	24–40 hours
Stain, destain, and view gel	1–2 hours
Screening a YAC Library by Hybridization	
Boil salmon sperm DNA and set up prehybridization	20 minutes
Prehybridize colony filters[1]	overnight
Set up probe labeling reaction	15–30 minutes
Label probe[1]	overnight
Extract labeled probe	40 minutes
Boil probe and set up hybridization	20 minutes

STEPS	***TIME REQUIRED***
Screening A YAC Library by Hybridization (continued)	
Hybridize colony filters with labeled probe	overnight
Wash filters	1 hour
Set up X-ray cassette	10–15 minutes
Expose filters to X-ray film	usually overnight

[1]These steps should be performed simultaneously.

Yeast DNA Minipreparation

This procedure is a scaled-down version of a protocol described by Philippsen et al. (1991). Yeast cells are grown to saturation in a medium that selects for the presence of the YAC and the cell walls are then removed by digestion with β-1,3 glucanase (Zymolyase® 20T). The cells are lysed with the detergent sodium dodecyl sulfate and proteins are removed from the solution by precipitation with potassium acetate. Extraction with organic solvents is then carried out to remove further proteins from the solution and this is followed by treatment with RNase A to remove RNA. Finally, the DNA is precipitated with ethanol and again with isopropanol to concentrate it and to remove contaminating low-molecular-weight compounds.

YEAST CLONES CONTAINING YACs

YACs are maintained in yeast strain AB1380 (*Mat-a, ade2-1, can1-100, lys2-1, trp1, ura3, his5 [psi+]*) (Burke et al. 1987). YAC libraries of *Arabidopsis* are available from the *Arabidopsis* Biological Resource Center at Ohio State University (see Appendix 2).

MATERIALS

Sterile distilled H_2O
Sorbitol (0.9 M)/EDTA (0.1 M)/dithiothreitol (DTT) (50 mM) (pH 7.5)
Zymolyase® 20T (50 mg/ml in H_2O) (Seikagaku America, Inc.)
Sodium dodecyl sulfate (SDS) (10%)
Potassium acetate (5 M) (not buffered)
Phenol/chloroform (1:1)
Ethanol (100% and 50%) (ice-cold)
RNase (20 mg/ml) (DNase-free)
Isopropanol (100% and 50%) (ice-cold)

MEDIA

AHC medium
(For recipe, see Preparation of Media, pp. 425–426)

REAGENTS

TE
(For recipe, see Preparation of Reagents, pp. 427–429)

SAFETY NOTES

- Wear a mask when weighing SDS.
- Phenol is highly corrosive and can cause severe burns. Wear gloves, protective clothing, and safety glasses when handling it. All manipulations should be carried out in a chemical fume hood. Any areas of skin that come in contact with phenol should be rinsed with a large volume of water or PEG 400 and washed with soap and water; do not use ethanol!
- Chloroform is irritating to the skin, eyes, mucous membranes, and respiratory tract. It should only be used in a chemical fume hood. Gloves and safety glasses should also be worn. Chloroform is a carcinogen and may damage the liver and kidneys.

PROCEDURE

1. Grow a yeast clone to saturation in 3 ml of AHC medium at 30°C. Depending on the clone, this may take 2–3 days.

2. Collect the cells by spinning the culture in a table-top centrifuge at 5000 rpm for 5 minutes at room temperature.

3. Resuspend the cells in 1 ml of sterile distilled H_2O and transfer them to a 1.5-ml microfuge tube. Spin in a microfuge for 10 seconds at room temperature.

4. Resuspend the cells in 0.3 ml of 0.9 M sorbitol/0.1 M EDTA/50 mM DTT (pH 7.5).

   Note: DTT is a reducing agent that acts upon the cell wall, making it more susceptible to digestion by β-1,3 glucanase.

5. Add 1 μl of 50 mg/ml Zymolyase® 20T and incubate with occasional shaking for 30 minutes at 37°C.

   Note: Thirty minutes is sufficient to completely digest the cell walls of strain AB1380 using this procedure. Overdigestion with Zymolyase® 20T causes premature lysis of the cells, which results in red pigment remaining in the supernatant after step 6.

6. Spin the resultant spheroplasts in a microfuge at low speed for 10 seconds at room temperature and remove the supernatant. Resuspend the spheroplasts in 0.3 ml of TE by gently drawing them into a pipette tip.

   Note: The spheroplasts are relatively robust, but centrifugation at high speed in a microfuge will cause them to rupture. If this has occurred, red pigment will be visible in the supernatant as it is aspirated off.

7. Add 30 μl of 10% SDS and incubate for 30 minutes at 65°C.

   Note: The SDS, together with high temperature, solubilizes the cell membranes. Heating at 65°C also helps to inactivate contaminating nucleases.

8. Add 0.1 ml of 5 M potassium acetate and place on ice for at least 1 hour.

9. Spin in a microfuge for 15 minutes at 4°C to pellet the precipitate and transfer the supernatant to a new tube.

   Note: The large flocculent precipitate contains mostly proteins complexed with SDS and SDS itself. Small particles of the pellet are often carried over with the supernatant to the next step. These are removed by the phenol/chloroform extraction and do not seem to interfere with subsequent manipulation of the DNA.

10. Extract the supernatant once with an equal volume of phenol/chloroform (1:1). Spin to separate the phases.

11. Transfer the upper (aqueous) phase to a clean tube, and add an equal volume of ice-cold 100% ethanol.

    Note: The red pigment in the solution is a by-product of the adenine synthesis pathway produced as a result of the *ade* genotype of strain AB1380. This pigment precipitates with ethanol just as DNA does and its presence actually makes the DNA pellet more visible.

12. Spin for 15 minutes. Rinse the pellet with 0.5 ml of ice-cold 50% ethanol.

13. Dry the pellet under vacuum and resuspend the DNA in 0.3 ml of TE.

14. Add 1 μl of 20 mg/ml DNase-free RNase and incubate for 30 minutes at 37°C.

15. Add 0.3 ml of ice-cold 100% isopropanol, mix, and spin for 15 minutes. Wash the pellet with 0.5 ml of ice-cold 50% isopropanol, and dry the pellet under vacuum.

Note: If you spin the sample immediately after adding the 100% isopropanol, the DNA will precipitate but most of the pigment (and presumably other low-molecular-weight contaminants) will remain in solution.

16. Redissolve the DNA in 50 μl of TE. The usual yield is about 3–5 μg of DNA. Store the DNA at 4°C.

Recovery of YAC Insert Terminal Sequences by Plasmid Rescue in *E. coli*

Several uses of YACs, such as chromosome walking and the construction of physical maps, require that the sequences at the extreme ends of the insert be isolated from the rest of the YAC. The YAC vector is propagated as a plasmid and as such contains elements necessary for replication and selection in the bacterium *E. coli* (Burke et al. 1987) (for review, see Schlessinger 1990). As shown in Figure 1 (A–B), the circular plasmid is converted to a pair of vector arms by digestion with two restriction enzymes, and high-molecular-weight genomic DNA is cloned between the two arms. The left arm of the vector retains the ampicillin resistance gene and the bacterial replication origin, which allows recovery of the sequences at the left end of the insert as a plasmid. Digestion of DNA prepared from a YAC clone with *Xho*I results in cleavage of the left arm of the YAC at a *Xho*I restriction site distal to the ampicillin resistance gene. It is also expected that a *Xho*I site will exist in the YAC insert within a reasonable distance of the cloning site (Fig. 1, B–C). Thus, digestion with *Xho*I will yield linear molecules with compatible termini which, when ligated together, form a plasmid that can be transformed into and maintained stably in *E. coli* (Fig. 1, D).

This method depends on the fortuitous occurrence of a *Xho*I site in the insert. Virtually all YAC inserts will contain *Xho*I restriction sites, but one of these needs to be close enough to the vector arm so that self-ligation will yield a plasmid that can be transformed into *E. coli* with reasonable efficiency. Also, YACs are maintained in yeast at one copy per cell, and therefore the DNA, when digested with *Xho*I, will consist predominantly of yeast genomic DNA, with relatively few of the desired fragments. To self-ligate DNA fragments efficiently and to avoid artifacts arising from co-ligation, the DNA concentration should be lower than 1 ng/μl. However, to achieve this concentration and to also have enough of the desired vector-insert junction fragments would result in ligation reactions that are impractically large. In practice, a compromise is reached in which the ligation reactions are a reasonable volume and the yield of the desired plasmid is great enough so that transformants are routinely obtained by transformation using a high efficiency electroporation protocol. Occasionally, anonymous *Xho*I fragments are cloned adjacent to the genuine insert DNA, but these can be detected and eliminated by digestion of the plasmid with *Xho*I. The insert sequences can be separated from the plasmid by digestion with *Xho*I and *Eco*RI and preparative agarose gel electrophoresis. The insert band may then be purified for subcloning, radioactive labeling, sequencing, etc., using standard methods (Sambrook et al. 1989).

MATERIALS AND EQUIPMENT

DNA isolated from a yeast clone containing a YAC (see pp. 390–393)
*Xho*I (Boehringer Mannheim Biochemicals)
Reaction buffer appropriate for *Xho*I (10x) (Boehringer Mannheim Biochemicals)
Sodium acetate (3 M; pH 5.2)
Ethanol (100% and 70%) (ice-cold)
T4 DNA ligase (Boehringer Mannheim Biochemicals)
Ligase buffer (10x) (appropriate for T4 DNA ligase) (Boehringer Mannheim Biochemicals)
ATP (10 mM) (Pharmacia Biotech, Inc.)
Bovine serum albumin (BSA) (1 mg/ml; molecular biology grade) (Boehringer Mannheim Biochemicals)
High-efficiency electrocompetent *E. coli* cells. The highest efficiency cells are those supplied commercially by Bio-Rad Laboratories. We use strain HB101 (*supE44 hsdS20* $[r^-_B\ m^-_B]$ *recA13 ara-14 proA2 lacY1 galK2 rpsL20 xyl-5 mtl-1*) prepared according to protocols supplied by Bio-Rad Laboratories.
Electroporation apparatus (Bio-Rad Laboratories)

MEDIA

Liquid:
SOC medium

Plates:
LB agar supplemented with 100 μg of ampicillin/ml
(For recipes, see Preparation of Media, pp. 425–426)

PROCEDURE

Digestion

1. Digest the DNA isolated from a yeast clone containing a YAC with *Xho*I restriction endonuclease by preparing the following reaction mixture in a 1.5-ml microfuge tube:

DNA	45 μl
Reaction buffer appropriate for *Xho*I (10x)	10 μl
H_2O	43 μl
*Xho*I (20 units)	2 μl

Incubate for 2–3 hours at 37°C.

Note: When you are dealing with a YAC clone of which nothing is known, this is an appropriate amount of DNA to start with. If no bacterial transformants are obtained after the ligation and transformation steps, the reason is likely to be that the desired plasmid is very large and is transforming with very low efficiency. If this is the case, you can use more DNA in this initial reaction.

2. Inactivate the enzyme by incubating the sample for 15 minutes at 70°C.

3. Precipitate the DNA by adding 10 μl of 3 M sodium acetate (pH 5.2) and 300 μl of ice-cold 100% ethanol. Incubate the sample for 20 minutes at –20°C.

4. Spin the sample in a microfuge for 10 minutes at 4°C, aspirate the supernatant, and wash the pellet with 1 ml of ice-cold 70% ethanol.

5. Aspirate the 70% ethanol. Spin the sample briefly to collect the remainder of the liquid at the bottom of the tube, and remove it with an automatic pipettor.

6. Dry the sample under vacuum.

Ligation

1. Ligate the linear DNA fragments into covalently closed circular molecules by preparing the following reaction mixture in a 1.5-ml microfuge tube:

DNA (from step 6 above)	dry
Ligase buffer (10x)	10 μl
ATP (10 mM)	10 μl
BSA (1 mg/ml)	1 μl
H_2O	78 μl
T_4 DNA ligase	1 μl

Incubate overnight at 16°C.

2. Precipitate the DNA by adding 10 μl of 3 M sodium acetate (pH

5.2) and 300 μl of ice-cold 100% ethanol. Incubate the sample for 20 minutes at –20°C.

3. Spin the sample in a microfuge for 10 minutes, aspirate the supernatant, and wash the pellet with 1 ml of ice-cold 70% ethanol.

4. Aspirate the 70% ethanol. Spin the sample briefly to collect the remainder of the liquid at the bottom of the tube, and remove it with an automatic pipettor. Repeat this step once.

   Note: It is important to wash the DNA pellet thoroughly with 70% ethanol so as to remove all traces of salt from the sample. The electroporation step is sensitive to salt contamination, which increases the conductance of the sample and reduces the efficiency of transformation. In some cases, salt can cause arcing of electricity between the electrodes of the electroporation cuvette.

5. Dry the sample under vacuum.

Electrotransformation of E. coli

1. Dissolve the DNA in 8 μl of distilled H_2O.

2. Transfer 4 μl of this DNA to a microfuge tube and place the sample on ice.

3. Thaw an aliquot of high-efficiency electrocompetent *E. coli* cells and place on ice. Mix 40 μl of the competent cells with the DNA and transfer the sample to a 1-mm gap electroporation cuvette.

   Note: Thaw the cells by holding the tube between your fingers. As soon as the cells have thawed, place the tube on ice. Allowing the cells to warm up can reduce the transformation efficiency.

4. Electroporate the bacteria using the following parameters: 1.5 kV, 25 mF, and 200 Ω.

   Note: These are the parameters most recently recommended by Bio-Rad Laboratories for use with the Bio-Rad Gene Pulser®.

5. Immediately after the pulse is delivered, add 1 ml of SOC medium to the bacteria and transfer them to a sterile glass culture tube.

   Note: Adding the SOC as rapidly as possible after the pulse is delivered increases the transformation efficiency.

6. Incubate with shaking for 1 hour at 37°C.

7. Plate the transformed cells on LB agar supplemented with 100 μg of ampicillin/ml to select for the plasmid of interest. Plate 100 μl

of the bacteria on one petri dish and the remainder on another.

Note: The number of transformants is often low, and it is usually acceptable to plate the entire sample on one 8-cm petri dish. If difficulty is encountered plating a sample of this volume, the bacteria may be collected by spinning them briefly in a microfuge and then resuspended in 100 μl of LB broth before plating again. We routinely plate 1-ml volumes using a plating wheel and incubate the petri dishes topside up until the liquid evaporates or is absorbed by the agar.

8. Incubate the plates overnight at 37°C. Colonies should be visible the next morning.

Note: The number of transformants varies greatly, and is influenced by many variables. However, provided the DNA is of the expected quantity and quality and the digestion and ligation steps were successful, these can be narrowed down to the size of the desired plasmid and the efficiency of transformation. The transformation efficiency can be monitored by electroporating control plasmids; we routinely use 1 ng each of pBluescript (3.0 kb) and pYAC4 (11.4 kb). pYAC4 usually gives efficiency two orders of magnitude lower than pBluescript. If the efficiency with pYAC4 is less than 10^7 transformants/μg, zero or very few rescued colonies can be expected. If the transformation efficiency with the control plasmids is high and still no colonies are obtained, the most likely reason is that the positioning of the nearest *Xho*I site in the insert results in a very large plasmid that is difficult to transform into *E. coli*. This problem can sometimes be cured by using more DNA in the initial digestion reaction.

Recovery of YAC Insert Terminal Sequences by Inverse PCR

As shown above, the presence of the ampicillin resistance gene and the replication origin on the left arm of the YAC offers a convenient way of cloning the sequences adjacent to the left end of the insert as a plasmid in *E. coli*. However, the right arm has no such elements and an alternative method must be used if the right-end sequences are to be recovered. Inverse PCR (IPCR) (Ochman et al. 1988) can be used to amplify insert sequences immediately adjacent to either arm of the YAC as shown in Figure 2 (Arveiler and Porteous 1991). The method depends on the presence of *Pst*I and *Hin*dIII restriction sites in the right arm of the vector and the random occurrence of the same sites in the insert within an amplifiable distance (<4 kb) of the right arm (Fig. 2, A). Total DNA from a yeast clone containing a YAC is digested with either of these enzymes and self-ligated at low concentration so that the formation of circular molecules is favored (Fig. 2, B–C). PCR primer pairs, originally pointing in opposite directions (5′ to 5′, hence *inverse* PCR) are now positioned so that amplification of insert DNA between them is possible (Fig. 2, D). Two pairs of primers are used. One primer, PVR, which is adjacent to the *Eco*RI cloning site of the vector with its 3′ end pointing towards the insert, is common to both pairs. Primers PVH and PVP are adjacent to the *Hin*dIII and *Pst*I sites, respectively, in the right arm of the vector. Amplification using appropriate thermal cycling results in specific amplification of insert DNA along with short pieces of vector DNA. If necessary, these vector sequences can be removed by digestion of the PCR product with the appropriate combination of enzymes and agarose gel purification of the insert DNA.

MATERIALS AND EQUIPMENT

DNA isolated from a yeast clone containing a YAC (see pp. 390–393)
*Pst*I (Boehringer Mannheim Biochemicals)
Reaction buffer appropriate for *Pst*I (10x) (Boehringer Mannheim Biochemicals)
*Hin*dIII (Boehringer Mannheim Biochemicals)
Reaction buffer appropriate for *Hin*dIII (10x) (Boehringer Mannheim Biochemicals)

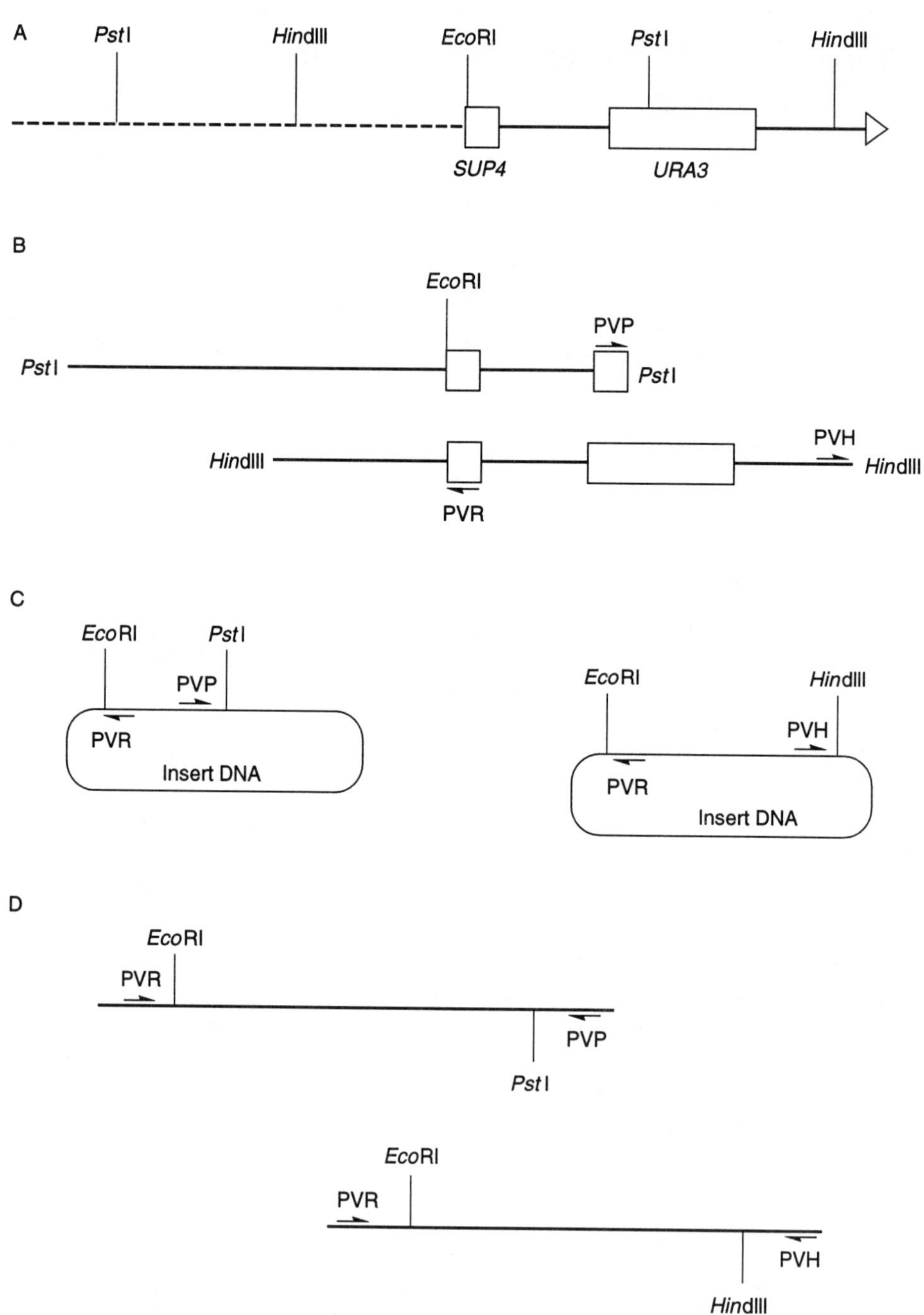

Figure 2 *(See facing page for legend)*

Phenol/chloroform (1:1)
T_4 DNA ligase (Boehringer Mannheim Biochemicals)
Ligase buffer (10x) (appropriate for T_4 DNA ligase) (Boehringer Mannheim Biochemicals)
ATP (10 mM) (Pharmacia Biotech, Inc.)
Mineral oil
Taq polymerase (Promega Corporation)
Reaction buffer appropriate for *Taq* polymerase (10x) (Promega Corporation)
PCR thermal cycler (e.g., Perkin Elmer Corp.)
Agarose (low electroendosmosis [EEO]; electrophoresis grade) (Boehringer Mannheim Biochemicals)

PRIMER SEQUENCES

PVR: 5'-ATATAGGCGCCAGCAACCGCACCTGTGGCG-3'
PVP: 5'-GGAAGAACGAAGGAAGGAGC-3'
PVH: 5'-AAACTCAACGAGCTGGACGC-3'

REAGENTS

dNTP mixture for PCR
Loading dye (2x)
(For recipes, see Preparation of Reagents, pp. 427–429)

Figure 2
Isolation of insert sequences adjacent to the right arm of the YAC by inverse PCR. (*A*) Structure of the right vector/insert junction. The right arm of the YAC contains *Pst*I and *Hin*dIII sites at the indicated positions. In many YACs, the insert will also contain *Pst*I and *Hin*dIII sites close to the junction and these hypothetical sites are indicated. (*B*) Structure of the junction fragments after digestion with the appropriate enzyme showing the positions of the primer sequences PVR, PVP, and PVH. (*C*) Circular molecules formed by religation of the *Pst*I and *Hin*dIII digestion products. (*D*) PCR products from amplification using the appropriate primer pairs: PVR/PVP in the case of *Pst*I or PVR/PVH in the case of *Hin*dIII.

SAFETY NOTES

- Phenol is highly corrosive and can cause severe burns. Wear gloves, protective clothing, and safety glasses when handling it. All manipulations should be carried out in a chemical fume hood. Any areas of skin that come in contact with phenol should be rinsed with a large volume of water or PEG 400 and washed with soap and water; do not use ethanol!
- Chloroform is irritating to the skin, eyes, mucous membranes, and respiratory tract. It should only be used in a chemical fume hood. Gloves and safety glasses should also be worn. Chloroform is a carcinogen and may damage the liver and kidneys.

PROCEDURE

Digestion

1. Digest genomic DNA isolated from a yeast clone containing a YAC with *Hin*dIII and *Pst*I (one sample for each enzyme) by preparing the following reaction mixture in a 1.5-ml microfuge tube:

DNA	1 μl
Reaction buffer appropriate for restriction enzyme (10x)	5 μl
H_2O	42 μl
*Hin*dIII or *Pst*I (20 units)	2 μl

   Incubate for 2–3 hours at 37°C.

   Note: A relatively large reaction volume (50 μl) is used for the digestion because 1/50 of 1 μl of the yeast DNA minipreparation (pp. 390–393) provides an appropriate amount of DNA for the subsequent ligation. A volume of 50 μl can also be extracted with organic solvents without the addition of further liquid, thereby reducing the number of manipulations required.

2. Extract once with phenol/chloroform (1:1) to remove the enzyme from the solution. Spin to separate the phases and transfer the upper (aqueous) phase to a clean tube. Use 1 μl of this for the ligation.

   Note: The restriction enzyme must be removed before the ligation step otherwise it may work against the ligase by re-digesting the desired circular molecules. Enzymes that are heat sensitive can be deactivated by incubation for 15 minutes at 65°C.

Ligation

1. Self-ligate the products of digestion into covalently closed circles by preparing the following reaction mixture in a 1.5-ml microfuge tube:

DNA (<10 ng) (digested with either *Hin*dIII or *Pst*I)	1 µl
Ligase buffer (10x)	1 µl
ATP (10 mM)	1 µl
H_2O	6 µl
T_4 DNA ligase (1 unit)	1 µl

Incubate overnight at 16°C.

Note: The critical part of this step is to have a sufficiently low concentration of DNA so that the ends of individual fragments only ever encounter their own other end, with the result that virtually all of the products of ligation are circular molecules. Higher concentrations will increase the risk of co-ligating random small fragments of DNA from the yeast genome. The exact amount of DNA in the ligation is unknown, and will vary since different clones yield different amounts of DNA. The sodium chloride carried over from the restriction buffer does not inhibit the ligase.

Amplification

1. Prepare the following reaction mixture in a 0.5-ml microfuge tube:

Template DNA (from the ligation above)	1 µl
Primer PVR (5 pmol/µl)	1 µl
Primer PVP or PVH (5 pmol/µl)	1 µl
Reaction buffer appropriate for *Taq* polymerase (10x)	2 µl
dNTP mixture for PCR	3.2 µl
H_2O	10.8 µl

Note: When analyzing multiple samples, it is convenient to premix the required volumes of each of the above reagents *except the template* and dispense 18-µl aliquots into each microfuge tube. You should then add template to the appropriate tubes and proceed with step 2.

2. Cover the sample with a drop of mineral oil and place it in a heating block for 5 minutes at 92°C.

Note: This heat treatment ensures that the template is fully denatured before the enzyme is added.

3. Add 0.5–1.0 units of *Taq* polymerase in a volume of 1 μl and return the sample to the heating block until enzyme has been added to every sample.

   Note: Addition of the enzyme to a reaction mixture at high temperature is known as "hot-start " PCR and ensures that the primers never come into contact with the template at any temperature that will allow mismatch annealing to sequences that arbitrarily have some homology to the primers. This eliminates the background of nonspecific bands that sometimes occur when "cold-start" PCR is carried out on complex templates.

4. Subject the sample to the following amplification profile:

Process	Time	Temperature
Denature	1 minute	92°C
Anneal primers	1 minute	60°C
Polymerize	3 minutes	72°C

   Perform this cycle 35 times. After cycling, place the samples at 4°C.

5. Analyze the PCR product by running it out on a 1% agarose gel alongside size markers. Mix 3 μl of the PCR reaction mixture with 3 μl of 2x loading dye and load the sample onto a 1% agarose minigel (Sambrook et al. 1989).

   Note: If amplification has been successful, loading this amount of PCR product will give a clear band on the gel. As the size of the template increases, amplification becomes decreasingly efficient so that more of the sample must be loaded in order to see the product. Absence of a band usually means that there is no restriction site located near enough to the cloning site to allow amplification. Amplification of longer fragments can sometimes be achieved by using more complex amplification profiles (Arveiler and Porteous 1991) or by using reaction mixtures lacking potassium chloride and buffered by tricine rather than the usual Tris-HCl (Ponce and Micol 1992).

Screening a YAC Library for a Sequence-tagged Site by PCR

PCR screening is sensitive, less prone to false positive and negative results than hybridization screening, and does not require the use of radiation. It is also compatible with the latest genetic and physical mapping technology of simple sequence length polymorphisms (SSLPs) (Jeffreys et al. 1985; Weber and May 1989; Deitrich et al. 1992) by using primers that define microsatellite loci that are likely to form the backbones of future genetic maps. Clearly, to screen libraries containing thousands of clones it would be impractical to perform PCR amplification of each clone, thus it is necessary to devise a clone-pooling strategy that allows the isolation of positive clones with a minimum number of PCR reactions (Green and Olson 1990). For a YAC library of a small genome such as that of *Arabidopsis*, the pooling strategy is fairly simple and is illustrated in Figure 3. Twenty-four 96-well microtiter plates contain 2300 clones. If the DNA from each plate (96 samples) is pooled (Fig. 3, A), then 24 PCR reactions are required to determine which plates contain the sequence-tagged site (STS) of interest. Twenty PCR reactions are performed for each positive plate—12 column pools and 8 row pools—to determine which rows and columns contain the positive clones. The addresses of the positive clones can be inferred from this data (Fig. 3, B). If there is only one positive clone on the plate, its address is immediately obvious, but with more than one there are false positive addresses that can be eliminated by amplification of individual clones.

YEAST CLONES CONTAINING YACs

YACs are maintained in yeast strain AB1380 (*Mat-a, ade2-1, can1-100, lys2-1, trp1, ura3, his5 [psi+]*) (Burke et al. 1987). YAC libraries of *Arabidopsis* are available from the *Arabidopsis* Biological Resource Center at Ohio State University (see Appendix 2).

MATERIALS AND EQUIPMENT

Forward primer (appropriate for STS of interest) (5 pmol/μl)
Reverse primer (appropriate for STS of interest) (5 pmol/μl)

A

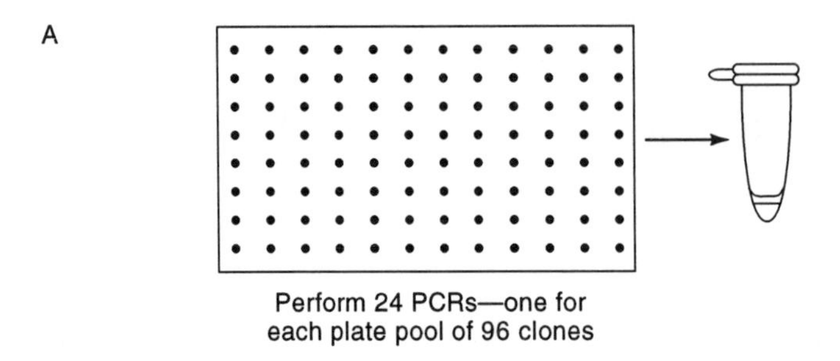

B

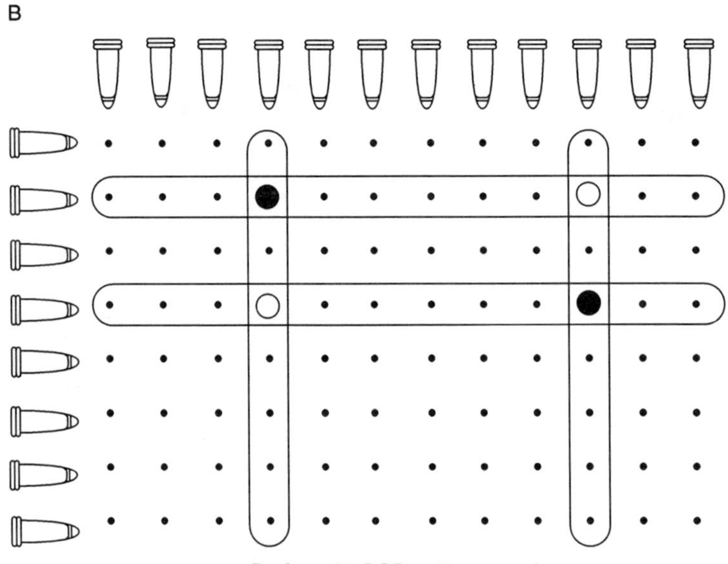

Figure 3
Screening a YAC library by PCR. (*A*) Sets of 96 DNA samples (each corresponding to one microtiter plate in the YAC library) are pooled, and PCR amplification is carried out on each of these pools with a pair of primers that identifies an STS. (*B*) Plates that give amplification products are further examined by amplifying DNA from 12 column pools and 8 row pools. Rows and columns that give amplification products can be cross-referenced to the addresses of the positive clones (*closed circles*). False positive clones (*open circles*) can then be eliminated by PCR on individual clones.

Reaction buffer appropriate for *Taq* polymerase (10x) (Promega Corporation)
Yeast DNA
Arabidopsis DNA
Mineral oil
Taq polymerase (Promega Corporation)
PCR thermal cycler (e.g., Perkin Elmer Corp.)

Agarose (low electroendosmosis [EEO]; electrophoresis grade)
(Boehringer Mannheim Biochemicals)
Ethidium bromide (1 µg/ml in TAE buffer)

REAGENTS

dNTP mixture for PCR
Loading dye (2x)
TAE buffer
(For recipes, see Preparation of Reagents, pp. 427–429)

SAFETY NOTE

- Ethidium bromide is a powerful mutagen and is moderately toxic. Gloves should be worn when working with solutions that contain this dye. After use, decontaminate and discard solutions according to local safety office regulations.

PROCEDURE

1. Prepare the PCR reaction mixture by multiplying the volumes in the following basic recipe by the number of samples you are analyzing:

Forward primer (5 pmol/µl)	1 µl
Reverse primer (5 pmol/µl)	1 µl
dNTP mixture for PCR	3.2 µl
Reaction buffer appropriate for *Taq* polymerase (10x)	2 µl
H_2O	10.8 µl

2. Dispense 18-µl aliquots of this reaction mixture into the required number of 0.5-ml microfuge tubes.

3. Add 1 µl of template DNA from the appropriate pools (see Fig. 3) to each tube. (Pool the cells and extract the DNA using the DNA minipreparation protocol.) Include yeast DNA and *Arabidopsis* DNA as controls.

   Note: The amount of DNA added from each pool need not be exact. In our experiments, each pool contains approximately 10 ng/µl of DNA, which is about 7 x 10^5 copies of the yeast genome per µl. The most complex pool (96 clones) therefore contains about 7 x 10^3 copies of each clone's genome per µl. Normally, each positive clone is confirmed by amplification of it alone in a PCR reaction. If there is no DNA sample available for the positive clone, PCR can be carried out using a small sample of yeast cells themselves as a template. This is done by mixing all the necessary reagents except the enzyme and simply touching a tooth pick to the yeast colony of interest then twirling it into the PCR mixture. Treat the sample as a standard PCR reaction: cover with mineral oil, heat for 5 minutes at 92°C to rupture the cells and liberate the nuclear DNA, add the enzyme, and carry out thermal cycling.

4. Cover each sample with a drop of mineral oil, close the tubes, and place them in a heating block for 5 minutes at 92°C.

5. Add one unit of *Taq* polymerase in a volume of 1 μl to each sample, and amplify for 35 cycles with the following profile:

Process	Time	Temperature
Denature	1 minute	92°C
Anneal primers	1 minute	55°C
Polymerize	1 minute	72°C

6. Mix 5 μl of the PCR reaction mixture with 5 μl of 2x loading dye and load onto a 1.5% agarose gel (Sambrook et al. 1989). Run the gel in TAE buffer until the bromophenol blue has migrated two-thirds of the length of the gel. Stain in ethidium bromide (1 μg/ml in TAE buffer) for 20–30 minutes, and take a Polaroid photograph under short-wave ultraviolet light.

Gridding a YAC Library for Colony Hybridization

The yUP YAC library (Ecker 1990) contains 2300 independently picked clones that are stored frozen in glycerol in 24 96-well microtiter plates. To make filters for colony hybridization, the microtiter plates are thawed and cells are transferred from the wells of each microtiter plate (i.e., 96 clones at a time) onto nylon filters, which are then placed on the surface of nutrient agar plates to allow the cells to proliferate into colonies. The yeast clones are transferred using a 96-prong replicating tool (ours was manufactured in the Instrument Shop at Washington University School of Medicine). The replicator consists of a platform with a countersunk space, milled to fit a microtiter plate, and a detachable metal frame in which an 8 x 12-cm nylon filter is secured. Between the microtiter plate and the frame is a rotating vertical column upon which sits a removable 96-prong array; this array is free to move up and down so that the prongs dip into the wells of the microtiter plate and can be rotated around its vertical axis, positioned over the metal frame, and lowered to deposit cells on the nylon filter. The metal frame can be moved to 8 offset positions so that 8 x 96 samples (768 YACs or eight microtiter plates) can be accommodated on one 8 x 12-cm filter (Fig. 4). Using this device, the entire library can be plated on three filters. After sufficient growth of the colonies (which takes 1–2 days, depending on the age of the microtiter plates, how often they have been thawed, and how much yeast remains in the wells), the cell walls are digested by incubating the filters on Whatman 3MM filter paper that has been soaked with a solution containing Zymolyase® 20T. Following cell wall digestion, the cells are lysed and DNA is immobilized on the filters by a standard series of steps used in colony hybridization: lysis with an alkaline solution that also denatures the DNA and neutralization with a strongly buffered Tris-HCl solution. The last treatment contains a high concentration of sodium chloride, which helps to immobilize the DNA on the filters. Following these treatments, much cellular debris remains on the filters and this must be eliminated by wiping with wet Kimwipes to prevent interference with nucleic acid hybridization experiments. Finally, the filters are washed briefly in 2x SSPE, air dried, and baked for 2 hours at 80°C to bind the DNA to the nylon. The filters can be used immediately or stored dry, sealed in plastic hybridization bags, for many months.

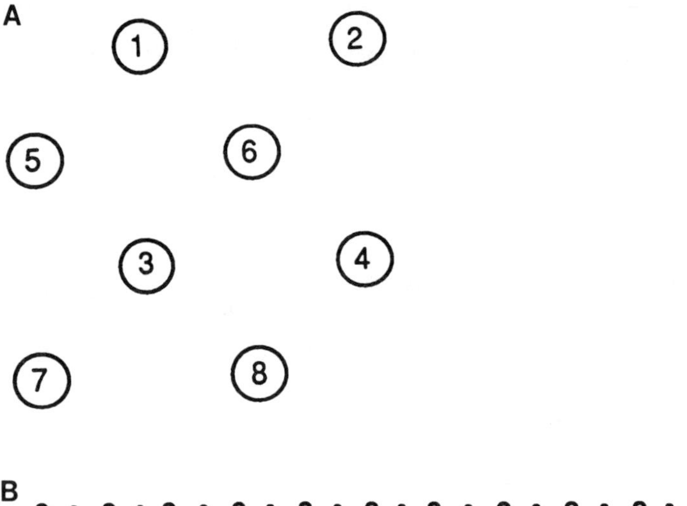

Figure 4
(*A*) Offset positions for replicating eight YAC library microtiter plates on one filter. The positions labeled 1–8 correspond to positions on the metal frames that are used to position the nylon filters correctly on the replicating device. (*B*) A representation of the final colony density on the nylon filters. Colonies derived from position 1 are shown as open circles.

MATERIALS AND EQUIPMENT

yUP YAC library (Ecker 1990)
Ethanol (95%)
96-prong replicator
Nylon filters (Magna Graph™ [Micron Separations, Inc.] or HYBOND™ [Amersham Life Science, Inc.])

0.5 M NaOH
Tris-HCl (1 M; pH 7.5)/NaCl (1.5 M)
Tris-HCl (0.1 M; pH 7.5)/NaCl (0.15 M)

MEDIA

Plates:
YPD nutrient agar
(For recipe, see Preparation of Media, pp. 425–426)

REAGENTS

SOE buffer containing 1% 2-mercaptoethanol and 1 mg of Zymolyase® 20T/ml (add the 2-mercaptoethanol and the Zymolyase® 20T [Seikagaku America, Inc.] just before use)
SSPE (2x)
(For recipes, see Preparation of Reagents, pp. 427–429)

SAFETY NOTE

- 2-Mercaptoethanol may be fatal if inhaled or absorbed through the skin and is harmful if swallowed. High concentrations are extremely destructive to the mucous membranes, upper respiratory tract, skin, and eyes. Use only in a chemical fume hood. Gloves and safety glasses should be worn.

PROCEDURE

Replicating the YAC Library onto Nylon Filters

1. Working in a laminar flow cabinet, remove the 96-prong array from the replicator and dip the prongs into a 95% ethanol bath. Use the lid of a pipette tip rack to contain the ethanol. Flame the prongs by passing the array through a gas flame. Working with the lights off so that the ethanol flame is clearly visible, repeat this step twice more and return the array to the replicator. Rotate the array to the rear of the replicator.

   Note: Effective sterilization of the prongs is a critical step; failure to do this will cause scrambling of the library by transferring cells between the plates.

SAFETY NOTE

- Exercise caution when flaming the prongs. Be absolutely certain that the ethanol flame has gone out before dipping the prongs back into the ethanol. If there is an accident and the ethanol bath is set alight, cover the ethanol bath with a steel tray, evacuate the flow bench taking the replicator with you if you can, and switch off the air flow. Put the fire out with a fire blanket.

2. Place microtiter plate #1 on the replicator platform. Row A on each microtiter plate is considered the top and should be positioned toward the back of the replicator.

3. Using sterile forceps, place an 8 x 12-cm nylon filter in the metal frame and position the frame on the replicator platform in position 1. Mark the bottom of the filter with a soft pencil so that it can be oriented correctly in subsequent manipulations.

4. Collect the cells on the 96-prong array. Dip the array into the top of the cell suspension and hold it there for a few seconds to allow any residual heat to disperse. Push the prongs to the bottom of the well where the cells have settled and move the plate in small circles to collect cells.

   Note: It is important that the prongs are correctly sterilized, but not still hot when they come in contact with the yeast cells.

5. Let the array rise out of the microtiter plate, rotate it round to above the metal frame, and lower it onto the nylon filter to transfer the cells. Touch the prongs to the filter several times. Repeat the transfer of cells until even wet dots are seen on the filter. This may take two or three transfers.

   Note: Try to standardize the replication process like a production line, treating each transfer exactly as the last. Differences in the number of transfers of cells, or differences in the method of picking the cells up on the prongs will cause variation in colony size, and later, differences in the strength of hybridization signals. When replicating, keep only the essentials in the laminar flow cabinet, i.e., the replicator, the current microtiter plate, and the water and ethanol baths. The library of microtiter plates represents a great investment of time and effort and should be kept away from possible accidents with burning ethanol.

6. Replace the lid of the microtiter plate, remove the array, and wash the tips of the prongs in a pipette rack lid containing deionized H_2O. Blot on a paper towel to remove liquid and excess remaining cells before flaming.

7. Flame the array three times as in step 1.

8. Replace microtiter plate #1 with plate #2, move the metal frame to position 2, and carry out steps 3–7 again.

   Note: The set of microtiter plates used to make the colony filters should be a sub-master set, which has been replicated from one set of a duplicated master copy that is kept permanently frozen at -80°C. A recently made sub-master set contains very dense suspensions of cells, which settle into thick deposits in the bases of the wells and a single transfer with the replicator prongs and one overnight incubation is enough to give a good colony. As the cells are used up and their viability decreases due repeated freeze-thaw cycles, it becomes necessary to make the transfer two or more times to inoculate enough cells to get even colony growth and it is also necessary to increase the time of incubation.

9. Continue plating, moving the frame to a new position for each plate until the first eight plates have been replicated onto one filter. Using sterile forceps, transfer the filter to the surface of a 15-cm YPD nutrient agar plate. Avoid trapping bubbles under the filter.

10. Using the same methods, transfer plates #9–#16 to a second filter and plates #17–#24 to a third.

11. Incubate the plates overnight at 30°C.

Cell Lysis and Transfer of DNA to the Filters

1. Cut three pieces of Whatman 3MM filter paper 9 x 13 cm in size and, using forceps, dip them into 4–6 ml of a solution of SOE buffer containing 1% 2-mercaptoethanol and 1 mg of Zymolyase® 20T/ml in a plastic food box. Place each piece of filter paper into a 150-mm petri dish.

   Note: Each piece of filter paper will absorb approximately 6 ml of buffer. Wetting the filter papers before putting them into the petri dishes avoids adding more buffer than the paper can hold and makes it much easier to avoid bubbles.

2. Place the nylon filters colony side up on the filter papers from step 1. Stack the petri dishes, seal them in a plastic bag with a heat sealer, and incubate overnight at 37°C.

3. Remove the petri dishes from the plastic bag and lyse the colonies by transferring the nylon filters colony side up to a series of sheets of Whatman 3MM filter paper soaked in the solutions shown below and placed on top of Saran Wrap:

Step	Time
0.5 M NaOH	20 minutes
Air dry on Whatman 3MM paper	5 minutes
1 M Tris-HCl (pH 7.5)/1.5 M NaCl	at least 5 minutes
0.1 M Tris-HCl (pH 7.5)/0.15 M NaCl	at least 5 minutes

Note: The NaOH causes the cells to break open, releasing their contents onto the nylon, and it also makes the two strands of the DNA duplexes separate. The colonies should glisten during this step and when the nylon filters are lifted there should be a pattern of pink dots on the filter paper where red pigment has leaked through after cell lysis. The precise function of the air drying step is unknown, but omitting this step greatly reduces the quality of the hybridization results. The treatment with Tris-HCl and NaCl neutralizes the high pH and binds DNA to the surface of the nylon.

4. Place the nylon filters on Saran Wrap and wipe off the cell debris with Kimwipes soaked in 0.1 M Tris-HCl (pH 7.5)/0.15 M NaCl. Change the Kimwipes frequently and use plenty of buffer. This step is especially important when the colonies are large.

   Note: The wiping of debris off the nylon filters is one of the most important steps, but the most difficult to control. Wiping should be just firm enough to eliminate all the cell debris, but not so harsh as to score or otherwise damage the surface of the nylon.

5. Wash the nylon filters briefly in 2x SSPE, dry them on Whatman 3MM filter paper for 30 minutes at room temperature, and then bake for 2 hours at 80°C.

   Note: After the treatment of the filters is complete, if the position of each colony is still visible as a very small shiny dot, this usually indicates a good filter set.

Preparation of High-molecular-weight Yeast DNA in Agarose Plugs

To immobilize high-molecular-weight DNA in agarose, Zymolyase® 20T is added to a suspension of yeast cells, which is then mixed with molten low-melting-point agarose at 45°C. The mixture is then pipetted into 100-μl slots in a plastic mold and allowed to solidify. The plugs are pushed from the mold into a high concentration of EDTA buffered with Tris and the cell walls are digested by incubating the plugs overnight at 37°C. The digestion buffer is replaced with a solution of 1% lithium dodecyl sulfate in 100 mM EDTA, which solubilizes the cell membranes during another overnight incubation. The chromosomes, including the YAC, are trapped in the agarose matrix while the remainder of the cellular components diffuse out of the agarose into the buffer. After a wash in EDTA, the plugs are ready to be loaded into the wells of a pulsed-field electrophoresis gel. The 100-μl plugs made in this way are ideal for rapid analysis of YACs, but only a few of these can be made from a 3-ml culture of yeast. When a need for more such plugs is anticipated, the procedure can be scaled up by using cells from a 25-ml culture and drawing the yeast/agarose mixture into a 1-ml syringe instead of pipetting it into molds. This procedure gives a cylindrical plug which is convenient to manipulate and can be stored inside a syringe from which the neck has been cut. 1–2-mm slices of the plug can be loaded into the wells of a gel.

YEAST CLONES CONTAINING YACs

YACs are maintained in yeast strain AB1380 (*Mat-a, ade2-1, can1-100, lys2-1, trp1, ura3, his5 [psi+]*) (Burke et al. 1987). YAC libraries of *Arabidopsis* are available from the *Arabidopsis* Biological Resource Center at Ohio State University (see Appendix 2).

MATERIALS

EDTA (50 mM and 100 mM; pH 8.0)
Zymolyase® 20T (20 mg/ml) (Seikagaku America, Inc.)
Low-melting-point (LMP) agarose (1.5% in 125 mM EDTA [pH 8.0]) (equilibrated at 45°C)

MEDIA

AHC medium
(For recipe, see Preparation of Media, pp. 425–426)

REAGENTS

LET buffer
Yeast lysis buffer
(For recipes, see Preparation of Reagents, pp. 427–429)

PROCEDURE

1. Grow a yeast clone to saturation in 25 ml of AHC medium for 2–3 days at 30°C.

   Note: With YACs maintained in yeast strain AB1380 and grown to saturation in AHC medium, 25 ml of culture provides enough cells so that a 1–2-mm slice of the plug gives ample YAC DNA to see as a band in an ethidium bromide stained agarose gel. Some clones yield fewer cells than others, but this protocol almost always gives useful plugs.

2. Collect the cells by spinning them at 3000 rpm for 15 minutes at room temperature.

3. Resuspend the cells in 20 ml of 50 mM EDTA (pH 8.0).

4. Repeat step 2 to wash the cells a second time.

5. Resuspend the cells in 225 μl of 50 mM EDTA, and transfer them to a 1.5-ml microcentrifuge tube. Add 12 μl of 20 mg/ml Zymolyase® 20T, quickly mix with 375 μl of 1.5% LMP agarose in 125 mM EDTA, which has been equilibrated at 45°C, and draw the mixture into a 1-ml syringe. To avoid bubbles, do not try to recover all of the mixture.

   Note: The best plugs are homogeneous, flexible, and slightly transparent. To avoid lumpy preparations that break easily, the agarose must be thoroughly melted, and the yeast/agarose mixture well mixed by pipetting up and down with a 1-ml Pipetman® before it is drawn into the syringe.

6. Cover the end of the syringe with Parafilm and push it into an ice bucket to solidify the agarose.

7. Cut the end of the syringe off with a razor blade and expel the cyl-

inder of agarose into 7 ml of LET buffer in a 15-ml disposable centrifuge tube and incubate overnight at 37°C.

8. Pour off the Zymolyase® 20T/LET solution and replace it with 7 ml of yeast lysis buffer and incubate overnight at 40–50°C.

   Note: If the plug is required quickly, the LET and yeast lysis buffers can each be replaced several times, reducing steps 7 and 8 to 4 hours each. As lysis proceeds, pigment leaking from the ruptured cells will turn the medium pink. This serves as a monitor of lysis and can be used to estimate when this step is complete.

9. Equilibrate the plug in 100 mM EDTA at room temperature, return it to a 1-ml syringe, seal the end with Parafilm, and store at 4°C.

   Note: Lithium dodecyl sulfate (LDS) is insoluble in 100 mM EDTA at 4°C, therefore it is important to equilibrate the plug thoroughly in 100 mM EDTA to remove all the LDS before storing it at 4°C. Plugs can be kept in the first wash of 100 mM EDTA at room temperature for many weeks without affecting the quality of the DNA.

Analysis of YACs by Contour-clamped Homogeneous Electric Field Electrophoresis

Until the advent of pulsed-field gel electrophoresis (PFGE) (Carle and Olson 1984; Schwartz and Cantor 1984), the upper limit of resolution of DNA molecules in an agarose gel was around 50 kb. By periodically reversing the field direction, it is now possible to resolve molecules up to 12,000 kb, and analysis of molecules in the size range of YACs (100–1000 kb) is a routine task. In this experiment, we use the contour-clamped homogeneous electric field electrophoresis (CHEF) configuration of electrodes in a hexagonal array (Chu et al. 1986), which has the advantage of causing the DNA to migrate in straight tracks along the gel, unlike some other electrode configurations. The resolution range of CHEF gels is principally dependent on the switching interval. In our experience, using CHEF apparatus designed by G. Chu and the Stanford University School of Medicine Biochemistry Department Instrument Shop, good separation between 50 and 600 kb is achieved with a 40-second switching interval, an applied voltage of 180 V (6.4 v/cm between opposite electrodes), and a running time of 24–40 hours.

MATERIALS AND EQUIPMENT

Agarose (low electroendosmosis [EEO]; electrophoresis grade)
(Boehringer Mannheim Biochemicals)
CHEF apparatus (e.g., CHEF-DR® II; Bio-Rad Laboratories)
YAC plugs (see pp. 415–417)
Concatamerized λ DNA molecules
Low-melting-point (LMP) agarose (1% in 0.5x TBE)
Ethidium bromide (10 mg/ml)

REAGENTS

TBE buffer (0.5x)
(For recipe, see Preparation of Reagents, pp. 427–429)

SAFETY NOTE

- Ethidium bromide is a powerful mutagen and is moderately toxic. Gloves should be worn when working with solutions that contain this dye. After use, decontaminate and discard solutions according to local safety office regulations.

PROCEDURE

1. Melt 0.8 g of electrophoresis grade agarose in 80 ml of 0.5x TBE buffer in a microwave oven, making sure it dissolves completely. Cool the agarose until it can be comfortably held in the palm of the hand and cast the gel in a 12 x 12 cm mold on the base of the CHEF tank. Position the comb before pouring the gel and weight the corners with the lead containers used to transport radioisotopes.

2. After the agarose has set, carefully remove the comb, cut around the edges of the gel with a scalpel, and remove the mold by holding one corner down and lifting up the opposite corner.

   Note: One of the most common problems is that the gel becomes detached from the base of the CHEF tank. Gentle removal of the comb and mold usually avoid this problem, but if the gel should move, it can often be sealed down with LMP agarose. Sometimes it is necessary to cast a new gel.

3. Load the gel with slices of YAC plugs. Remove the Parafilm from the syringe containing the plug, expose 1–2 mm of the plug, and slice it off cleanly with a razor blade. Transfer it gently to a well of the gel using a pipette tip. Load other samples as required, and in one lane, load a slice of a plug containing concatamerized λ DNA molecules as size markers.

4. After all the samples are loaded, seal the plugs in the wells with molten 1% LMP agarose.

5. Cover the gel with 0.5x TBE and equilibrate the buffer to 10–12°C by circulating it through a water bath set to 4°C.

6. Set the switching interval to 40 seconds, switch on the current, and adjust the voltage to 180 V. Make sure that the switching apparatus is working correctly and check that current is flowing.

7. Let the gel run for 24–40 hours, checking on the temperature occasionally.

8. After switching off the current, drain 500 ml of the running buffer into a suitable container, add 10 μl of 10 mg/ml ethidium bromide, and stain the gel for 20 minutes with gentle shaking.

9. Examine the gel over a ultraviolet transilluminator and take a Polaroid photograph. If the background staining is unacceptably high, destain the gel in H_2O for 1–2 hours.

Screening a YAC Library by Hybridization

There are two major factors that affect the results of yeast colony hybridization: the quality of the filter sets and the specific activity of the probe. To prepare probes, we use the random hexamer labeling method of Feinberg and Vogelstein (1983) with minor modifications. With high-quality filter sets (see pp. 409–414) and labeling that yields greater than 30 million counts per minute from 50–100 ng of input DNA, identification of positive signals should be routine, although screening may need to be repeated to ensure detection of all positives. False positive signals are also frequent.

MATERIALS

Random hexamer oligonucleotides (90 OD_{260} units/ml in distilled H_2O) (Pharmacia Biotech, Inc.)
DNA to be labeled (50–100 ng)
HEPES (1 M; pH 6.6)/bovine serum albumin (BSA) (0.2 mg/ml)
[α-^{32}P]dCTP (3000 Ci/mmol)
Klenow fragment of *E. coli* DNA polymerase I (Amersham Life Science, Inc.)
Phenol/chloroform (1:1)
Salmon sperm DNA (sonicated) (10 mg/ml in TE)
Ammonium acetate (2 M)
Ethanol (100%)
Colony filters (see pp. 409–414)
pBR322 DNA labeled with ^{35}S (10^6 cpm)

REAGENTS

TE
dNTP mixture for DNA labeling
Hybridization/prehybridization buffer
SSPE (2x) containing 1% sodium dodecyl sulfate (SDS)
SSPE (0.1x) containing 1% SDS
(For recipes, see Preparation of Reagents, pp. 427–429)

SAFETY NOTES

- Wear gloves when handling radioactive substances. Consult the local safety office for further guidance in the appropriate use of radioactive materials.
- Phenol is highly corrosive and can cause severe burns. Wear gloves, protective clothing, and safety glasses when handling it. All manipulations should be carried out in a chemical fume hood. Any areas of skin that come in contact with phenol should be rinsed with a large volume of water or PEG 400 and washed with soap and water; do not use ethanol!
- Chloroform is irritating to the skin, eyes, mucous membranes, and respiratory tract. It should only be used in a chemical fume hood. Gloves and safety glasses should also be worn. Chloroform is a carcinogen and may damage the liver and kidneys.
- Wear a mask when weighing SDS.

PROCEDURE

Probe Preparation

1. Add 1 µl of 90 OD_{260} units/ml random hexamers to 50–100 ng of the DNA to be labeled in a volume of 8 µl of TE and boil for 10 minutes. Let it cool slowly.

2. Add 5 µl of 1 M HEPES (pH 6.6)/0.2 mg/ml BSA and 5 µl of dNTP mixture.

3. Add 5 µl of [α-^{32}P]dCTP.

4. Add 1 unit of Klenow fragment and incubate overnight at room temperature.

5. Add 75 µl of TE and extract once with 100 µl of phenol/chloroform (1:1). Spin to separate the phases.

6. Transfer the upper (aqueous) phase to a clean tube. Add 50 µg of salmon sperm DNA (5 µl of a 10 mg/ml solution) and precipitate the DNA with 100 µl of 2 M ammonium acetate and 600 µl of ice-cold 100% ethanol on ice for 10 minutes.

   Note: With a short incubation on ice, ammonium acetate causes the DNA to precipitate but leaves the unincorporated nucleotides in solution. Longer incubations cause the nucleotides to precipitate also.

7. Spin for 10 minutes in a microfuge, dispose of the supernatant, and resuspend the DNA in 200 μl of TE.

   Note: In normal ethanol precipitation procedures, the pellet is washed in 70% ethanol and dried under vacuum. However, under the conditions described in step 7 above, the bulk of the unincorporated isotope is removed in the supernatant, and the small amount of ethanol that remains does not prevent the pellet dissolving in TE.

8. Measure the activity of 2 μl of the probe by Cerenkov counting.

Colony Hybridization

There are many variations on the basic colony hybridization method, but the one we have had most success with is that of Larin et al. (1991), which is described below.

1. Place the colony filters in a plastic hybridization bag.

2. Boil 150 μl of salmon sperm DNA (10 mg/ml) for 10 minutes, mix with 15 ml of hybridization/prehybridization buffer, and pour this mixture into the hybridization bag.

3. Taking care to eliminate as many bubbles as possible, close the bag with a heat sealer and incubate on a shaking platform overnight at 42°C.

4. Cut open the bag, discard the buffer, and replace it with another 15 ml of identical buffer.

   Note: A problem often encountered with this procedure is excessive background on the X-ray film. Discarding the prehybridization buffer appears to reduce this background.

5. Boil the probe (>30 x 10^6 cpm) together with 150 μl of salmon sperm DNA (10 mg/ml) for 10 minutes. Include 10^6 cpm of pBR322 DNA labeled with ^{35}S. Add this to the hybridization buffer.

   Note: The filters have an array of very closely spaced colonies and in the absence of any background the precise identification of positive signals is difficult. The plasmid pBR322 contains sequences specific to the YAC vector and, when labeled with ^{35}S, causes a faint background signal on every colony that helps identify colonies that are positive with the ^{32}P probe. The ^{35}S pBR322 is prepared just like any other probe (see above), except that [α-^{32}P]dCTP is replaced with [^{35}S]dCTP.

6. Close the bag with a heat sealer, seal it inside a second bag to guard against leakage, and incubate on a shaking platform at 42°C overnight.

7. Remove the colony filters from the hybridization buffer and wash them briefly in 2x SSPE containing 1% SDS.

8. Wash the filters twice in 0.1x SSPE containing 1% SDS for 20 minutes each at 65°C. Blot them with Whatman 3MM filter paper and expose to Kodak XAR film at –80°C. The appropriate length of exposure is determined empirically for each probe and is usually between 1 day and 1 week.

• *PREPARATION OF MEDIA*

LIQUID:

AHC medium

0.17% Yeast nitrogen base without amino acids and $(NH_4)_2SO_4$
0.5% $(NH_4)_2SO_4$
1% Acid-hydrolyzed casein
2% Glucose
0.002% Adenine hemisulfate

Dissolve all the components except the glucose in 900 ml of deionized H_2O. Transfer to a 1-liter media bottle and autoclave. Cool, then add 50 ml of sterile 40% glucose.

SOC medium

NaCl	0.5 g
Bacto-yeast extract	5 g
Bacto-tryptone	20 g

Dissolve the components in 900 ml of deionized H_2O. Add 5 ml of 0.5 M KCl and adjust the volume to 1 liter with deionized H_2O. Sterilize by autoclaving. After the medium has cooled, add 10 ml of sterile 1 M $MgCl_2$ and 20 ml of sterile 1 M glucose.

PLATES:

LB agar supplemented with 100 μg of ampicillin/ml

NaCl	10 g
Bacto-tryptone	10 g
Bacto-yeast extract	5 g

Dissolve the components in 900 ml of deionized H_2O. Adjust the volume to 1 liter with deionized H_2O, add 15 g of Bacto-agar, and autoclave. After the medium has cooled to below 60°C, add 1 ml of 100 mg/ml ampicillin, and swirl to mix. Pour the plates, allowing 25 ml of medium per 8-cm plate.

YPD nutrient agar plates

1% Bacto-yeast extract
2% Bacto-peptone
2% Dextrose
2% Bacto-agar

Dissolve the yeast extract and peptone in 900 ml of deionized H_2O. Transfer to a 1-liter media bottle containing 20 g of agar and autoclave. After the medium has cooled to below 60°C, add 50 ml of sterile 40% dextrose. Pour the plates, allowing 50–60 ml of medium per 15-cm plate.

• PREPARATION OF REAGENTS

dNTP mixture for DNA labeling

100 mM dATP
100 mM dGTP
100 mM dTTP

Make this solution up in 50 mM $MgCl_2$. Store at –20°C. The dNTPs can be obtained from Pharmacia Biotech, Inc.

dNTP mixture for PCR

1.25 mM dATP
1.25 mM dCTP
1.25 mM dGTP
1.25 mM dTTP

Make this solution up in distilled H_2O. Store at –20°C. The dNTPs can be obtained from Pharmacia Biotech, Inc.

Hybridization/prehybridization buffer

50% Formamide
8% Dextran sulfate
1% Sodium dodecyl sulfate (SDS)
10x Denhardt's solution
5x SSPE

Make fresh and keep at 42°C until ready to use.

Denhardt's solution (100x)

Ficoll (type 400, Pharmacia Biotech, Inc.)	10 g
Polyvinylpyrrolidone	10 g
Bovine serum albumin (Fraction V, Sigma)	10 g

Dissolve in H_2O and then adjust the volume to 500 ml with deionized H_2O. Filter sterilize. Store at –20°C.

LET buffer

10 mM Tris-HCl (pH 7.5)
0.5 M EDTA

Store at room temperature.

Loading dye (2x)

Dilute 6x loading dye 1 in 3 with TE buffer. Store at room temperature.

Loading dye (6x)

0.25% Bromophenol blue
0.25% Xylene cyanol
40% Glycerol

SOE buffer

1 M Sorbitol
20 mM EDTA
10 mM Tris-acetate (pH 8.0)

Store at room temperature.

SSPE (0.1x)

Dilute 20x SSPE 1 in 200 with deionized H_2O. Store at room temperature.

SSPE (20x)
(1 liter)

NaCl	175.3 g
$NaH_2{\cdot}PO_4$	27.6 g
EDTA	7.4 g

Dissolve the components in 800 ml of H_2O. Adjust the pH to 7.4 with 10 M NaOH and adjust the volume to 1 liter with deionized H_2O.

SSPE (2x)

Dilute 20x SSPE 1 in 10 with deionized H_2O. Store at room temperature.

TAE buffer (50x)
(1 liter)

Tris base	242 g
Glacial acetic acid	57.1 ml
0.5 M EDTA (pH 8.0)	100 ml

Adjust the volume to 1 liter with distilled H_2O (Sambrook et al. 1989). 1x TAE buffer is 40 mM Tris-acetate/1 mM EDTA. Store at room temperature.

TBE buffer (0.5x)
(45 mM Tris-borate/1 mM EDTA)

Dilute 5x TBE 1 in 10 with distilled H_2O. Store at room temperature.

TBE buffer (5x)
(1 liter)

Tris base	54 g
Boric acid	27.5 g
0.5 M EDTA (pH 8.0)	40 ml

Adjust the volume to 1 liter with distilled H_2O.

TE

10 mM Tris-HCl (pH 8.0)
1 mM EDTA

Store at room temperature.

Yeast lysis buffer

0.1 M EDTA
10 mM Tris-HCl (pH 8.0)
1% Lithium dodecyl sulfate

Store at room temperature.

Appendices

Appendix 1
Regulation of Experiments with Transgenic Plants

Experiments with transgenic plants may be carried out only in contained laboratory and greenhouse facilities, and require approval from the biosafety committee of the institution. Growing plants outside is considered field testing and requires a permit from the U.S. Department of Agriculture (USDA). The National Institutes of Health guidelines for research involving recombinant DNA molecules were published in the Federal Register (vol. 51, no. 88, pp. 16958–16985 [May 7, 1986]). Import of *Agrobacterium* strains (even if disarmed), transgenic plants, and transgenic seeds into the United States also requires a permit, as does interstate transportation of these regulated materials (see Federal Register, vol. 58, no. 60, pp. 17044–17059 [March 31, 1993] and the USDA Animal and Plant Inspection Service Technical Bulletin no. 1783). However, mailing of isolated DNA is not regulated. The interstate movement of genetically engineered *Arabidopsis thaliana*, even when containing genetic material derived from a plant pest, is exempt from permit requirements, provided it does not present a risk of plant pest introduction and dissemination. The interstate movement of *Escherichia coli* genotype K12 (K12 strain and its derivatives) and sterile strains of *Saccharomyces cerevisiae* is also exempt from permit requirements, provided specific conditions are met. Importation and interstate movement of transgenic corn, cotton, potato, soybean, tobacco, and tomato is now feasible using the faster notification procedures (for details, consult the USDA Animal and Plant Inspection Service Technical Bulletin no. 1783). Intrastate movement of regulated biomaterials is subject to state law. For a permit or an update of the rules of transportation of regulated biomaterials within the United States, contact Animal and Plant Inspection Service, USDA, Biotechnology Permits, 4700 River Road, Unit 147, Riverdale, Maryland 20737-1237; telephone 301-734-7612.

Appendix 2
Sources of Materials and Equipment

ACADEMIC RESOURCES

Tobacco (*Nicotiana tabacum*) seeds; Wisconsin 38 variety

Small amounts of transgenic seeds harboring the pED32 construct and parental, nontransformed seeds can be obtained from the laboratory of Dr. David W. Ow, Plant Gene Expression Center, U.S. Department of Agriculture and University of California at Berkeley, 800 Buchanan Street, Albany, CA 94710. Telephone 510-559-5909. Fax 510-559-5678. e-mail address: ow@mendel.Berkeley.EDU

Maize cell suspension culture

A maize cell suspension culture established from immature, embryo-derived type II embryogenic callus can be obtained from the laboratory of Dr. Indra K. Vasil, Horticultural Sciences Department, University of Florida, Gainesville, FL 32611. Telephone 904-392-1193. Fax 904-392-9366.

C1-Sh-GUS, 35S-CAT, and 35S-Sh-Vp1 plasmids

These plasmids can be obtained from Dr. Donald R. McCarty, Horticultural Sciences Department, University of Florida, Gainesville, FL 32611. Telephone 904-392-1928. Fax 904-392-6479.

Tomato *RbcS3B* primers I and II

These primers can be synthesized using the published sequence or obtained from Dr. Wilhelm Gruissem, Department of Plant Biology, 111 Koshland Hall, University of California, Berkley, CA 94720. Telephone 510-642-1079. Fax 510-642-4995. e-mail address: gruissem@nature.berkeley.edu

YAC libraries of *Arabidopsis*

YAC libraries of *Arabidopsis* are available from the *Arabidopsis* Biological Resource Center at Ohio State University, 1735 Neil Avenue, Columbus, Ohio 43210. Telephone 614-292-2116. Fax 614-292-5379. e-mail address: arabidopsis+@osu.edu

COMMERCIAL SUPPLIERS

Amersham Life Science, Inc., 2636 S. Clearbrook Dr., Arlington Heights, IL 60005. Telephone 800-323-9750 or 708-593-6300. Fax 800-228-8735.

Amicon, Inc., 72 Cherry Hill Dr., Beverly, MA 01915. Telephone 800-426-4266 or 508-777-3622. Fax 508-777-6204.

Analytical Luminescence Laboratory, 11760 Sorrento Valley Rd., Ste. E, San Diego, CA 92121. Telephone 800-854-7050 or 619-455-9283. Fax 619-455-9204.

Beckman Instruments, Inc., 2500 Harbor Blvd., P.O. Box 3100, Fullerton, CA 92634. Telephone 800-742-2345 or 714-871-4848. Fax 800-643-4366.

Bio-Rad Laboratories, Life Science Group, 2000 Alfred Nobel Dr., Hercules, CA 94547. Telephone 800-4BIORAD or 510-741-1000. Fax 800-879-2289.

Boehringer Mannheim Corporation, Biochemical Products Division, 9115 Hague Rd., P.O. Box 50414, Indianapolis, IN 46250. Telephone 800-262-1640 or 317-849-9350. Fax 317-576-2754.

Burton-Rogers Co., Calibron Instruments Division, 220 Grove St., Waltham, MA 02154. Telephone 800-225-4678 or 617-894-6440. Fax 617-893-8393.

Calbiochem-Novabiochem International, Inc., 10394 Pacific Center Ct., San Diego, CA 92121. Telephone 800-854-3417 or 619-450-9600. Fax 800-776-0999.

Ciba-Geigy Agrochemicals AG, CH-4002 Basle, Switzerland. Telephone 41-0-61-6971111. Fax 41-0-223-835211.

CLONTECH Laboratories, Inc., 4030 Fabian Way, Palo Alto, CA 94303. Telephone 800-622-CLON or 415-424-8222. Fax 415-424-1064.

Corex. *See* Corning, Inc.

Corning, Inc., Science Products Division, P.O. Box 5000, Corning, NY 14831. Telephone 800-222-7740. Fax 607-974-0345.

Difco Laboratories, Inc., P.O. Box 331058, Detroit, MI 48232. Telephone 800-521-0851 or 313-462-8500. Fax 313-462-8594.

DuPont Medical Products, Sorvall® Centrifuges, P.O. Box 80024, Wilmington, DE 19880. Telephone 800-555-2121 or 800-551-2121.

DuPont NEN®, 549 Albany St., Boston, MA 02118. Telephone 800-551-2121 or 617-542-9595. Fax 617-350-9562.

Eastman Kodak Co., Scientific Imaging Systems, 343 State St., Rochester, NY 14650. Telephone 800-225-5352 or 716-588-2572. Fax 716-588-8368.

Fisher Scientific Co., 711 Forbes Ave., Pittsburgh, PA 15219. Telephone 800-766-7000 or 412-562-8300. Fax 800-926-1166.

GIBCO, GIBCO/BRL. *See* Life Technologies, Inc.

Hamamatsu Photonic Systems, Division of Hamamatsu Corporation, 360 Foothill Rd., P.O. Box 6910, Bridgewater, NJ 08807. Telephone 908-231-1116. Fax 908-231-0852.

Hasleton Biologics Co., P.O. Box 14848, Lenexa, KS 66215. Telephone 800-255-6032 or 913-469-5580. Fax 913-469-5584.

Hoefer Scientific Instruments, 654 Minnesota St., P.O. Box 77387, San Francisco, CA 94107. Telephone 800-227-4750 or 415-282-2307. Fax 415-826-8755.

Kenwood. *See* Burton-Rogers Co., Calibron Instruments Division (a distributor of this brand name)

Kipp & Zonen, 390 Central Ave., Bohemia, NY 11716. Telephone 516-589-2885. Fax 516-589-2068.

Kodak. *See* Eastman Kodak Co.

Kontes Glass, 1022 Spruce St., P.O. Box 729, Vineland, NJ 08360. Telephone 800-223-7150 or 609-692-8500. Fax 609-692-3242.

L.L. Olds Seeds Co., P.O. Box 7790, Madison, WI 53707. Telephone 608-249-9291.

Life Technologies, Inc. (GIBCO BRL), 8400 Helgerman Ct., P.O. Box 6009, Gaithersburg, MD 20884. Telephone 800-828-6686 or 301-840-8000. Fax 800-331-2286.

Magenta Corporation, 3800 N. Milwaukee Ave., Chicago, IL 60641. Telephone 313-777-5050. Fax 312-777-4055.

Micron Separations, Inc., 135 Flanders Rd., P.O. Box 1046, Westboro, MA 01581. Telephone 800-444-8212 or 508-366-8212. Fax 508-366-5840.

Millipore Corporation, Process Membrane Division, 80 Ashby Rd., Bedford, MA 01730. Telephone 800-MILLIPORE or 617-275-9200. Fax 617-553-8873.

Nalgene. *See* Nalge Co.

Nalge Co., a subsidiary of Sybron Corporation, 75 Panorama Creek Dr., P.O. Box 20365, Rochester, NY 14602-0365. Telephone 716-586-8800. Fax 716-586-8987.

Narishige USA, Inc., 404 Glen Cove Ave., Sea Cliff, NY 11579. Telephone 516-676-0044. Fax 516-676-1480.

Nikon, Inc., Biomedical Instrument Group, 1300 Walt Whitman Rd., Melville, NY 11747. Telephone 516-547-8500. Fax 516-547-0306.

Out Patient Services, Inc., 1320 Scott Street, Petaluma, CA 94952. Telephone 707-763-1581.

The Perkin-Elmer Corporation, Applied Biosystems Division, 850 Lincoln Centre Dr., Foster City, CA 94404. Telephone 800-345-5224 or 415-570-6667. Fax 415-572-2743.

Pipetman®. *See* Rainin Instrument Co., Inc.

Pharmacia Biotech, Inc., 800 Centennial Ave., P.O. Box 1327, Piscataway, NJ 08855. Telephone 800-526-3593 or 908-457-8000. Fax 800-329-3593.

Polysciences, Inc., 400 Valley Rd., Warrington, PA 18976. Telephone 800-523-2575 or 215-343-6484. Fax 215-343-0214.

Promega Corporation, 2800 Woods Hollow Rd., Madison, WI 53711. Telephone 800-356-9526 or 800-356-9526. Fax 800-356-1970.

Rainin Instrument Co., Inc., Mack Rd., P.O. Box 4026, Woburn, MA 01888. Telephone 800-4-RAININ or 617-935-3050. Fax 617-938-5152.

Research Products International Corporation, 410 N. Business, Mt. Prospect, IL 60056-2190. Telephone 800-323-9814 or 708-635-7330. Fax 708-635-1177.

Savant Instruments, Inc., 100 Colin Dr., Holbrook, NY 11741. Telephone 800-634-8886 or 516-249-4600. Fax 516-249-4639.

Schleicher and Schuell, Inc., 10 Optical Ave., P.O. Box 2012, Keene, NH 03431. Telephone 800-245-4024 or 603-352-3810. Fax 603-357-3627.

Seikagaku America, Inc., 30 W. Gude Dr., Ste. 260, Rockville, MD 20850. Telephone 800-237-4512 or 301-424-0546. Fax 301-424-6961.

Seishin Pharmaceuticals, Inc., 4-13, Koami-Aho Nihonbashi, Tokyo, Japan.

Serva Biochemicals, Inc., 200 Shamus Dr., Westbury, NY 11590. Telephone 516-333-1575. Fax 516-333-1582.

Sigma Chemical Co., P.O. Box 14508, St. Louis, MO 63178. Telephone 800-325-8070 or 314-771-5750. Fax 800-325-5052.

Small Parts, Inc., 13980 N.W. 58th Ct., Miami Lakes, FL 33014. Telephone 305-558-1038. Fax 800-423-9009.

Stratagene, 11011 N. Torrey Pines Rd., La Jolla, CA 92037. Telephone 800-424-5444 or 619-535-5400. Fax 619-535-0034.

Sutter Instrument Co., 40 Leveroni Ct., Novato, CA 94949. Telephone 415-883-0128. Fax 415-883-0572.

Tetko, Inc., 333 S. Highland Ave., Briarcliffe Manor, NY 10510. Telephone 914-941-7767. Fax 914-941-2446.

Turner Designs, 920 W. Maude Ave., Sunnyvale, CA 94086. Telephone 408-749-0994.

Whatman, Inc., 9 Bridewell Pl., Clifton, NJ 07014. Telephone 800-631-7290 or 201-773-5800. Fax 201-773-6138.

World Precision Instruments, Inc., 175 Sarasota, Sarasota, FL 34240-9258. Telephone 813-371-1003. Fax 813-377-5428.

Yakult Pharmaceuticals, Inc., 1-1-19, Higashi-Shinba, Sinato-ku, Tokyo 105, Japan. Telephone 03375746766.

Index